Stoichiometric Asymmetric Synthesis

Postgraduate Chemistry Series

A series designed to provide a broad understanding of selected growth areas of chemistry at postgraduate student and research level. Volumes concentrate on material in advance of a normal undergraduate text, although the relevant background to a subject is included. Key discoveries and trends in current research are highlighted, and volumes are extensively referenced and cross-referenced. Detailed and effective indexes are an important feature of the series. In some universities, the series will also serve as a valuable reference for final year honours students.

Titles in the series:

Catalysis in Asymmetric Synthesis
Jonathan M.J. Williams

Protecting Groups in Organic Synthesis
James R. Hanson

Organic Synthesis with Carbohydrates
Geert-Jan Boons and Karl J. Hale

Organic Synthesis using Transition Metals
Roderick Bates

Stoichiometric Asymmetric Synthesis
Mark Rizzacasa and Michael Perkins

Stoichiometric Asymmetric Synthesis

MARK A. RIZZACASA
School of Chemistry
The University of Melbourne
Victoria
Australia

and

MICHAEL PERKINS
Department of Chemistry
Flinders University
Adelaide
Australia

**Blackwell
Science**

First published 2000
Copyright © 2000 Sheffield Academic Press

Published by
Sheffield Academic Press Ltd
Mansion House, 19 Kingfield Road
Sheffield S11 9AS, England

ISBN 1-84127-111-X

Published in the U.S.A. and Canada (only) by
Blackwell Science, Inc.
Commerce Place
350 Main Street
Malden, MA 02148-5018, U.S.A.
Orders from the U.S.A. and Canada (only) to Blackwell Science, Inc.

Printed on acid-free paper in Great Britain by
Bookcraft Ltd, Midsomer Norton, Bath

British Library Cataloguing-in-Publication Data:
A catalogue record for this book is available from the British Library

Library of Congress Cataloging-in-Publication Data: $\alpha C \ast$
A catalog record for this book is available from the Library of Congress

Foreword

by Dr Ian Paterson, Department of Chemistry, University of Cambridge, UK

The control of relative and absolute stereochemistry is an intellectually stimulating and challenging aspect of modern synthesis design. To understand the origins of stereoinduction, reaction pathways generally need to be evaluated in three dimensions, where competing transition states are energetically responsive to steric and electronic influences. Over the last three decades, powerful methods and strategies have been developed to synthesise organic compounds containing multiple stereocentres with high levels of diastereoselectivity and in enantiomerically pure form. Through appropriate use of substrate and reagent-based control of stereochemistry, chiral molecules having diverse functionality can now be synthesised in a practical and predictable fashion.

This book by Mark Rizzacasa and Michael Perkins does an admirable job of presenting, logically and systematically, the underlying stereochemical and mechanistic aspects associated with asymmetric synthesis. Both authors are active and skilled practitioners in stereocontrolled organic synthesis, so it is good to see their combined expertise and knowledge being made available to a wider audience. Emphasis is given to practical, stoichiometric methodology and the analysis of reaction transition states. There is a wealth of illustrative examples of stereocontrolled synthesis taken from the recent literature (including those involving radical reactions), accompanied by a useful bibliography for each chapter, including references to specialist reviews and research publications. The coverage is eminently suitable for graduate and advanced undergraduate courses concentrating on the synthesis of chiral organic compounds, and in addition it provides an up-to-date and authoritative overview of stereocontrolled transformations for synthetic chemists working in academia and industry.

Preface

This book focuses on the use of a stoichiometric amount of chirality to induce asymmetry in carbon-carbon bond-forming reactions. Stoichiometric asymmetric synthesis is widely used in the academic and industrial sectors for the synthesis of chiral molecules of biological importance. Although catalytic asymmetric synthesis is an alternative, the use of equimolar amounts of chirality usually provides high selectivities over a wider range of substrates, without extensive modifications of reaction conditions. This book is aimed primarily at graduate students, but postdoctoral researchers and teaching staff may also find it useful. It is not intended as a review of all the literature. However, the majority of the methods included are widely used and provide high selectivities. Those that are experimentally simple and well documented are highlighted.

Chapter 2 details additions to carbonyl compounds by simple nucleophiles, beginning with a description of the models used to predict facial selectivity. A section on the use of such additions in the synthesis of some important molecules is included. Chapter 3 discusses the asymmetric alkylation of enolates, including methods for stereoselective enolate formation. Chapters 4 and 5 show the strategies for one of the most important asymmetric C-C bond-forming reactions, the aldol condensation, while chapter 6 details the allyl and crotylmetal alternatives. Chapter 7 covers pericyclic reactions, while chapter 8 deals with asymmetric reactions of alkenes. The book concludes with chapter 9, which presents examples of asymmetric radical reactions.

To show the origin of stereoselectivity, attention has been paid to the facial selectivities of reactions, and detailed transition states are included where appropriate. We hope that this text provides the researcher with a source of asymmetric reactions that can be utilised for the synthesis of chiral compounds.

We wish to thank the following for their invaluable assistance with the preparation of this volume: Dr Ian Paterson (Cambridge University), Ivona Czuba for indexing, Anthony Cuzzupe, Dr Mike Lilly, Mariana El Sous, Georgina Holloway and Robert Mann (all from the University of Melbourne).

Mark A. Rizzacasa
Michael Perkins

Contents

8 Reactions of alkenes 188

9 Free radical processes 205

1 Introduction

1.1 Background

The past 20 years have been a renaissance period for asymmetric organic synthesis. The development of efficient methods for the enantiospecific synthesis of chiral molecules has intensified as more complex chiral natural products have been targeted.[1] The requirement that all chiral drugs must be produced in optically pure form has also driven research in asymmetric synthesis.[2,3] Currently, the most widely used general method for the production of chiral compounds is the stoichiometric approach where the chiral information is present in either the substrate or in the reagent. This book will outline the major methods of stoichiometric asymmetric synthesis, with the main focus being on the formation of carbon–carbon bonds.

Some key types of asymmetric carbon–carbon bond-forming reactions are shown in Tables 1.1 and 1.2. The reactions in Table 1.1 involve the generation of a new C—C bond along with the formation of an asymmetric centre at one or both of the carbon atoms involved in the process. The first example is the simple addition of carbon-based nucleophiles to aldehydes and ketones (Chapter 2) which can generate a new asymmetric centre. The nucleophile can attack either face of the sp^2 hybridised carbonyl carbon (trigonal planar) as shown in Scheme 1.1. These two faces can be described as either Re (if the

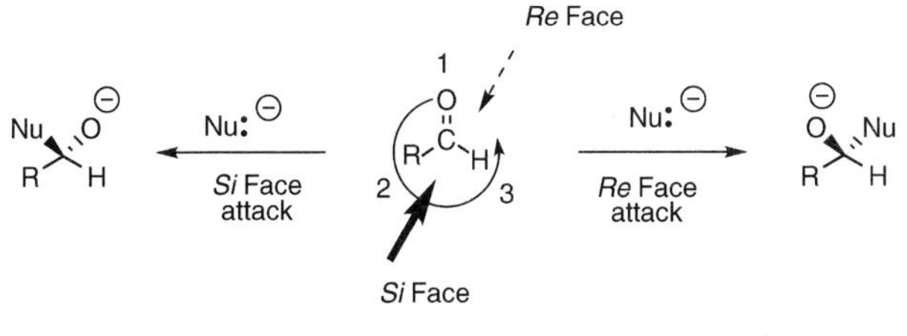

Scheme 1.1

three groups attached to the carbonyl are clockwise according to decreasing priority according to the Cahn–Ingold–Prelog system) or Si (if the three groups are anticlockwise).[4,5] The facial selectivity or topicity of the reaction must be controlled in order to produce the desired asymmetry at the newly formed stereogenic centre.

In the reactions of enolates and crotylmetal reagents, two new asymmetric centres can be formed simultaneously. One can now consider the topicity of

Table 1.1 Carbonyl additions

Simple carbon nucleophiles

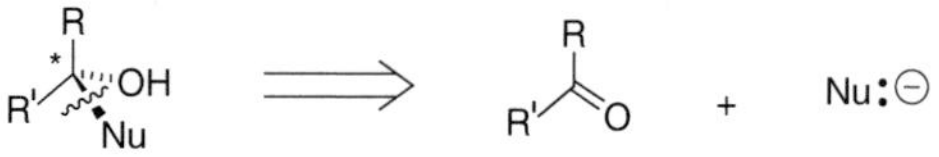

Enolates

  akylations

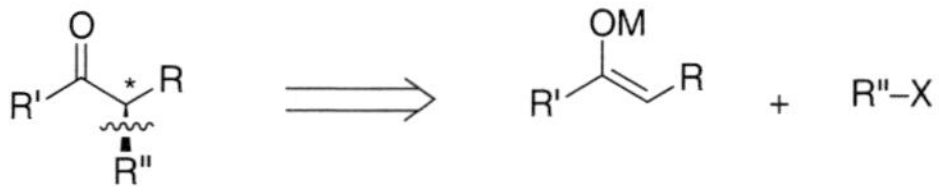

  aldol reactions

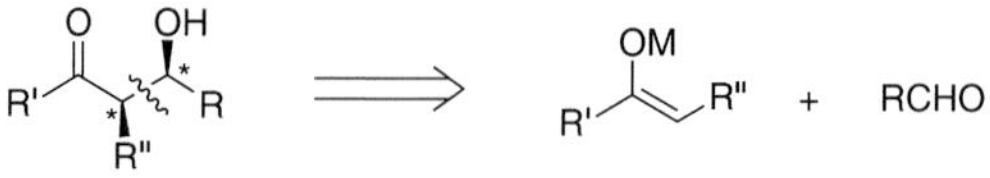

Crotylmetal and allyl nucleophiles

  crotylmetallation

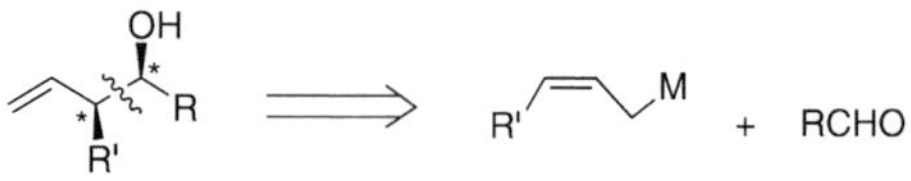

  allylmetallation

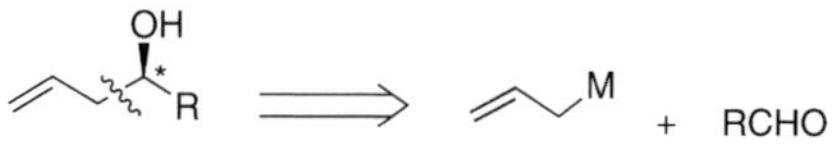

Conjugate additions

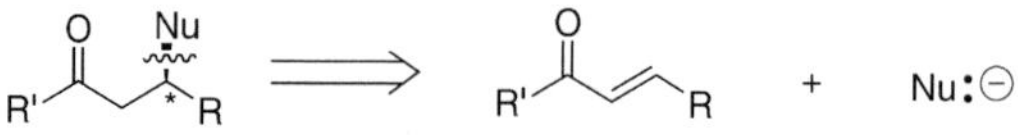

*Ene reactions (pericyclic reactions)

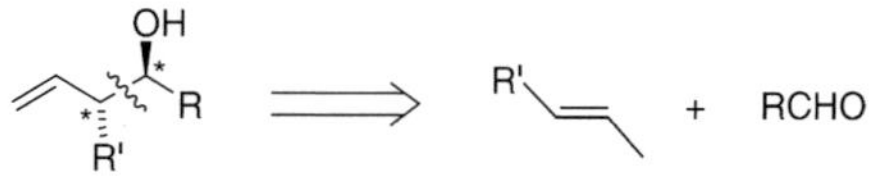

Table 1.2 Pericyclic reactions

[4 + 2] Diels–Alder reactions

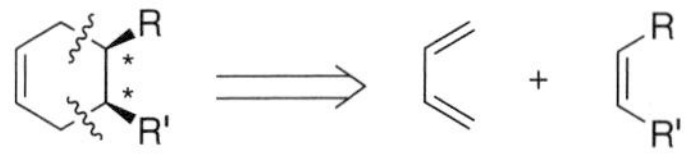

[3 + 2] Dipolar cycloadditions

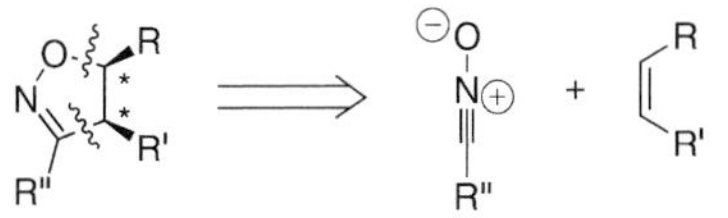

[3,3] Rearrangements

  Cope rearrangements

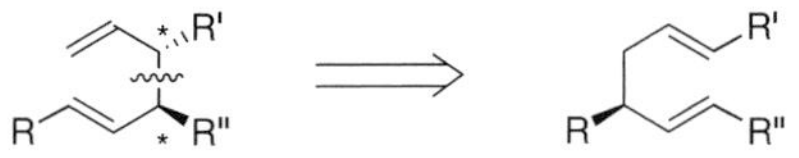

  Claisen rearrangements

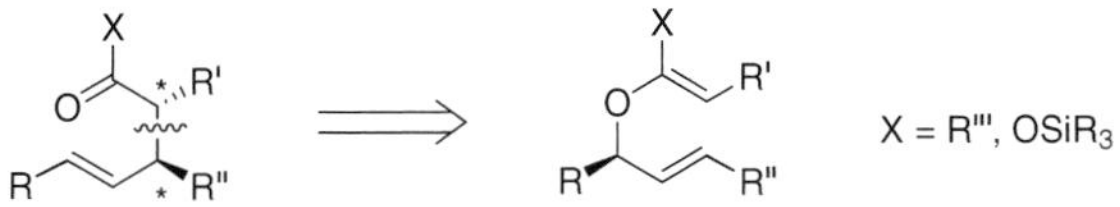

X = R''', OSiR$_3$

2,3-Wittig rearrangements

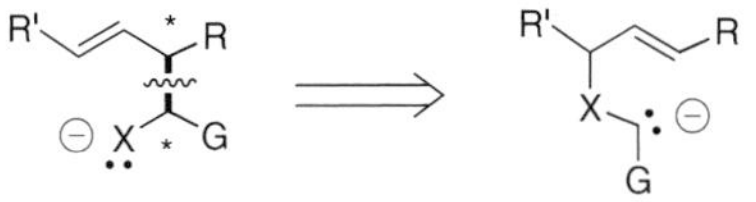

the reaction of the enolate and designate *Re* and *Si* faces of the sp^2 hybridised enolate carbon (R = alkyl group) by the orientation of the three substituents, as is shown in Scheme 1.2 for the alkylation of an ester enolate (Chapter 3).

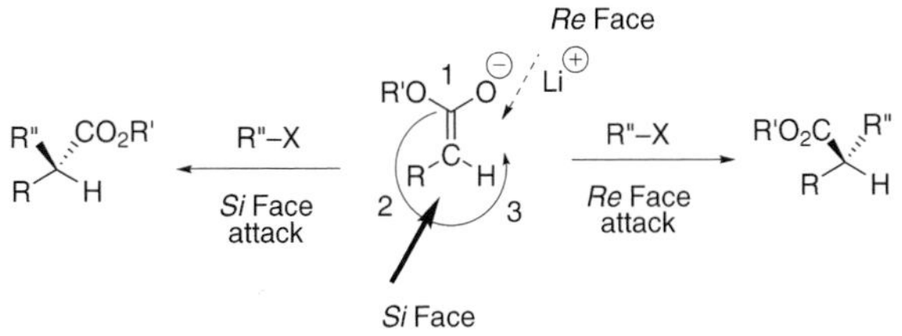

Scheme 1.2

In the case of a propionate type aldol reaction (see Chapter 4) the topicity of the enolate and aldehyde must both be considered and there are several stereochemical outcomes, an enantiomeric *syn* pair (OH and Me on the same side of the 'zigzag' carbon chain) and an *anti* pair (OH and Me on opposite sides) (Figure 1.1).[5,6] This type of nomenclature will be used throughout this book

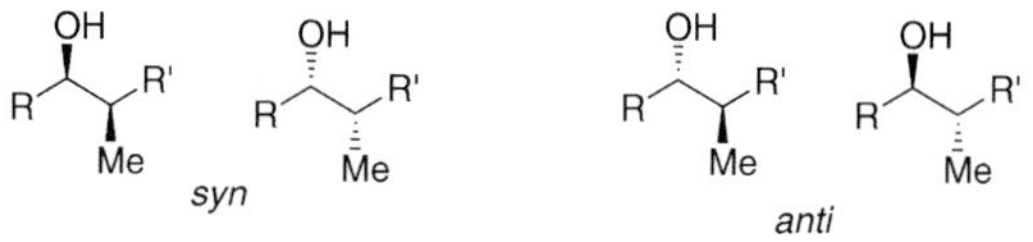

Figure 1.1

in preference to the older *threo* and *erythro* designation of the stereochemical relationship between two asymmetric adjacent centres. Asymmetric crotyl-metallation (Chapter 6) and ene type reactions (Chapter 7) can also provide either *syn* or *anti* type products with stereocontrol similar to that for aldol reactions. Conjugate additions to α,β-unsaturated carbonyl compounds (Chapter 8) is the other method of asymmetric carbon–carbon bond formation presented where both simple and enolate type nucleophiles can be utilised.

In some pericyclic reactions (Table 1.2), formation of two σ-bonds results in the introduction of several stereocentres simultaneously. Stereocontrol in these reactions can be high since they are concerted processes that proceed via highly ordered transition states (Chapter 7). Cycloaddition reactions can often result in the formation of more than two new asymmetric centres while rearrangements involve the formation of a new σ-bond and the 'transfer' of chirality from one stereogenic centre to another.

1.2   Strategies

There are a number of stoichiometric strategies that can be utilised to generate new asymmetric centres in molecules and these can be divided into three main types: (1) substrate control, (2) auxiliary control and (3) reagent control.[7] These strategies are explained in detail below and will be the main ones covered in this volume. A representative substrate-controlled asymmetric reaction is shown in Scheme 1.3. Otherwise known as a 'first generation method',[7] this approach involves the use of an achiral substrate or portion of a molecule (S) with a chiral group or asymmetric centre (*A*) covalently attached nearby. The chiral influence is close enough so that the subsequent reaction with an achiral reagent (R) is effectively controlled by *A* which induces asymmetry in the substrate portion (S) of the molecule to provide the product (*P*) with the new stereogenic

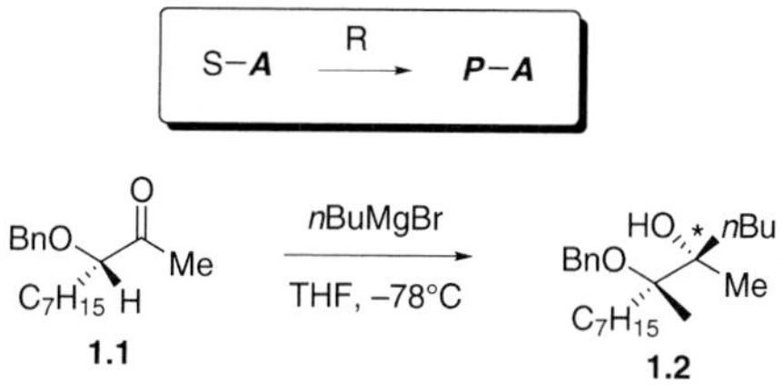

Scheme 1.3

centre(s) present. The chiral group *A* is retained in this process and, in most cases, throughout the synthesis. In the example shown in Scheme 1.3, the chiral ketone **1.1** is converted into the alcohol **1.2** by a simple nucleophilic addition to a carbonyl group (Chapter 2). In this reaction, the stereogenic centre marked is introduced with good control.

The second method, auxiliary control, is shown in Scheme 1.4. In this case a temporary chiral influence or 'chiral auxiliary' (*Aux*) is covalently bound to the

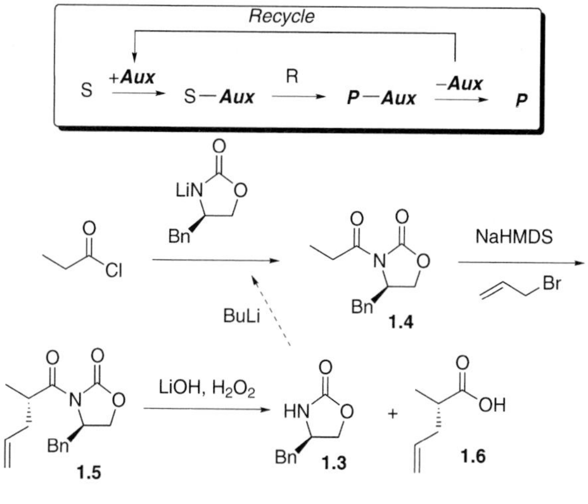

Scheme 1.4

substrate (S), normally by a weaker C—O or C—N bond which can be cleaved in a later step. The substrate–auxiliary (S—*Aux*) molecule is then treated with an achiral reagent, and an asymmetric reaction occurs, controlled by the auxiliary, to give the product (*P*) which now contains asymmetry and is still attached. The chiral auxiliary (*Aux*) is then removed in a subsequent step to give the desired product and the auxiliary, which can be recycled by attachment to more of the substrate. In the example shown in Scheme 1.4, the chiral auxiliary is the oxazolidinone **1.3** derived from an amino acid. The achiral substrate is attached to give the substrate–auxiliary intermediate **1.4** which undergoes an asymmetric alkylation to provide product **1.5**; the auxiliary is then removed by hydrolysis to give the acid product **1.6** and neutral auxiliary **1.3** ready to recycle.

The final type of method is reagent control where a chiral reagent (**R**) is allowed to react with an achiral substrate (S) to produce a chiral product (**P**) (Scheme 1.5). In this method, the chirality of the reagent is transferred to the

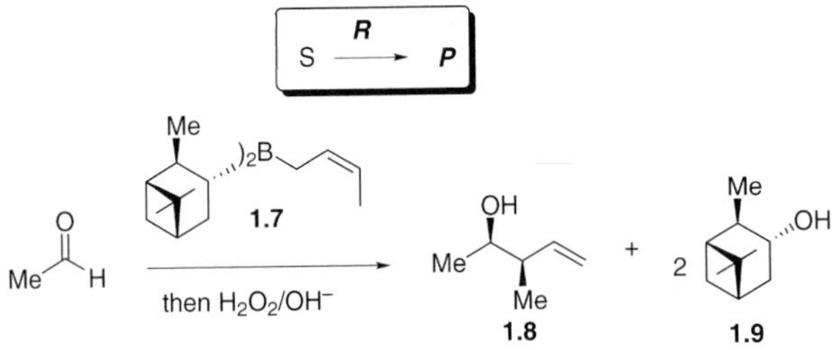

Scheme 1.5

substrate. The chiral crotyl metal reagent **1.7** (Chapter 7) attacks acetaldehyde exclusively from the *Si* face while the double bond geometry dictates the relative stereochemical outcome (Scheme 1.5). This results in the production of the *syn*-propionate product **1.8** as well as two molar equivalents of alcohol **1.9** derived from the chiral auxiliary part of the reagent.

1.3 Chiral sources

The chiral sources from which auxiliaries and reagents can be synthesised for use in asymmetric synthesis can be divided into a number of main natural product groups (Table 1.3). Terpenes are inexpensive and readily available chiral precursors which have been used extensively in both auxiliary-based and reagent-based methods. Amino and hydroxy acids are useful as chiral starting materials for the production of heterocyclic chiral auxiliaries. Amino acids, in particular, have a wide range of hydrocarbon side-chains which are effective bulky substituents. Carbohydrates are also useful chiral materials and are relatively inexpensive. However, the large number of similar hydroxy groups and stereogenic centres in these compounds makes their use as chiral auxiliaries somewhat limited. Alkaloids and amines are effective chiral sources and can also be used for the classical resolution of racemic mixtures to provide non-natural optically pure compounds.

Table 1.3 Natural product groups

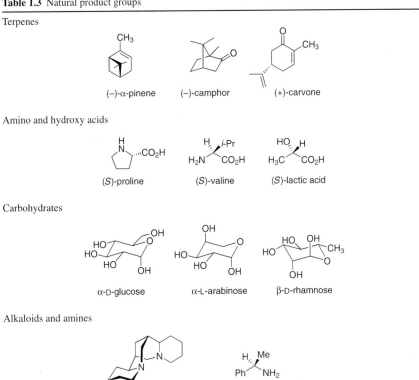

Terpenes

(−)-α-pinene   (−)-camphor   (+)-carvone

Amino and hydroxy acids

(S)-proline   (S)-valine   (S)-lactic acid

Carbohydrates

α-D-glucose   α-L-arabinose   β-D-rhamnose

Alkaloids and amines

(−)-sparteine   (R)-1-phenylethylamine

1.4 Strategies in the total synthesis of natural products

The strategies outlined in section 1.2 have been extensively developed, as will be shown in this book, and can be applied to the total synthesis of complex chiral natural products. The first example is the total synthesis of zaragozic acid C (**1.10**) by Evans *et al.* (Scheme 1.6).[8] Several key steps in this synthesis involved the use of stoichiometric methods to generate asymmetry. A chiral auxiliary controlled aldol reaction (Chapter 4) between the oxazolidinone **1.11**

1.10

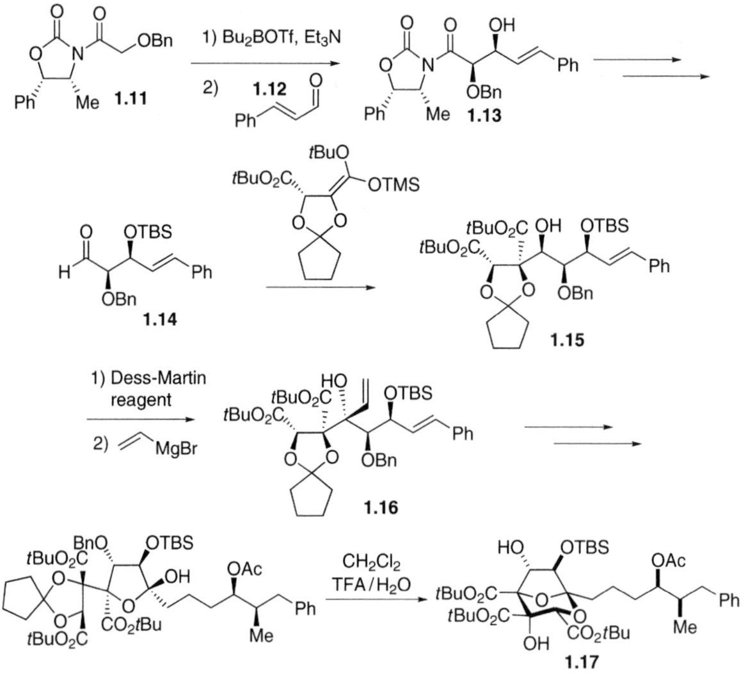

Scheme 1.6

and aldehyde **1.12** generated the alcohol **1.13** which was converted into the aldehyde **1.14**. A substrate and reagent controlled aldol reaction then generated the diester **1.15** which, upon oxidation and subsequent nucleophilic addition to the resultant ketone (Chapter 2), provided the highly functionalised alkene **1.16**. This was eventually carried on to the key zaragozic acid intermediate **1.17** which possesses the bicyclic ring system.

In the total synthesis of sanglifehrin A (**1.18**), reported by Nicolaou *et al.*, a number of stoichiometric methods were utilised to synthesise the spirolactam

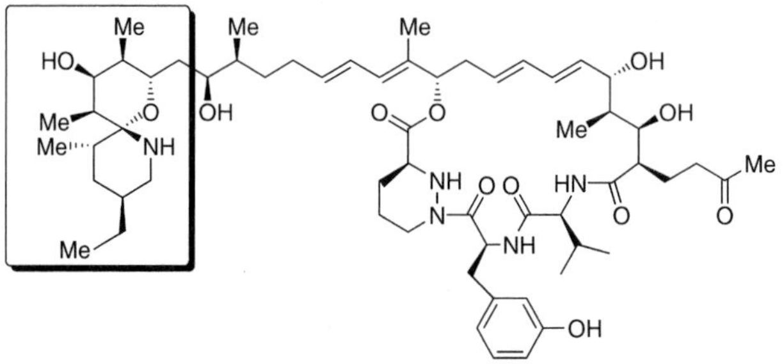

1.18

fragment (the boxed region).[9] A reagent controlled aldol reaction (Chapter 5) between the chiral enolate generated from 3-pentanone with methacrolein gave the *syn* product **1.19** which was enolised and condensed with aldehyde **1.20** (Scheme 1.7). Subsequent substrate controlled asymmetric reduction

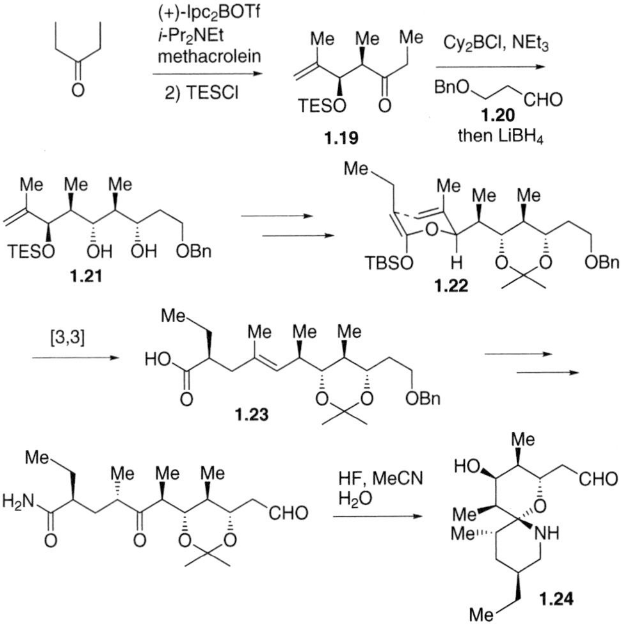

Scheme 1.7

(Chapter 2) *in situ* then generated diol **1.21** which was converted into the silylketene acetal **1.22**. Compound **1.22** underwent a stereoselective [3,3] sigma-tropic rearrangement (Chapter 7) to give acid **1.23** which was carried through to the bicyclic lactam **1.24**.

A substrate controlled aldol reaction (Chapter 4) was also utilised by Paterson *et al.*, in their total synthesis of elaiolide (**1.25**) (Scheme 1.8).[10] Treatment of the boron enolate derived from chiral ketone **1.26** with the aldehyde **1.27** resulted in the formation of adduct **1.28**. Stereoselective reduction then gave a diol which

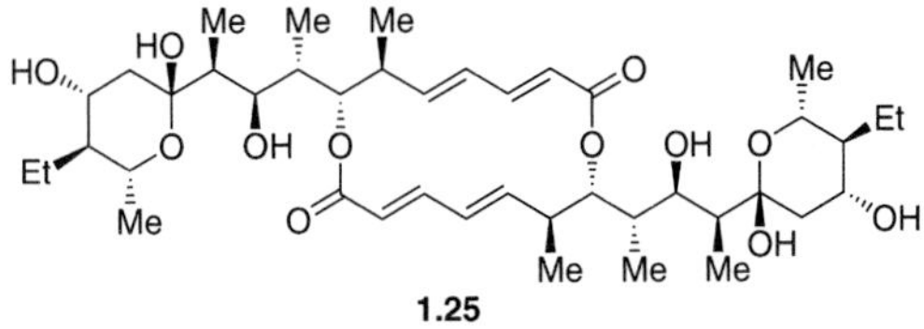

1.25

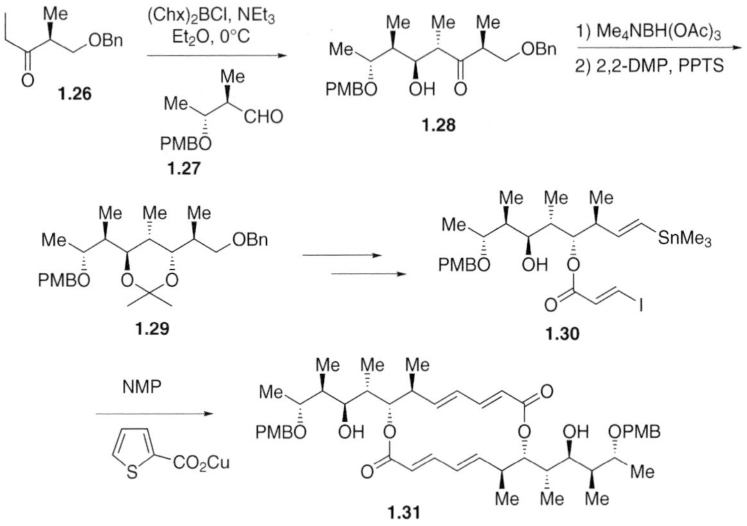

Scheme 1.8

was protected as the acetonide **1.29**. This intermediate was eventually converted into the vinylstannane **1.30** which underwent a novel copper-promoted Stille cyclodimerisation reaction to yield the macrodiolide **1.31**.

1.5 Selectivity

The selectivity of a stoichiometric asymmetric reaction can be expressed in a number of ways depending on the strategy used.[2,7] If one new asymmetric centre is created using either a substrate or an auxiliary controlled method, two new diastereoisomers are possible and the selectivity can be expressed as a ratio or as a percentage excess of one diastereoisomer (the *diastereoisomeric excess*, or percentage d.e.). A ratio can simply be reported as the amount of one diastereoisomer obtained as a ratio of the other. For example, a reaction that provides a ratio of diastereoisomers of 90:10 or 9:1 is one in which 9 times the amount of one diastereoisomer is formed over the other, and the d.e. of such a reaction is 80% (or 90%–10%). In the example shown in Scheme 1.9,

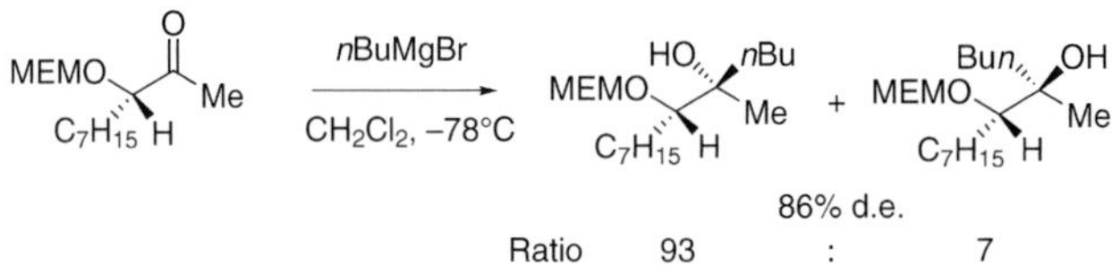

Scheme 1.9

the reaction proceeds to give the isomers shown in a 93:7 ratio which can be expressed as a d.e. of 86%.

For a reagent controlled method the selectivity can be expressed as either a ratio, percentage d.e. (if a chiral substrate is converted into one major diastereo-isomer) or percentage excess of one enantiomer (*enantiomeric excess*, or per-centage e.e.) if an achiral substrate is converted into a chiral product in which one (or more) asymmetric centre(s) has (have) been introduced. An asymmetric reaction between an achiral substrate and chiral reagent which provides a ratio of enantiomers of 90:10 or 9:1, where 9 times the amount of one enantiomer is formed over the other, proceeds with an e.e. of 80% (90% − 10%). Scheme 1.10

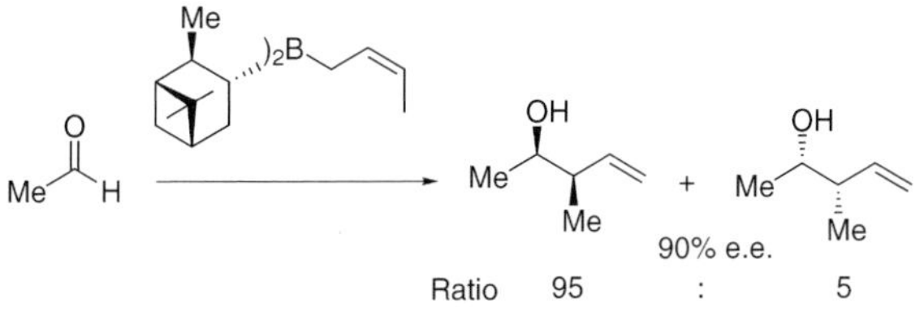

Scheme 1.10

shows an example where the two possible isomeric outcomes are enantiomers. The ratio obtained (95:5) can be represented as an e.e. of 90%.

If two new asymmetric centres are created by using either a substrate or auxiliary controlled method, such as in the aldol reaction of an α-substituted enolate, up to *four* new diastereomeric products are possible. In these cases the idea of an excess of one diastereoisomer is less well defined and it is clearer simply to report the diastereoselectivity (percentage d.s.; that is, the percentage of the major isomer). That means if a reaction produced four isomeric products in the ratio 88:8:3:1, then the selectivity could be reported as 88% d.s.

Throughout this book, the stereochemical outcomes of reactions will be expressed either as ratios or as percentage d.e., d.s. or e.e.

References

1. Nicolaou, K.C. and Sorensen, E.J., *Classics in Total Synthesis*, VCH, Weinheim, **1996**.
2. Ager, D.J. and East, M.B., *Asymmetric Synthetic Methodology*, CRC, Boca Raton, FL, **1996**.
3. Crosby, J., *Tetrahedron*, **1991**, *47*, 4789.
4. Hanson, K.R., *J. Am. Chem. Soc.*, **1966**, *88*, 2731.
5. Eliel, E.L., Wilen, S.H. and Mander, L.N., *Stereochemistry of Organic Compounds*, John Wiley, New York, **1994**.
6. Masamune, S., Ali, A., Snitman, D.L. and Garvey, D.S., *Angew. Chem., Int. Ed. Engl.*, **1980**, *19*, 557.

7. Aitken, R.A. and Kilényi, S.N., *Asymmetric Synthesis*, Chapman & Hall, London, **1994**.
8. Evans, D.A., Barrow, J.C., Leighton, J.L., Robichaud, A.J. and Sefkow, M.J., *J. Am. Chem. Soc.*, **1994**, *116*, 12 111.
9. Nicolaou, K.C., Xu, J., Murphy, F., Barluenga, S., Baudoin, O., Wei, H.-X., Gray, D.L.F. and Ohshima, T., *Angew. Chem., Int. Ed. Engl.*, **1999**, *38*, 2447.
10. Paterson, I., Lombart, H.-G. and Allerton, C., *Org. Lett.*, **1999**, *1*, 19.

2   Additions to carbonyl compounds

2.1   Nucleophilic addition to chiral carbonyl compounds

2.1.1   *Background*

The addition of nucleophiles to acyclic chiral carbonyl compounds is a widely used method for the stereoselective synthesis of alcohols (Scheme 2.1).[1,2]

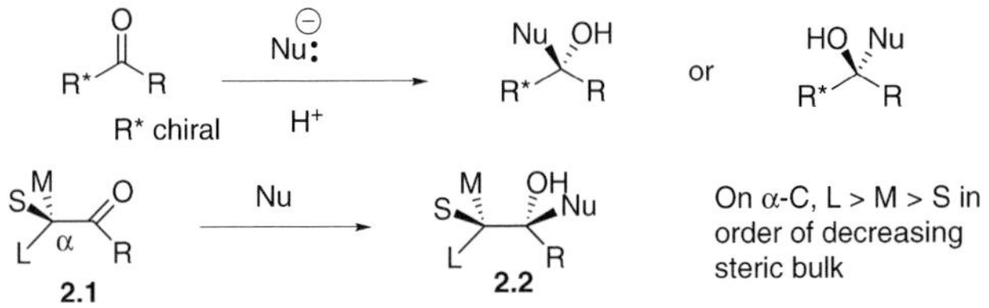

Scheme 2.1

Relative asymmetric control has been studied in detail and an early model for the prediction of the stereochemical outcome of additions to acyclic α-chiral aldehydes and ketones based on a large amount of experimental data was proposed by Cram and Elhafz in 1952.[3] In this early paper, it was suggested (emphasis in original) that

> 'In non-catalytic reactions of the type shown [Scheme 2.1], that dia-stereomer will predominate which would be formed by the approach of the entering group from the *least hindered side* of the double bond when the rotational conformation of the C—C bond is such that the double bond is flanked by the two least bulky groups attached to the adjacent asymmetric centre'.[3]

This can be represented by the approach of the nucleophile to the carbonyl group on the same side as the S (small) group in the conformer **2.1** to give the alcohol **2.2** as the major product.

Another view of this model is shown in Scheme 2.2 where the nucleophile approaches the carbonyl from the same side as the S group in the Newman projection of conformer **2.1**. This provides an alkoxide which on protonation gives the product **2.2** shown in the 'zigzag' format in Scheme 2.2. Alternatively, a two-conformer model was also proposed by Cram and Kopecky where stereodifferentiation results from the different *gauche* interactions in conformers

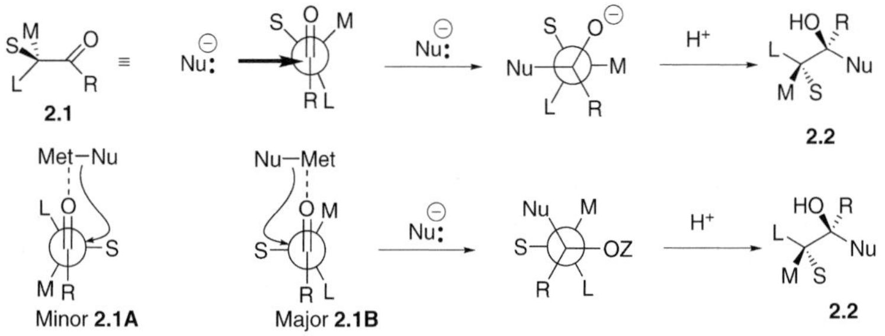

Scheme 2.2

2.1A and **2.1B** (Scheme 2.2).[4] In this case, the metal atom Met chelates to the carbonyl oxygen atom, effectively making this the bulkiest group, and this would tend to orientate itself between the two least bulky groups (S and M) on the adjacent carbon atom. Attack would then occur preferentially from the same side as the S group.

In 1959, Cornforth *et al.* proposed a polar model for nucleophilic addition to α-chlorocarbonyl compounds based on the Cram postulate.[5] In this model, the polarised C—Cl bond prefers to be *anti* to the C=O bond in the transition state because of favourable dipole–dipole interactions which increase the C=O polarisation and therefore the reactivity (Scheme 2.3). Addition then

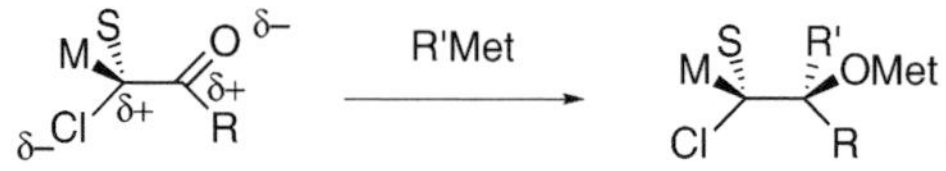

Scheme 2.3

occurs from the least hindered face of the carbonyl group in a similar manner to that described for the Cram model.

Karabatsos later suggested that the rule proposed by Cram had several shortcomings.[6] For example, it was noted that the diastereomeric ratio of products resulting from nucleophilic additions to simple α-chiral aldehydes decreases with R varying from methyl to *iso*-propyl (Scheme 2.4). If the Cram model

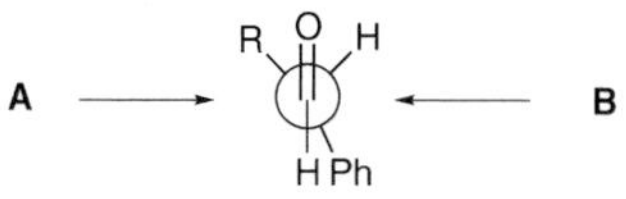

R	ratio **A:B**
Me	2.4:1
Et	2.5:1
i-Pr	1.9:1

Scheme 2.4

is applied, this would imply that an *iso*-propyl group would be effectively smaller than a methyl substituent.

A new model was then proposed which was based on the known minimum energy conformations of aldehydes and ketones where a group on the adjacent carbon atom is eclipsed by the C=O bond.[6] It was argued that bond breaking and making had not occurred to any great extent in the transition states and the arrangement of the groups on the α-carbon atom was such that the medium-sized group was eclipsed by the C=O bond in the major conformer (Scheme 2.5).

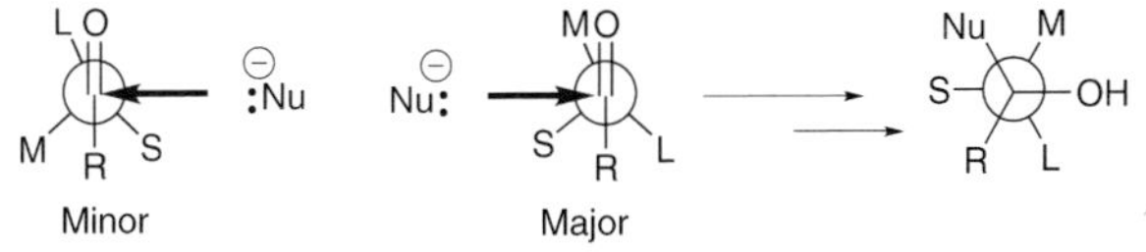

Scheme 2.5

Good correspondence was obtained between the size of the interactions in the major conformer and the product ratios obtained experimentally.

2.1.2   The Felkin–Anh model

It was eventually noted by Felkin *et al.* that the above models could not explain all experimental results.[7] A major problem with the Cram and Karabatsos proposals arose when it was observed that hydride addition to carbonyl compounds becomes more stereoselective when the R group attached to the carbonyl carbon increases in steric size (Scheme 2.6). Each of these early models would predict

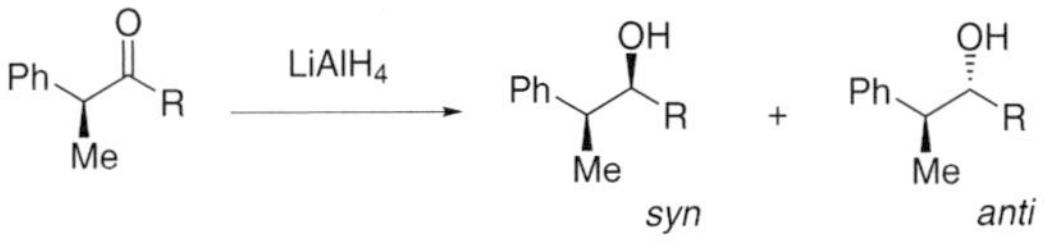

R	ratio *syn*:*anti*
Me	2.8:1
Et	3.2:1
i-Pr	5:1
t-Bu	49:1

Scheme 2.6

that the bulkier the R group the *less* stereoselective the addition would be, owing to the destabilising R↔L interactions in the transition state.

Felkin proposed that the transition states in these reactions are product-like rather than reactant-like and that torsional strain involving partial bonds was important. This assumption led to the suggestion that the important steric interactions involve the incoming nucleophile and R group on the carbonyl carbon rather than the carbonyl oxygen. It was also proposed that polar effects are important and these stabilise the transition states where the separation between the incoming nucleophile and an electronegative group is the greatest. Based on these premises, a model was proposed where the incoming nucleophile attacks the carbonyl group antiperiplanar to the L group where the L ligand is perpendicular to the C=O bond (Scheme 2.7).[7] From the two choices where

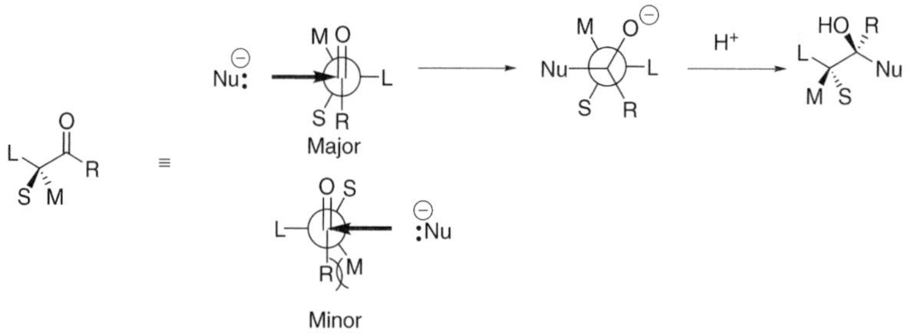

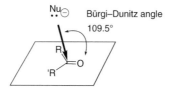

Scheme 2.7

the L group is perpendicular to the C=O bond, the one with the gauche M↔R interaction is destabilised leading to the preference for the other conformation.

Further support for the Felkin model came from calculations conducted by Anh and Eisenstein where the Felkin conformers shown in Scheme 2.7 were much lower in energy than the Cram or Karabatsos conformers.[8] More importantly, Anh and Eisenstein proposed that the nucleophile approaches the carbonyl at the Bürgi–Duniz angle of 109.5° to the C=O plane (Figure 2.1),

Nu⊖
·· Bürgi–Dunitz angle
109.5°
R
'R ═O

Figure 2.1

and this modification explains the stereoselectivities observed without considering the M↔R or O↔R interactions.[9,10] The preferred conformation would therefore be **2.1C** where the attacking nucleophile and the S ligand are interacting, as shown in Scheme 2.8. The so-called Felkin–Anh model is now the accepted method for the prediction of stereoselectivity in nucleophilic additions to acyclic α-chiral aldehydes and ketones.

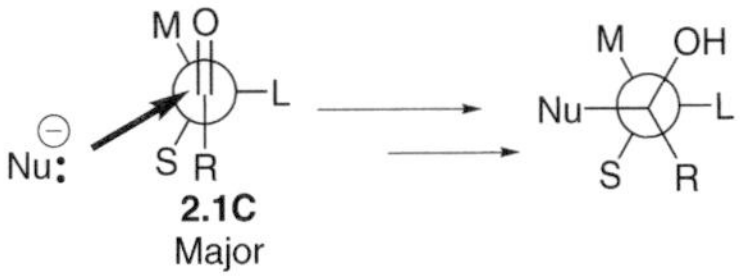

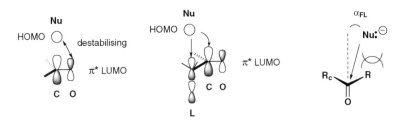

Scheme 2.8

The Bürgi–Dunitz mode of attack can be explained by considering frontier orbital theory where the destabilising interaction between the highest occupied molecular orbital (HOMO) of the nucleophile and the lowest unoccupied molecular orbital (LUMO) of the carbonyl group is minimised by increasing the angle of approach (Figure 2.2). Futhermore, there is a stabilising n-σ^*

Figure 2.2

interaction that develops between the nucleophile and the σ^* orbital of the adjacent L—C bond. The increase in stereoselectivity observed as the R group attached to the carbonyl carbon increases in size (Scheme 2.6) can be explained by considering the deviation of the approach of the nucleophile from the plane that bisects the carbonyl group along the C=O bond. The nucleophile will approach along the Bürgi–Dunitz angle and, as the size of R increases compared with R_c, the angle α_{FL} (the Flippin–Lodge angle)[11,12] decreases bringing the incoming nucleophile closer to the R_c substituent thereby increasing the effect of the ligands attached to the α-carbon (Figure 2.2). The combination of the angle of attack of the nucleophile and the orientation of the ligands of the adjacent asymmetric centre therefore explains the stereochemical outcome in these additions.

Anh and Eisenstein also proposed that the ligand with the lowest lying σ^* orbital was the one that would be orientated perpendicular to the C=O bond.[8] Lodge and Heathcock examined this proposal and found that steric effects and σ^* orbital energies were both important.[12] Based on experimental results obtained from additions of lithium enolates to various acyclic α-chiral aldehydes, a four-conformer model was suggested (Figure 2.3) for an α-methoxy aldehyde where the C—O bond has the lowest lying σ^* orbital followed by an sp^2—sp^3

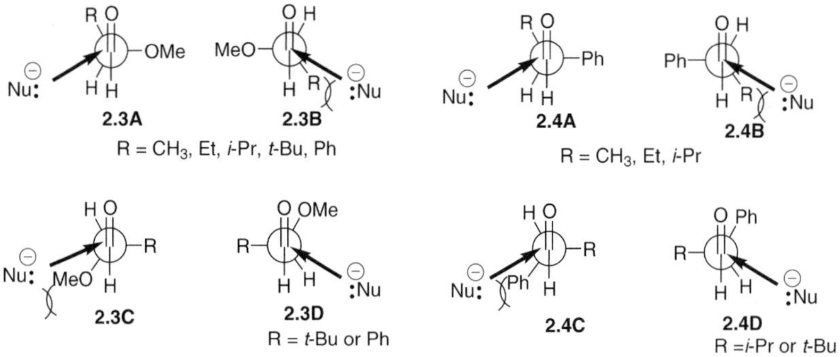

Figure 2.3

bond, with an sp^3—sp^3 bond having the highest σ^* orbital energy of the three. In the case of α-methoxy aldehydes, the methoxy group behaves as the L group when pitted against a methyl, ethyl or *iso*-propyl group (conformers **2.3A** and **2.3B**). However, the bulk of a *tert*-butyl group becomes important, making the Anh–Eisenstein postulate only partially correct, and the other conformers **2.3C** and **2.3D** need to be considered. Furthermore, a phenyl group behaves somewhat like the methoxy group in that its σ^* orbital is lower than that for an sp^3—sp^3 bond while its steric bulk is between that of an *iso*-propyl and *tert*-butyl group, leading again to some reaction via the alternative conformers **2.3C** and **2.3D**.

For α-phenyl aldehydes, when R is small (methyl or ethyl), conformers **2.4A** and **2.4B** are preferred because of the greater steric bulk of the phenyl group and the lower lying energy of the sp^2—sp^3 bond (Figure 2.3). When the R ligand is an *iso*-propyl or *tert*-butyl group, conformers **2.4C** and **2.4D** come into play. The effect of the *tert*-butyl substituent is the greatest and a preference for conformer **2.4D** is suggested in this case. These observations led to the proposal that the Felkin–Anh model can be used for a qualitative prediction of the stereochemical outcome of nucleophilic additions to acyclic α-chiral aldehydes with the order of preference for the L group being: MeO > *t*-Bu > Ph > *i*-Pr > Et > Me > H.[12]

In general, 1,2-asymmetric induction by the Felkin–Anh mode of attack (sometimes called Cram attack) proceeds with modest stereoselectivity. However, some modifications have allowed for higher selectivities. For example, addition of organometallics to aldehydes in the presence of crown ethers has resulted in some excellent results (Scheme 2.9).[13] It was proposed that the complexation of metal ions by the crown ethers reduces the complexation of the metal to the carbonyl oxygen of the aldehyde thus leading to increased Felkin–Anh selectivity.

Felkin–Anh selectivity can also be increased by having a temporary large R group, such as a trimethylsilyl (TMS) group, attached to the carbonyl carbon (Scheme 2.10).[14] As cited above, this increases the selectivity in Felkin attack

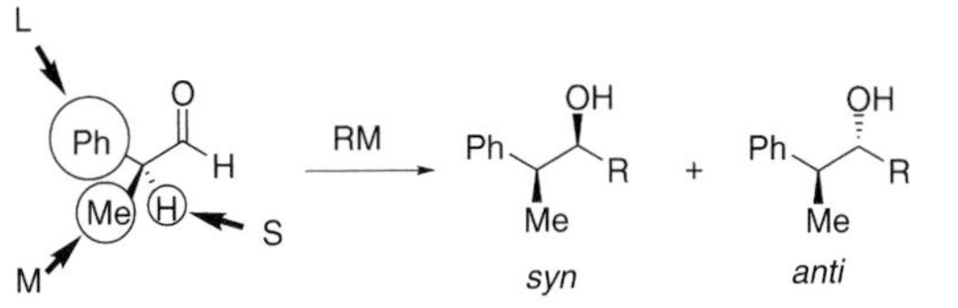

RM	ratio *syn:anti*
BuLi	5:1
MeLi	4:1
EtMgBr	4:1
Et$_2$Mg	8:1
BuLi/15-C-5	> 30:1

Scheme 2.9

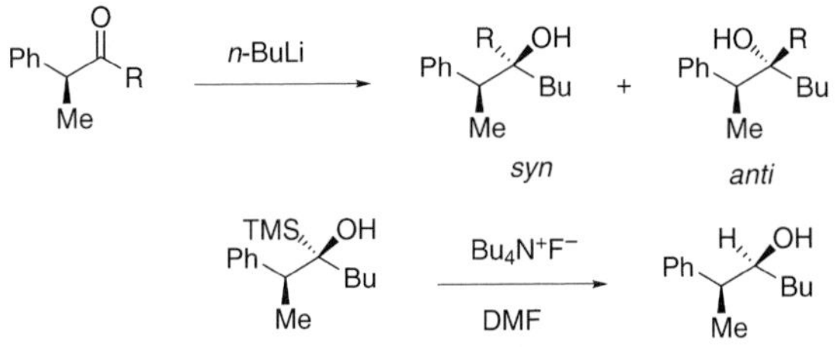

R	ratio *syn:anti*
H	5:1
TMS	100:1

Scheme 2.10

by reducing the Flippin–Lodge angle and increasing the M↔R interaction. The TMS group is then removed by treatment with fluoride to give the desired product.

2.1.3 The Cram chelate model

Cram also proposed another model for 1,2-asymmetric induction in carbonyl compounds where the asymmetric carbon atom bears a group such as OR or NR$_2$ which is capable of chelating to a metal atom.[3,4] In this model, the adjacent chelating group (X) and the carbonyl oxygen each complex to the metal atom to form a tight five-membered ring, and the nucleophile then approaches from the less-hindered side (Scheme 2.11). The so-called Cram chelate model often leads

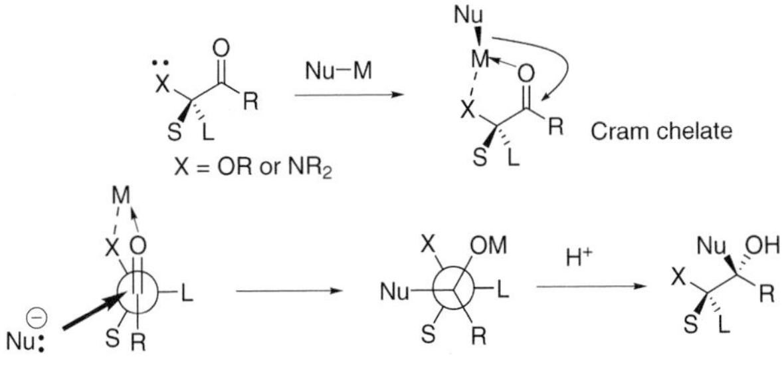

Scheme 2.11

to high stereoselectivity as a result of the rigid cyclic complex that is formed prior to addition of the nucleophile.

In 1980, Still and co-workers studied the effects of solvent and oxygen protecting group on the stereoselectivity of additions to α- and β-alkoxyketones (Scheme 2.12).[15,16] It was found that magnesium was more effective than

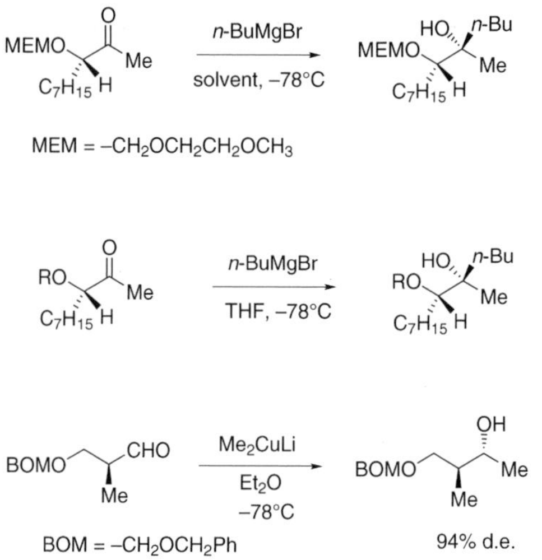

Solvent	d.e.
Pentane	80
CH_2Cl_2	86
Et_2O	80
THF	> 98

MEM = $-CH_2OCH_2CH_2OCH_3$

R	d.e.
MEM	> 98
MOM	> 98
Bn	> 98
BOM	> 98
THP	50

BOM = $-CH_2OCH_2Ph$

94% d.e.

Scheme 2.12

lithium as the metal and that tetrahydrofuran (THF) was the best solvent. Groups such as methoxymethyl (MOM), methoxyethoxymethyl (MEM), benzyl (Bn) and benzyloxymethyl (BOM) can be placed on the oxygen with no effect on the selectivity; however, a tetrahydropyranyl (THP) ether gave poor selectivity and this was attributed to steric hindrance in one of the two diastereomeric THP ethers. In the case of addition of simple nucleophiles to β-alkoxyaldehydes, lithium dialkylcuprates proved most effective in providing excellent selectivities.[16]

Another example of high diastereoselectivity via chelation controlled addition was reported by Eliel et al. (Scheme 2.13).[17] The 2-acyl-1,3-oxathiane 2.5 was synthesised from the corresponding 1,3-oxathiane by anion formation, and condensation with benzaldehyde gave the equatorially substituted product which upon oxidation provided 2-acyl derivative 2.5. Addition of methylmagnesium iodide to the 2-acyl-1,3-oxathiane 2.5 gave the alcohol 2.6 as the only detectable isomer. It was proposed that Cram chelate addition had occured via the conformer shown in Scheme 2.13. Methylation and removal of the oxathiane moiety then gave the chiral aldehyde 2.7.

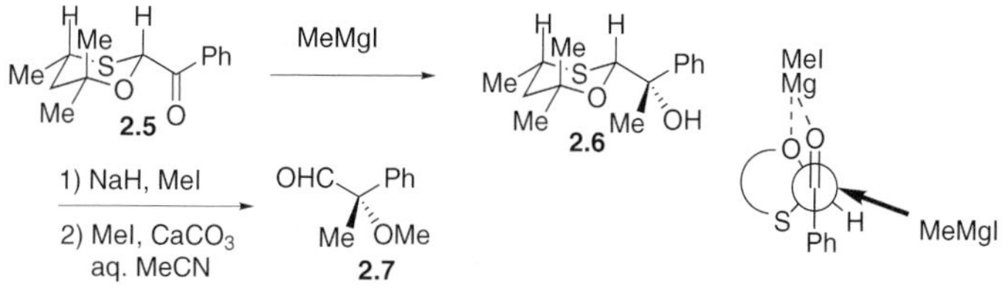

Scheme 2.13

Direct evidence for a Cram chelate intermediate was obtained by nuclear magnetic resonance (NMR) analysis of a reaction involving the ketone **2.8** and MeTiCl$_3$ which gives the diastereoisomer **2.9** as the only product (Scheme 2.14).[18] It had been observed earlier that chiral β-hydroxycarbonyl

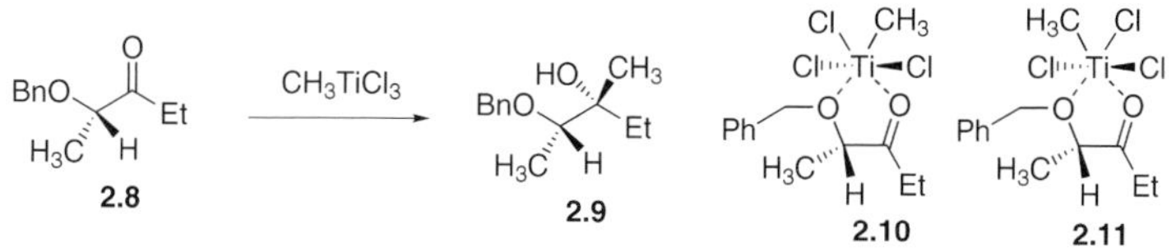

Scheme 2.14

compounds form a 1:1 complex with TiCl$_4$ which can be monitored by NMR spectroscopy.[19]

When the ^{13}C NMR spectrum was recorded immediately after addition of MeTiCl$_3$ to a solution of the ketone at low temperature, two new carbonyl signals were observed downfield to the carbonyl signal for the starting material. These signals were attributed to the two diastereoisomeric titanium complexes **2.10** and **2.11** which slowly disappeared to give the product. It was also found using deuterium labeling experiments that methyl transfer did not occur by an intramolecular process.

An early example of the importance of the oxygen protecting group in nucleophilic additions to α-alkoxy ketones is shown in Scheme 2.15.

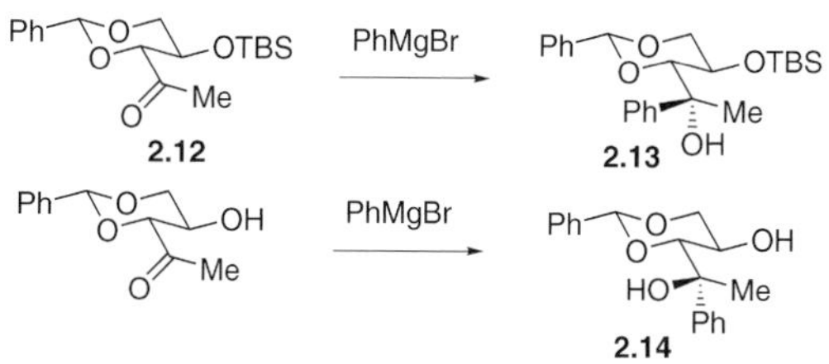

Scheme 2.15

Phenylmagnesium bromide addition to the ketone **2.12** gives alcohol **2.13** via
α-chelation since the β-oxygen is protected as a silyl ether, preventing
chelation.[20] In the other reaction shown, the β-oxygen can easily chelate giving
diastereoisomer **2.14** as the major product.

Another example of the importance of the protecting group on oxygen was
reported by Overman and McCready during synthetic studies on pumiliotoxin
B (Scheme 2.16).[21] Nucleophilic addition to the ketone **2.15** proceeded

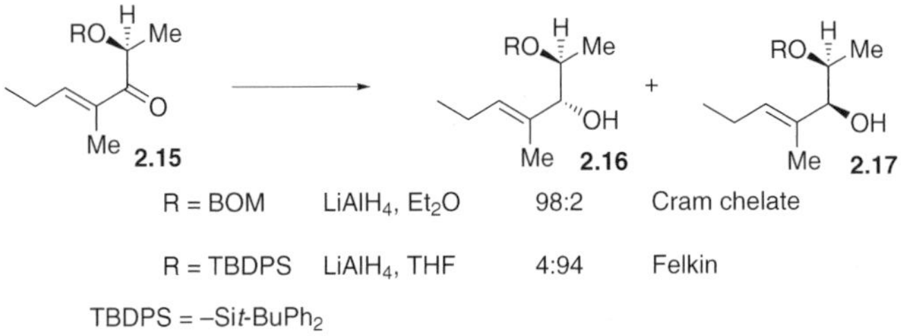

R = BOM   LiAlH₄, Et₂O   98:2   Cram chelate

R = TBDPS   LiAlH₄, THF   4:94   Felkin

TBDPS = –Si*t*-BuPh₂

Scheme 2.16

via chelation control to give the alcohol **2.16** in high d.e. However, replace-
ment of the benzyl ether with a silyl ether reversed the stereochemical outcome
of the reduction, providing the other isomer **2.17** with good stereoselectivity by
the Felkin–Anh mode of addition. The unusually high selectivity observed for
the Felkin mode of attack may arise from the destabilisation of the minor Felkin
conformer as a result of an interaction between the methyl group and β-hydrogen
of the enone, as shown in Figure 2.4.

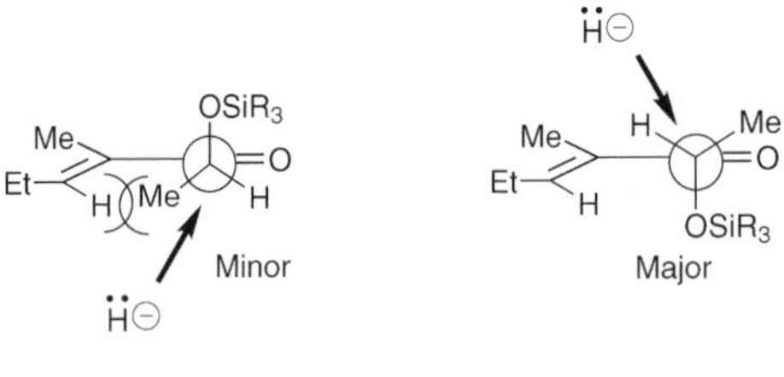

Figure 2.4

Frye and Eliel also observed that chelation to an oxygen function could be
prevented by protection with a trialkylsilyl group (Scheme 2.17).[22] Addition
of methyl Grignard reagent to the benzyl ether analogue of a 1,3-oxathiane
proceeded in low d.e. because of competing chelation between the oxygen of
the 1,3-oxathiane and the β-alkoxy group. Some improvement was observed
on changing the β-alkoxy group to a trityl ether because of an increase in steric

Scheme 2.17

R	% d.e.
Bn	17
Trityl	72
TIPS	95

bulk; however, the best results were obtained when a triisopropylsilyl (TIPS) ether was used.

Studies have revealed that the ability of trialkylsilyl groups to prevent chelation to the oxygen by Lewis acids can be attributed to electronic and steric effects. Metal–oxygen–carbon (M—O—C) bond angles in alkoxy metal complexes are mostly linear and this is because of the donation of the lone pairs on oxygen into empty p- or d-type orbitals on the metal. It was therefore suggested that a deviation from the tetrahedral bond angle in ethers would cause a reduction in the π-donor ability of oxygen. Calculations showed that the Si—O—C bond angle was larger than C—O—C, therefore resulting in a reduction in the ability of the oxygen atom to chelate.[23] In addition, direct evidence for the absence of chelation of Lewis acids with β-silyloxy aldehydes was observed by NMR spectroscopy.[24] No complexation was observed for a β-silyloxy aldehyde with $MgBr_2·OEt_2$ which was a very effective agent for forming complexes with the corresponding β-alkoxy aldehyde.

^{1}H NMR investigation of Lewis acid complexes of β-alkoxy aldehydes aided in explaining why aldehyde **2.18** reacts with certain nucleophiles, in

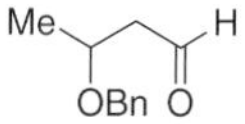

2.18

the presence of a bidentate ligand, on the face of the carbonyl opposite to the methyl group, with high selectivity.[19,25] In the case of $TiCl_4$, a tight 1:1 complex is formed and the methyl group occupies an axial position effectively blocking attack by a nucleophile from one face (Figure 2.5). It was proposed that relief

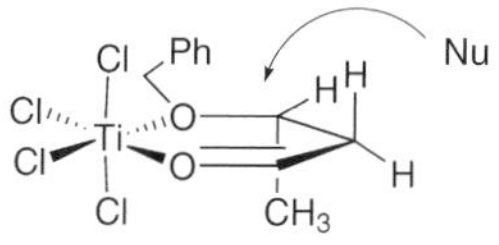

Figure 2.5

of steric strain between the methyl group and benzyl group on the oxygen was responsible for the preferred conformation of the chelate.

Additions to chiral carbonyl compounds with α- and β-nitrogen functional-ities are much less common than the corresponding oxygen counterparts, and some early examples are shown in Scheme 2.18. Addition of benzylmagnesium

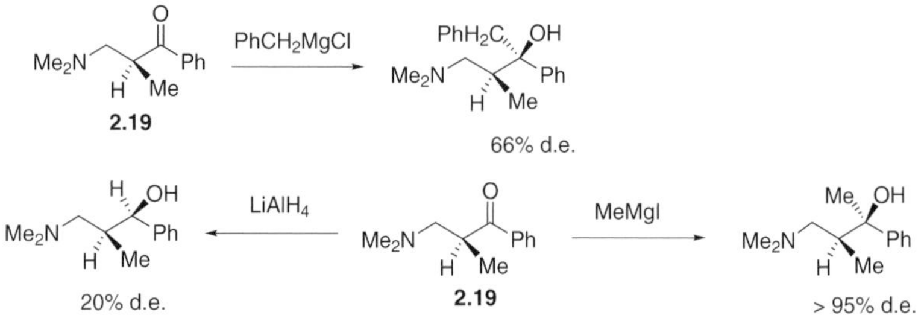

Scheme 2.18

chloride to the chiral β-aminoketone **2.19** proceeded with modest diastereo-selectivity.[26] In contrast to this, addition of methyl Grignard reagent to com-pound **2.19** gave only one detectable diastereoisomer,[27] and hydride reduction gave poor selectivity.[28] In the last case, Felkin–Anh attack may also be occurring where the dimethylaminomethyl group acts as the large substituent.

The choice of solvent and metal can often have a profound effect on stereo-selectivity, as shown in Scheme 2.19. Addition of lithium TMS acetylide to

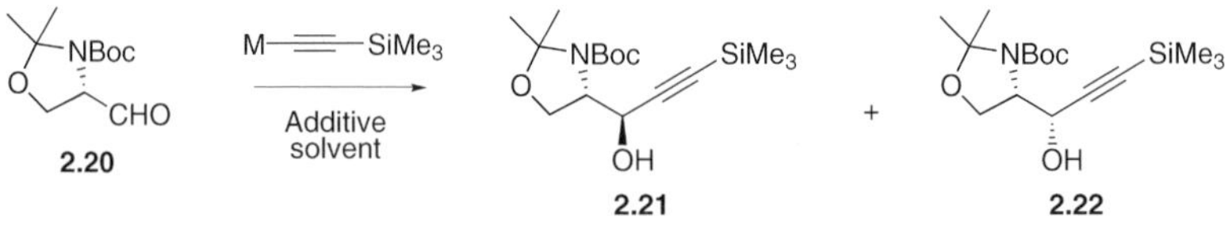

M	Solvent	Additive	2.21:2.22
Li	THF	HMPA	20:1
MgBr	THF/SMe$_2$	CuI	1:20

Scheme 2.19

aldehyde **2.20** in the presence of cation-complexing agents such as HMPA gave good selectivity for the *anti* isomer **2.21**, whereas addition of the corresponding Grignard reagent in the presence of a Lewis acid gave the *syn*-isomer **2.22** in high d.e.[29]

These results can be explained by assuming that Felkin–Anh mode of attack is preferred with lithium as the metal in the presence of coordinating solvents whereas the Cram-chelate mode of attack is preferred in the cases where the metal M is zinc or copper (Figure 2.6).

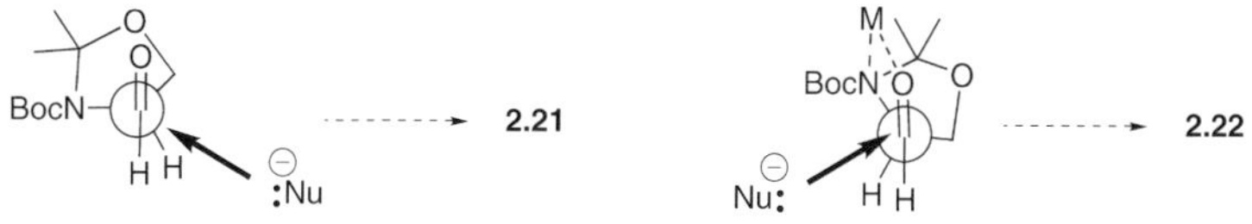

Figure 2.6

Mukaiyama[30] has reported an efficient and highly stereoselective synthesis of chiral α-hydroxyaldehydes which is analogous to that devised by Eliel *et al.*[17] (see Scheme 2.13). Treatment of the proline derivative **2.23** with methyl glyoxylate provides the *cis*-fused diazabicyclopentane **2.24** as one isomer (Scheme 2.20). Since the condensation is presumably under thermodynamic

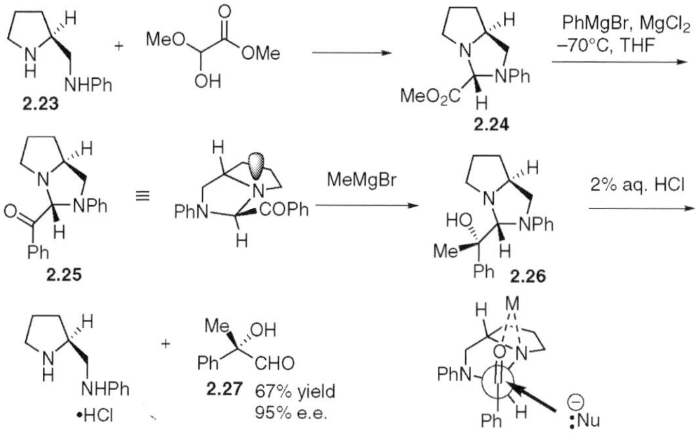

Scheme 2.20

control, the carboxymethyl group is in the *exo* position. Grignard addition then provides the ketone **2.25** which upon treatment with another Grignard reagent gives the alcohol **2.26**. Acid hydrolysis then yields the optically active α-hydroxyaldehyde **2.27** in high e.e. The stereoselective Grignard addition to ketone **2.25** proceeds by chelation control where the metal chelates preferentially to the more basic pyrrolidine nitrogen rather than the anilino nitrogen (Scheme 2.20).

2.1.4   *Synthetic applications*

An early and impressive example of the use of chelation controlled addition
to carbonyls is the total synthesis of polyether antibiotic monensin reported by
Still and co-workers in 1980.[31–33] Some of the stereochemical problems in monensin were reduced to the formation of vicinal stereocentres in a stereoselective
manner by chelation controlled additions.

The two key examples of the additions utilised in the total synthesis of
monesin are shown in Schemes 2.21 and 2.22. In the first, the C(12) stereocentre

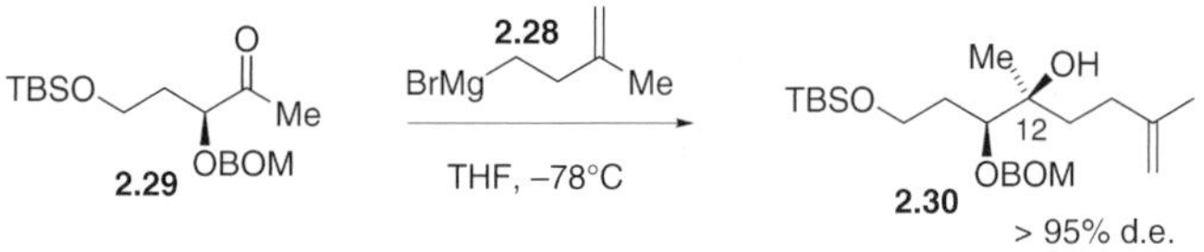

Scheme 2.21

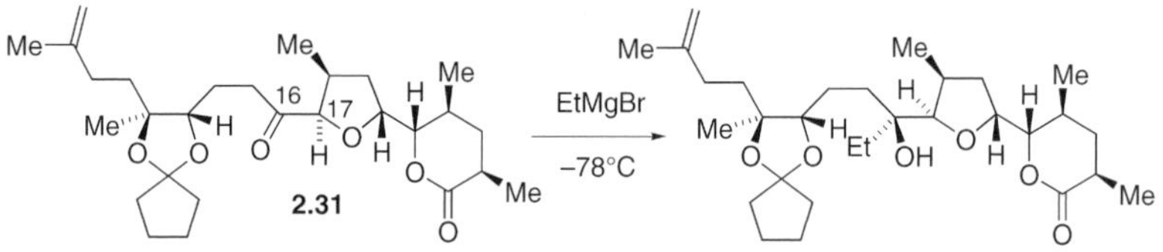

Scheme 2.22

was introduced by an addition of the Grignard reagent **2.28** to ketone **2.29** which
produced alcohol **2.30** in high d.e.[15] In this reaction, the benzyloxymethyl group
on the adjacent oxygen chelates to the magnesium and the nucleophile attacks
the ketone from the *Si* face (Figure 2.7).

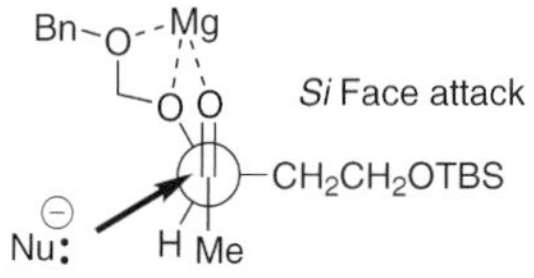

Figure 2.7

In the more complex example shown in Scheme 2.22, the C(16) stereocentre
of monensin was set by a chelation controlled addition to ketone **2.31**.[33] In this

case, the tetrahydrofuran oxygen chelates to the metal, forcing nucleophilic attack to occur from the *Re* face of the carbonyl as indicated in Figure 2.8.

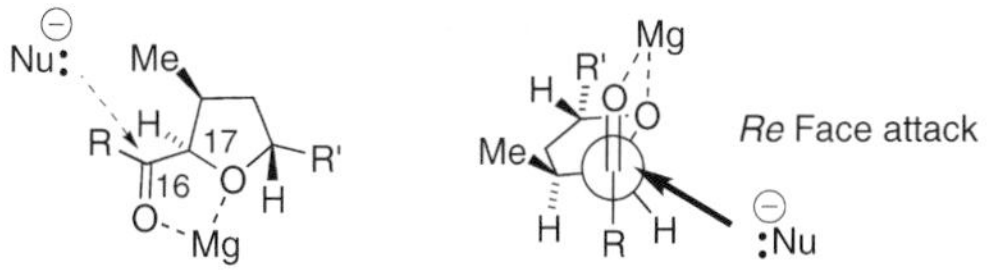

Figure 2.8

Nicolaou *et al.* also utilised chelation control in the total synthesis of racemic zoapatanol (**2.32**).[34] The stereochemistry at C(2) in **2.32** was set by a

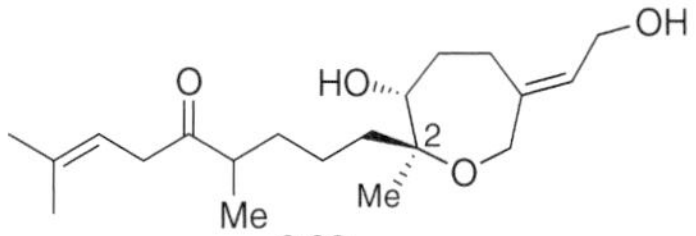

2.32

chelation controlled addition of methylmagnesium chloride to ketone **2.33** (Scheme 2.23). *Re* face attack by the methyl anion (Figure 2.9) provided the alcohol **2.34** as the only detectable isomer.

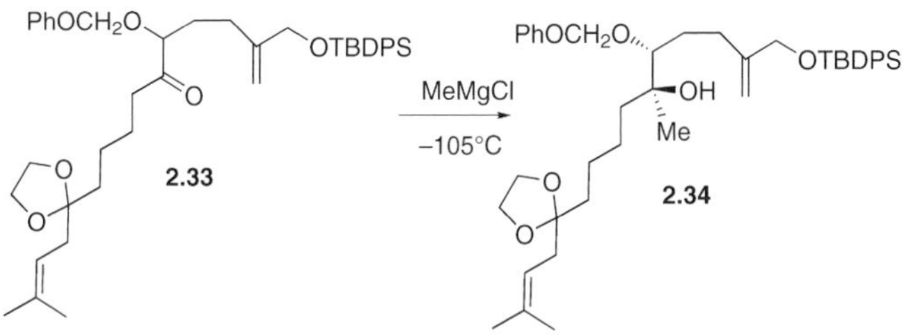

Scheme 2.23

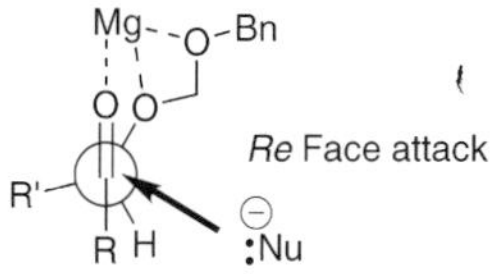

Figure 2.9

The use of chelation control for the production of polyether antibiotics was further extended to the synthesis of the bis-spiroketal-containing ionophore salinomycin by Yonemitsu and co-workers.[35–37] An early step in the synthesis of the E ring involved the addition of ethylmagnesium bromide to the ketone **2.35**. This reaction proceeded in a stereoselective manner (Figure 2.10) providing intermediate **2.36** (Scheme 2.24).[35,38]

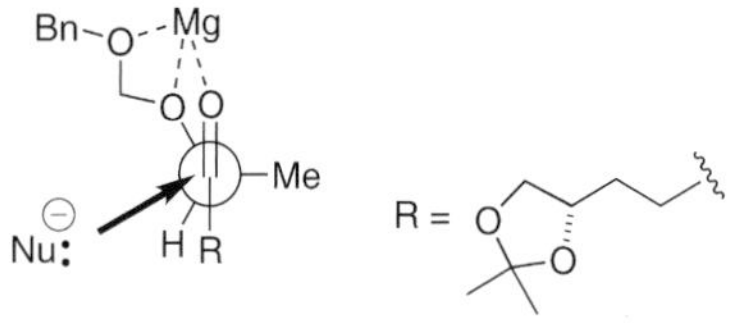

Figure 2.10

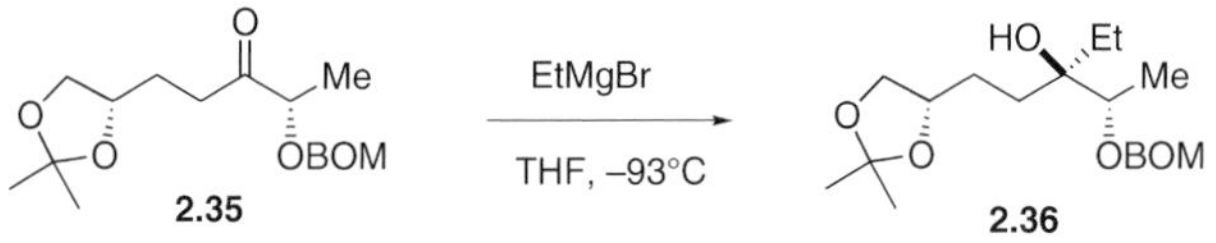

Scheme 2.24

Another key step in the synthesis of salinomycin by Yonemitsu involved the addition of methyl lithium to the ketone **2.37** (Scheme 2.25). When conducted

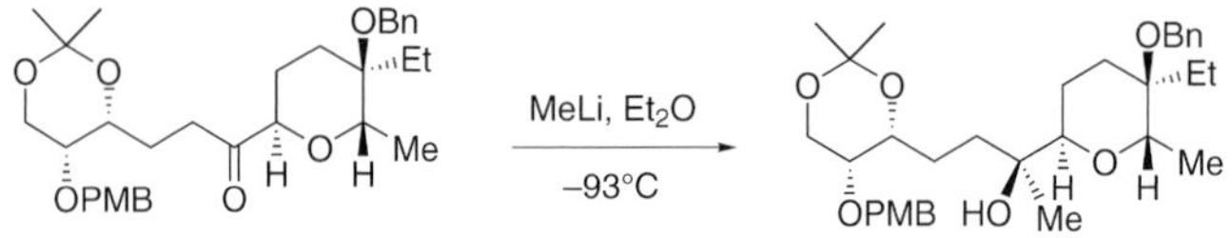

Scheme 2.25

at low temperature, this reaction proceeded with high stereoselectivity favouring the alcohol **2.38** (33:1).[38]

In a total synthesis of isolasalocid, addition of the lithium acetylide to the ketone **2.39** yielded the alcohol **2.40** as the only isomer in nearly quantitative yield (Scheme 2.26). Similar treatment of ketone **2.41** gave the lasalocid intermediate **2.42** again with exceptional stereoselectivity.[39]

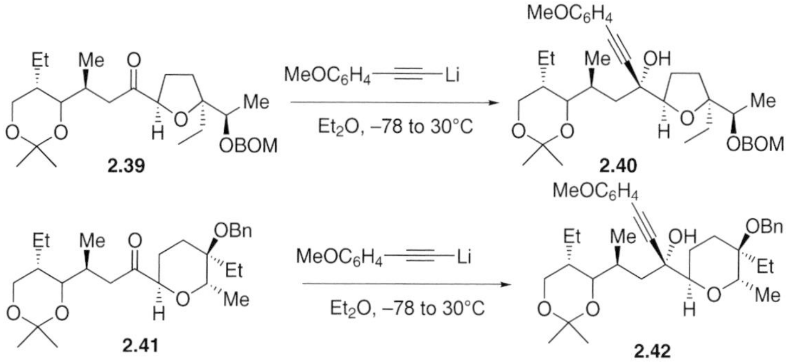

Scheme 2.26

An elegant example of a stereocontrolled addition to a complex ketone was reported by Evans *et al.* 'en route' to the macrolide antibiotic cytovaricin (Scheme 2.27).[40] A coupling between the Grignard reagent **2.43** and ketone

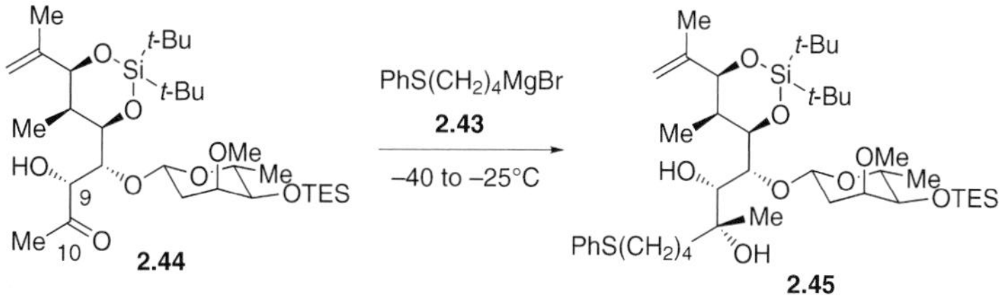

Scheme 2.27

2.44 proceeded stereoselectively via a chelated intermediate to give the alcohol **2.45** which was the only isomer detected.

The preference for the formation of 1,2-chelates over 1,3-chelates was exploited by two groups in separate total syntheses of zaragozic acid C (Figure 2.11). In considering a stereoselective synthetic approach to zaragozic

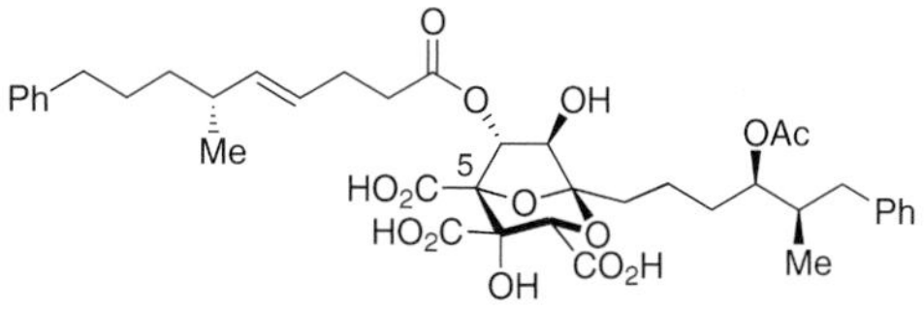

Figure 2.11 Zaragozic acid C.

STOICHIOMETRIC ASYMMETRIC SYNTHESIS

acid C, the highly oxygenated bicyclic core provides for a number of possiblities as far as Cram chelate addition is concerned.

In the Carreira synthesis, the crucial C(5) stereocentre (zaragozic acid numbering) was set by a chelation-controlled addition of TMS acetylide anion to the ketone **2.46** (Scheme 2.28).[41] The product was obtained in a diastereoisomeric

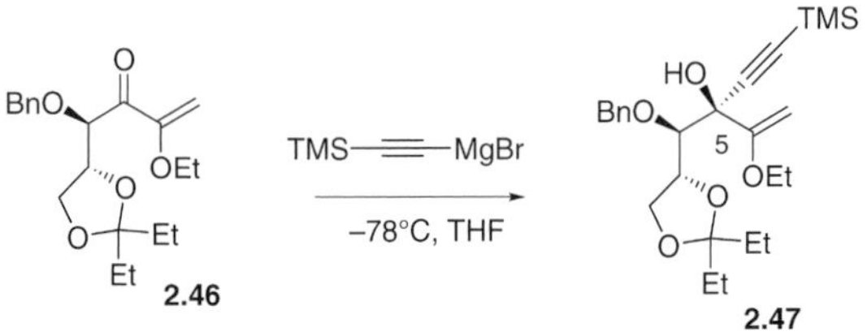

Scheme 2.28

ratio of 20:1 favouring the alcohol **2.47**. Three possible magnesium chelates can be formed in this reaction, **2.48–2.50**, however, only the 1,2-chelate **2.48**

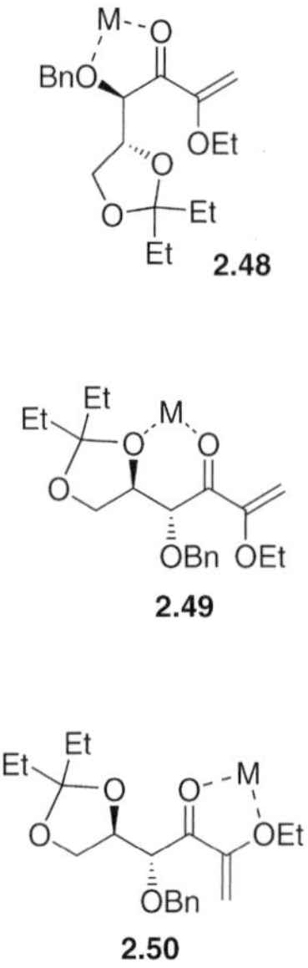

would provide the desired product (Figure 2.12). The 1,3-chelate **2.49** and the alternative 1,2-chelate **2.50** would give the unwanted stereochemistry at C(5).

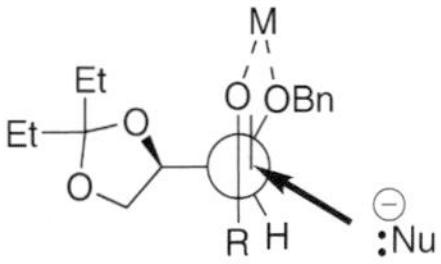

Figure 2.12 Structure **2.48**, Newman projection and attack of nucleophile.

It should be noted that a Felkin–Anh mode of addition to ketone **2.46** (where the OBn is the L group) would also yield the incorrect stereochemistry at C(5).

Evans *et al.* also utilised chelation control to introduce the C(5) stereocentre in their total synthesis of zaragozic acid C (Scheme 2.29).[42] In this example,

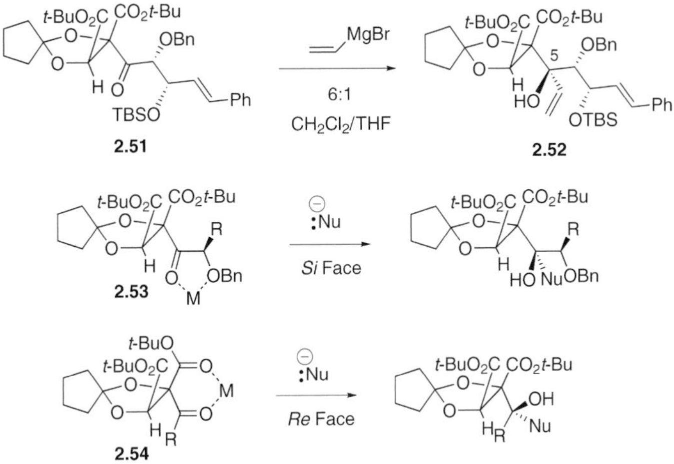

Scheme 2.29

addition of vinylmagnesium bromide to ketone **2.51** proceeded with at least 10:1 selectivity to provide allylic alcohol **2.52** as the major product. This result again suggests a preference for the 1,2-chelate, **2.53**, over the 1,3-chelate, **2.54**, which would yield the undesired C(5) isomer as the major product.

In the final example presented in Scheme 2.30, Kuwajima, Kusama and co-workers in their recent total synthesis of taxol utilised a chelation-controlled addition reaction to introduce the C ring and C(2) stereochemistry in one step.[43] Condensation of the magnesium-chelated aldehyde **2.55** with the lithium anion

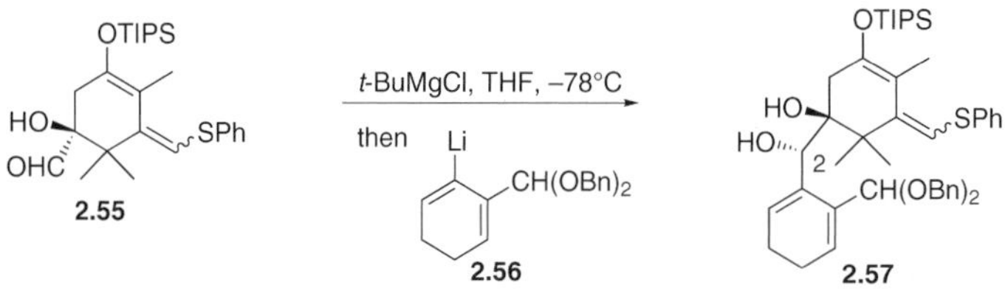

Scheme 2.30

2.56 gave the alcohol **2.57** with the desired C(2) stereochemistry as the only isomer.

2.2   Addition of chiral nucleophiles to carbonyl compounds

The addition of simple chiral nucleophiles or achiral nucleophiles in the presence of a stoichiometric amount of chiral additive to carbonyl compounds is not as common as addition to chiral carbonyl compounds (Scheme 2.31).[44]

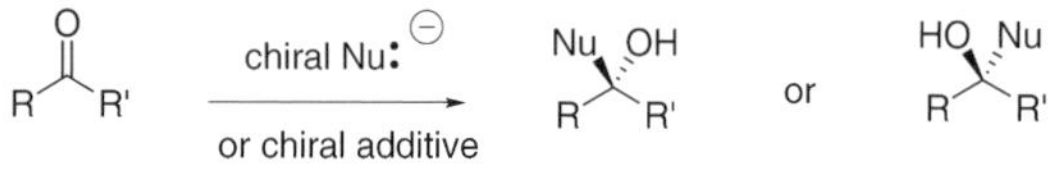

Scheme 2.31

However, the use of other types of nucleophile, such as chiral enolates and chiral crotyl or allylmetal reagents, is widespread. These types of chiral nucleophile will be discussed in detail in Chapters 4, 5 and 6 and only the addition of simple chiral nucleophiles will be discussed here.

An early example of the addition of an achiral nucleophile to an achiral aldehyde in the presence of a chiral additive was reported by Noyori and co-workers in 1971 (Scheme 2.32).[45] Since it was known that organolithium reagents

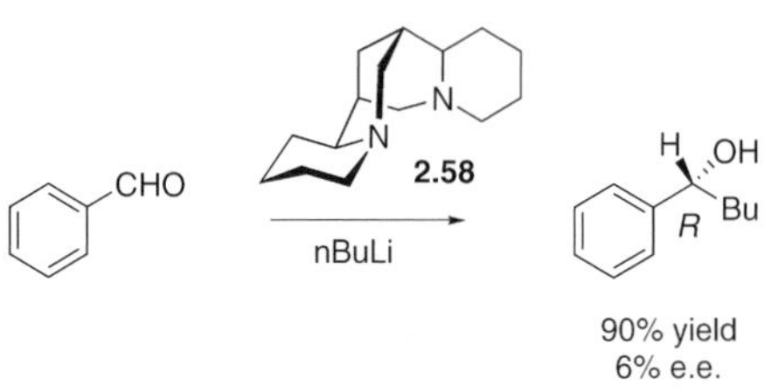

90% yield
6% e.e.

Scheme 2.32

are activated by complexation with tertiary amines such as N, N, N', N'-tetramethylethylenediamine, the chiral inducer chosen for study was the bidentate ligand (−)-sparteine (**2.58**) which can chelate to the metal to create a chiral environment around the nucleophile, as well as enhance reactivity. Addition of butyllithium to benzaldehyde in the presence of **2.58** gave 1-phenyl-1-pentanol in very low e.e. favouring the R-enantiomer. This result can be explained by the preferred mode of attack by the chiral nucleophile on the Re face of the carbonyl as shown in Figure 2.13.

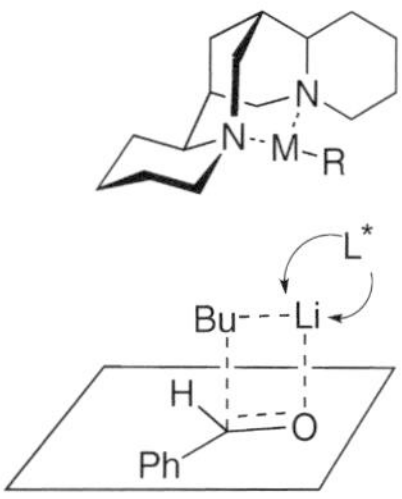

Figure 2.13

In 1979, Mukaiyama *et al.* reported a more effective chiral ligand, (2S, 2S')-2-hydroxymethyl-1-[(1-methylpyrrolidin-2-yl)methyl]pyrroline (**2.59**) which is readily synthesised from (S)-proline as shown in Scheme 2.33.[46] Coupling of

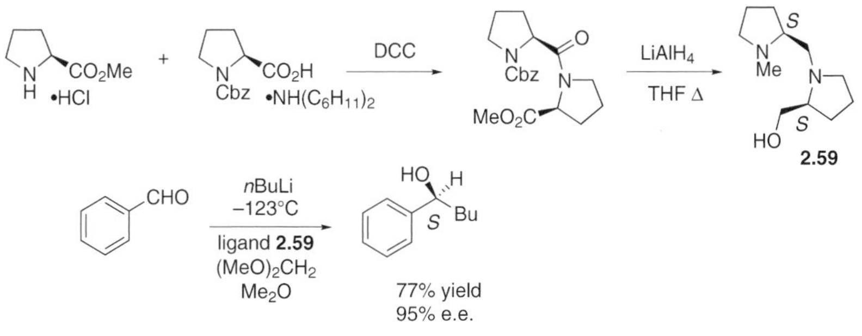

Scheme 2.33

(S)-proline methyl ester hydrochloride with (S)-N-benzyloxycarbonylproline mediated by DCC followed by reduction of the resultant amide-ester gave **2.59** in good yield. Addition of butyllithium to benzaldehyde in the presence of ligand **2.59** at low temperature gave (S)-1-phenyl-1-pentanol in high e.e.

A rigid complex such as **2.60** involving the ligand and organometallic reagent is probably formed where the coordination of the alkylmetal to two nitrogens and one oxygen atom provides an effective chiral environment.

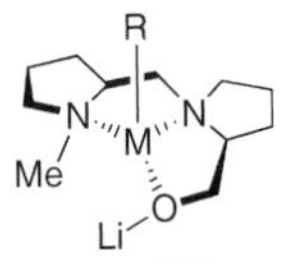

2.60

Ligand **2.59** was utilised by Johnson *et al.* to prepare the chiral alcohol **2.61** which is a precursor to a polyene utilised in the biomimetic synthesis of corticoids.[47] The addition of lithium trimethylsilylacetylide complexed with ligand **2.59** to aldehyde **2.62** gave alcohol **2.61** with excellent enantioselectivity (Scheme 2.34).

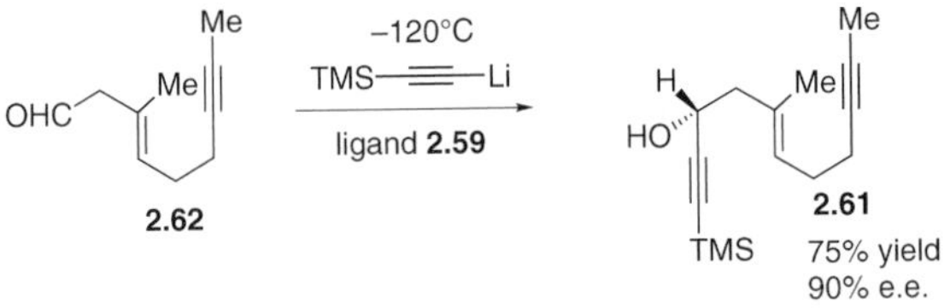

Scheme 2.34

The sulfonamidoalcohol **2.63** also proved to be an effective ligand in the asymmetric addition of methyltitanium to aromatic aldehydes (Scheme 2.35).[48] Treatment of **2.63** with tetramethyltitanium followed by addition of 2-propanol

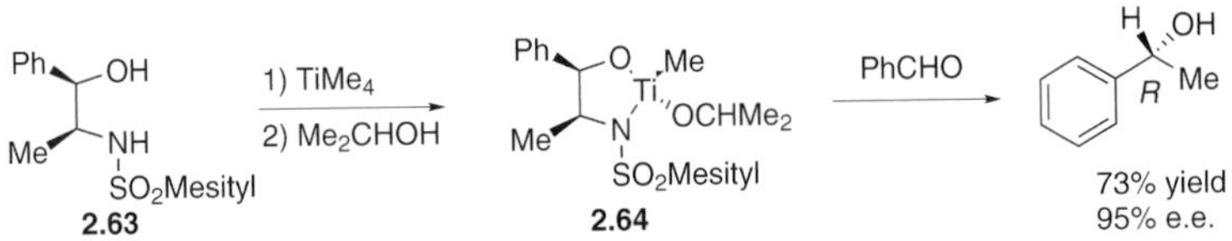

Scheme 2.35

to the intermediate dimethyltitanium species afforded the chirally modified methyltitanium reagent which can be represented by structure **2.64**. In fact, examination by NMR showed that there were a multitude of compounds present and so the exact nature of the chiral titanium species is unknown. Exposure of benzaldehyde to the mixture represented by **2.64** resulted in the production of (R)-1-phenylethanol with very high e.e. In contrast, aliphatic aldehydes gave lower yields and unacceptable e.e. values.

High enantioselectivities have also been observed for the addition of nucleophiles to benzaldehyde in the presence of diamine ligands which possess C_2 axes of symmetry. The binaphthyl ligand (R, R)-**2.65** was demonstrated to be an excellent chiral chelating agent for the asymmetric addition of nucleophiles to benzaldehyde (Scheme 2.36).[49] Aliphatic aldehydes such as pentanal also

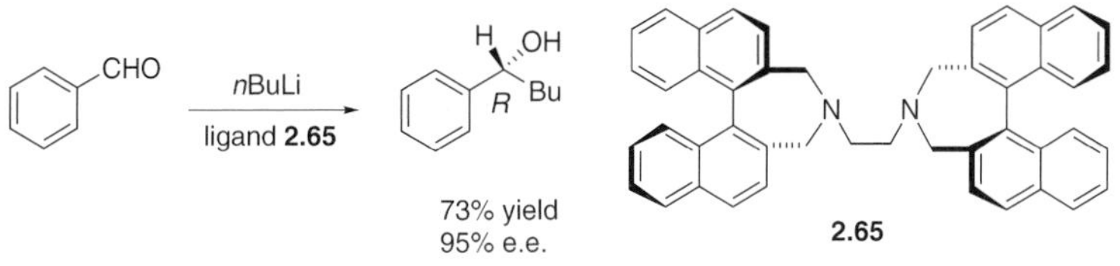

73% yield
95% e.e.

2.65

Scheme 2.36

underwent asymmetric addition but with less than half the e.e. observed for benzaldehyde.

The addition of α-chiral sulfoxide stabilised carbanions to carbonyl compounds is an effective way to induce asymmetry since chiral sulfoxides are easily synthesised.[44] In the example shown in Scheme 2.37 the chiral naphthyl

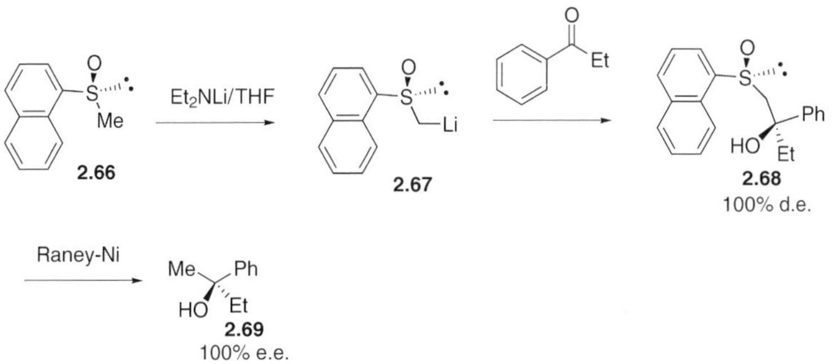

2.66

2.67

2.68
100% d.e.

Raney-Ni

2.69
100% e.e.

Scheme 2.37

sulfoxide **2.66** is first deprotonated with lithium diethylamide to provide the α-chiral anion **2.67**.[50] Subsequent addition of this anion to phenylethylketone gives the alcohol **2.68** with excellent diastereoselectivity. Removal of the sulfoxide chiral auxiliary was then effected with Raney-nickel to provide alcohol **2.69** in high e.e. The high level of stereocontrol observed in the addition to aryl ketones can be explained by the transition state model **2.70** (Figure 2.14) where there

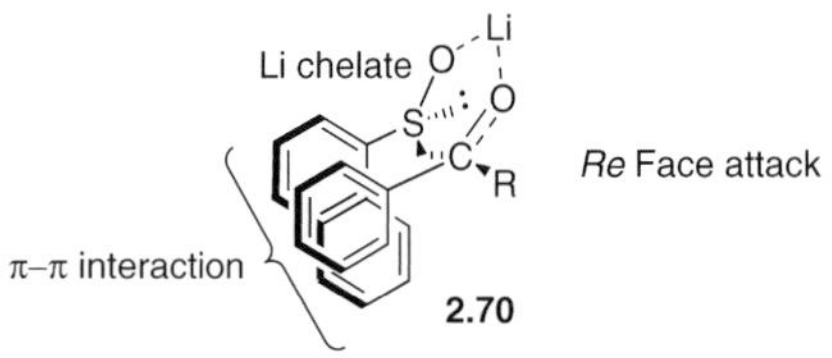

Figure 2.14

is a π–π interaction between the naphthyl group of **2.67** and the phenyl ring of the ketone, which would result in *Re* face attack to give diastereoisomer **2.68**. Further support for the model depicted came from the low diastereoisomeric excess observed for the addition of **2.67** to 2-butanone. In this case, no π–π interaction can be generated between **2.67** and 2-butanone, which explains the low selectivity observed.

Another example of addition of a chiral nucleophile is shown in Scheme 2.38. This time, the nucleophile is a configurationally stable carbanion which is

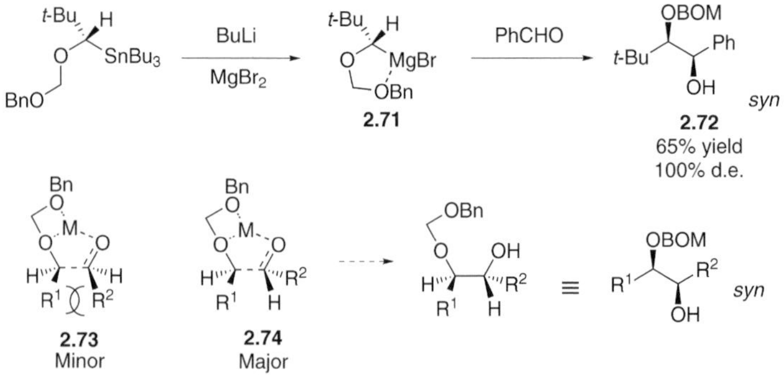

Scheme 2.38

generated from a chiral organotin compound as shown.[51] The addition of the chelated chiral anion **2.71** to benzaldehyde proceeds in 100% d.e. to give the *syn*-isomer **2.72**. The formation of isomer **2.72** as the major product can be explained by the transition states **2.73** and **2.74**. The steric interaction between

R^1 and R^2 in **2.73** is alleviated in **2.74** thus giving the *syn* isomer **2.72** as the preferred product.

Seebach and co-workers have shown that primary Grignard reagents can add enantioselectively to ketones in the presence of the chiral diol TADDOL[52] **2.75** ($\alpha, \alpha, \alpha', \alpha'$-tetraaryl-1,3-dioxolan-4,5-dimethanol). Treatment of acetophenone with various primary Grignard reagents in the presence of 1 equivalent of TADDOL **2.75** gave tertiary alcohols in very high e.e. (Scheme 2.39).[53,54]

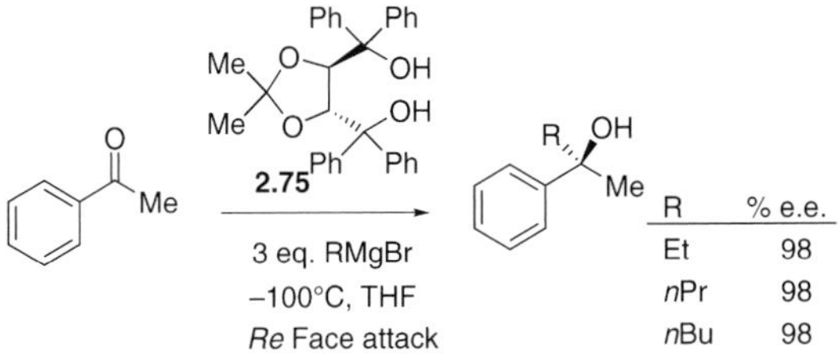

R	% e.e.
Et	98
nPr	98
nBu	98

Scheme 2.39

In the final example of a chirally modified nucleophile shown in Scheme 2.40, alkynylation of the ketone **2.78** with the novel chiral zinc aminoalkoxide **2.77**,

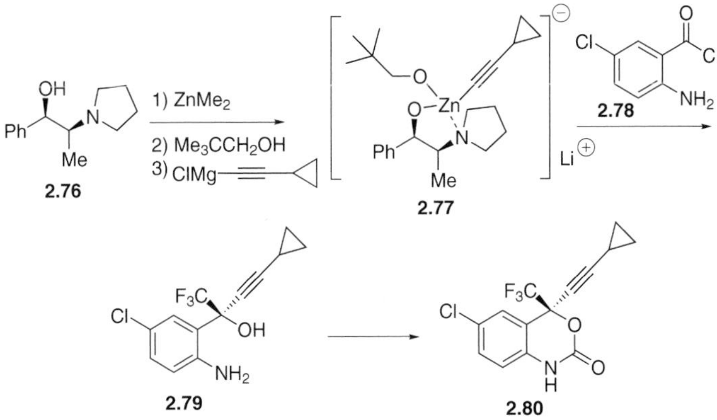

Scheme 2.40

prepared in three steps from aminoalcohol **2.76**, proceeds in over 97% e.e. to provide the alcohol **2.79**.[55] Compound **2.79** is an intermediate in the synthesis of Efavirenz (**2.80**), a potent non-nucleosidal HIV (human immunodeficiency virus) reverse transcriptase inhibitor used in the treatment of AIDS (acquired immunodeficiency syndrome). It is noteworthy that the addition is effective only in the presence of the unprotected *o*-amino group.

2.3   Asymmetric reduction

2.3.1   Relative diastereoselection

Relative diastereoselective reduction can be achieved by application of either Felkin–Anh or chelate control as discussed earlier in this chapter. In the example shown in Scheme 2.41, α-alkoxy ketones are reduced to *syn*-products under Cram-chelate control while Felkin control provides the *anti*-isomers. Examples of this type of reduction were presented in Schemes 2.6 and 2.16.

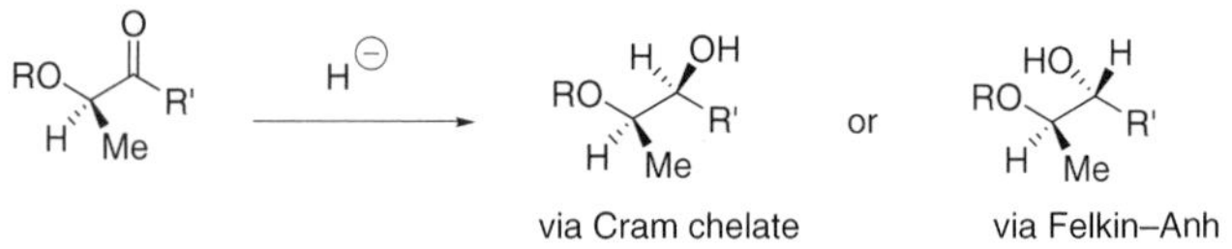

Scheme 2.41

1,3-Stereoinduction is possible in the reduction of β-hydroxy ketones **2.81** and the outcome is dependant on the hydride reagent utilised (Scheme 2.42).

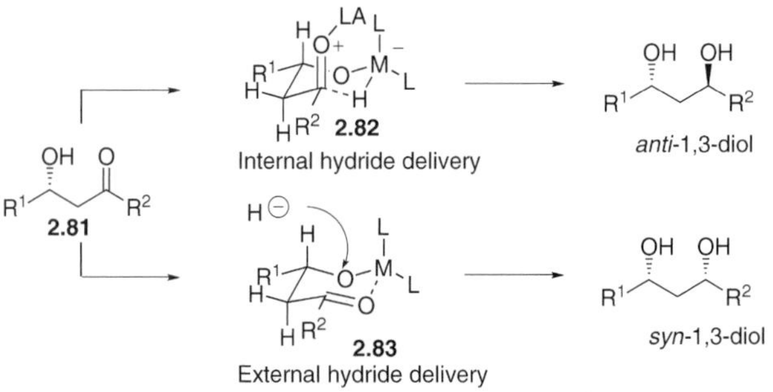

Scheme 2.42

For 'internal hydride delivery', the metal atom of the hydride is bound to the oxygen while the carbonyl is chelated to a Lewis acid or proton. The hydride is then delivered to the carbonyl via the transition state **2.82** to provide the *anti*-diol. Directed reduction to the *syn*-isomer can be achieved by 'external hydride delivery'. Initial chelation of **2.81** to a metal complexing agent forms the tight six-membered complex **2.83** and the hydride is then added. Hydride delivery then occurs from the face opposite the axial hydrogen on the α-carbon atom.

Narasaka and Pai showed that trialkylboranes can chelate effectively to β-hydroxy ketones to give dialkylborone complexes which are stereoselectively reduced with sodium borohydride to provide *syn*-diols.[56] Treatment of ketone **2.84** with tributylborane in the presence of a catalytic amount of oxygen gave the boron complex **2.85** which was selectively reduced to give the *syn*-diol **2.86** almost exclusively (Scheme 2.43).

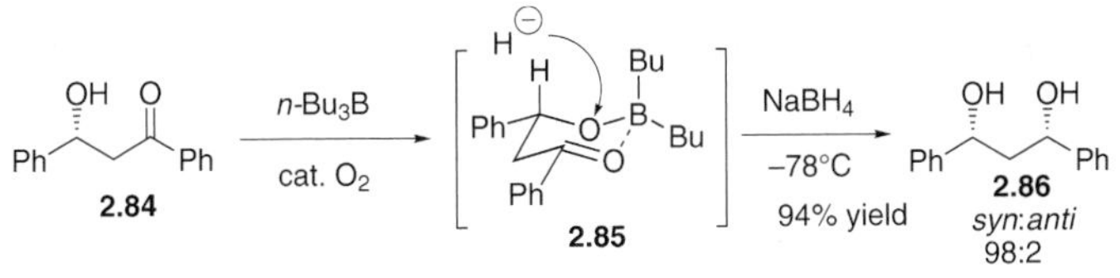

Scheme 2.43

Alkoxydialkylboranes can complex with hydroxyketones without air or acid activation to provide boron chelates and these can also be reduced stereoselectively to give *syn*-diols. For example, treatment of ketone **2.84** with diethylmethoxyborane gives diol **2.86** with excellent selectivity (Scheme 2.44).[57]

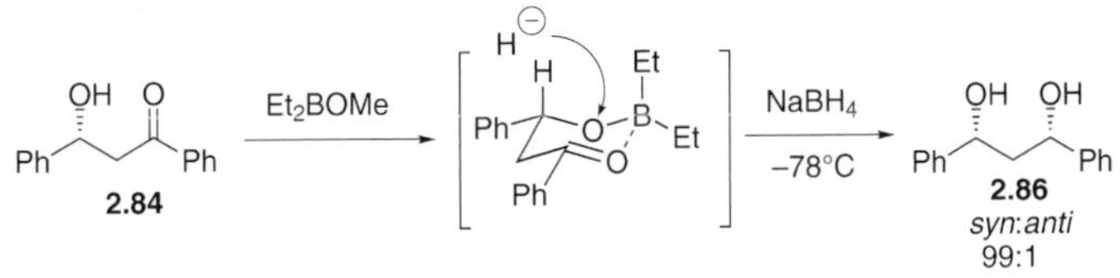

Scheme 2.44

Evans and Hoveyda showed that *syn*-diols are produced upon treatment of β-hydroxy ketones with at least 2 equivalents of catecholborane.[58] The first equivalent apparently forms the boron complex **2.87** and the second equivalent acts as the source of external hydride (Scheme 2.45). Evidence for intermediate **2.87** arose from treatment of the ketone **2.88** with 1.1 equivalents of catecholborane which resulted in less than 10% reduction.

Anti-selective reduction of β-hydroxy ketones by internal hydride delivery can be effected by reagents such as tetramethylammonium triacetoxyborohydride. Reduction of ketone **2.89** with this reagent in the presence of acetic acid occurred in a stereoselective manner to give the *anti*-diol as the major product (Scheme 2.46).[59]

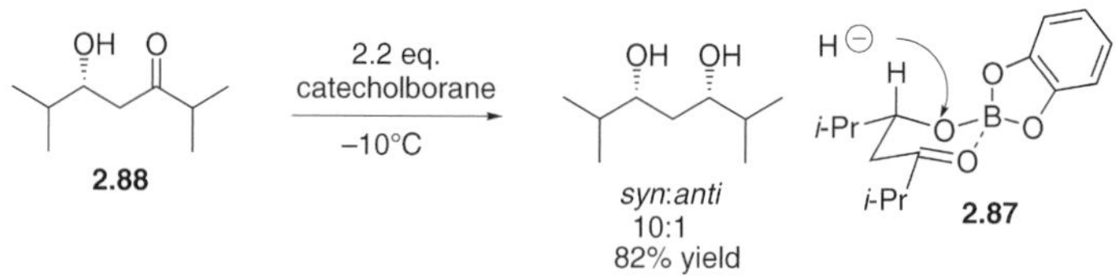

Scheme 2.45

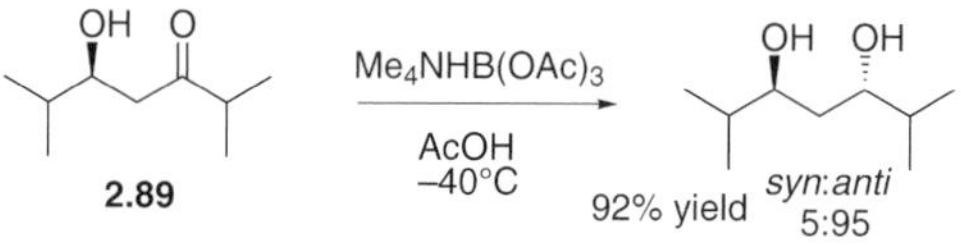

Scheme 2.46

The triacetoxyborohydride can exchange an acetoxy ligand for the hydroxy group of the substrate to give a boron complex. Internal delivery of hydride then occurs via intermediate **2.90** as the alternative complex **2.91** suffers from a destabilising steric interaction between an axially orientated alkyl substituent and an acetoxy group on the boron atom (Figure 2.15).

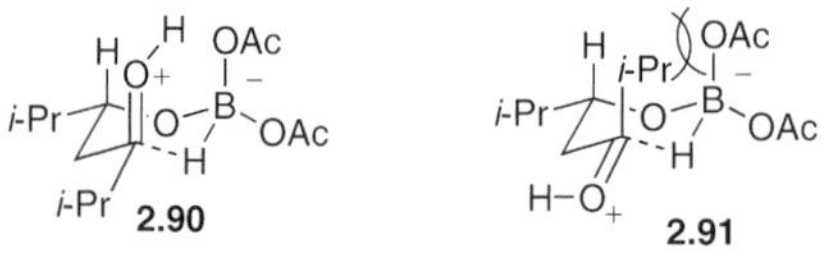

Figure 2.15

Chiral 2-ketosulfoxides are also reduced stereoselectively by various hydride reagents to provide both *syn*- and *anti*-type products. Reduction of the chiral 2-ketosulfoxide **2.92** with DiBALH provides alcohol **2.93**, and the diastereo-isomer **2.94** is produced on reduction with the same hydride in the presence of a chelating species such as zinc chloride (Scheme 2.47).[60]

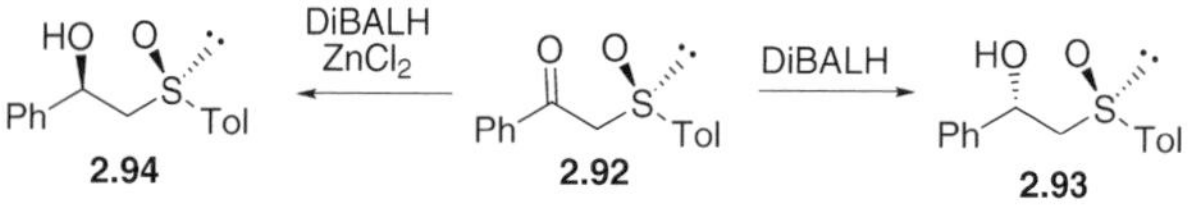

Scheme 2.47

In the former case, an intermediate ate complex such as **2.95** is probably formed and internal hydride delivery gives isomer **2.93**. The addition of zinc

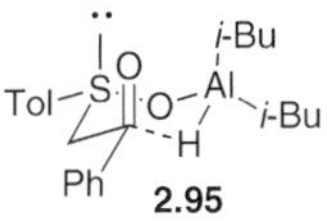

2.95

chloride causes the formation of the intermediate complex **2.96** which is reduced by external hydride delivery to provide diastereoisomer **2.94**.

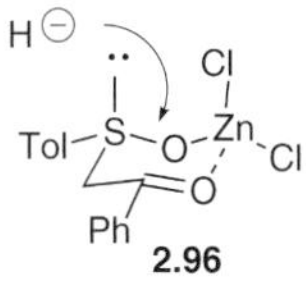

2.96

2.3.2 Chiral hydrides

There are a number of useful methods for the enantioselective reduction of unsymmetrical ketones (Scheme 2.48). Chirally modified hydrides are amongst

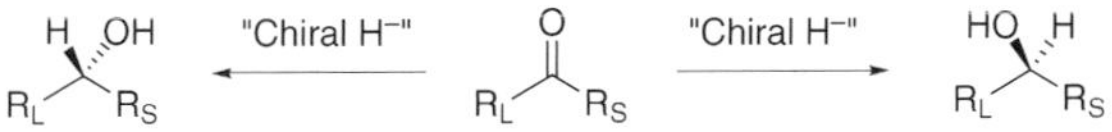

Scheme 2.48

the most common stoichiometric reducing agents and high enantioselectivies can be achieved.[61]

An early example of a chirally modified hydride was that derived from the amino alcohol **2.97** (also called Darvon alcohol) and lithium aluminium hydride (Scheme 2.49).[62] Reduction of acetophenone proceeded in 75% e.e.[62] but it was later found by Marshall and Wang that reduction of the α,β-acetylinic ketone **2.98** with the reagent derived from **2.97** and LiAlH$_4$ gave the alcohol **2.99** in over 90% e.e.[63]

Chiral reducing agents derived from axially dissymmetric molecules have been utilised by Noyori and co-workers to induce enantioselective reduction of ketones.[64] The chiral binaphthol hydride reagent **2.100**, reported in the early 1980s, reduces acetophenone with exceptionally high e.e. (Scheme 2.50).

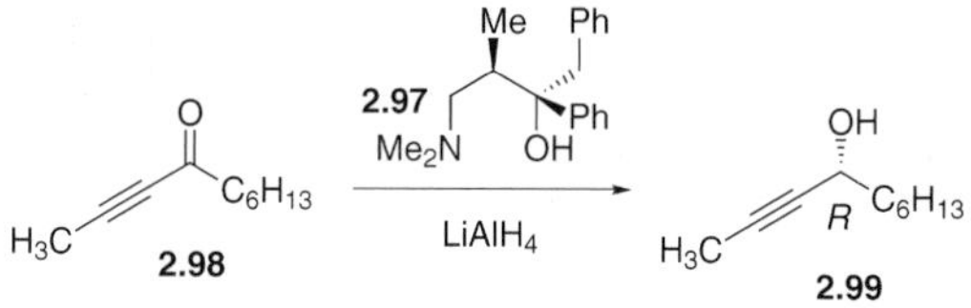

Scheme 2.49

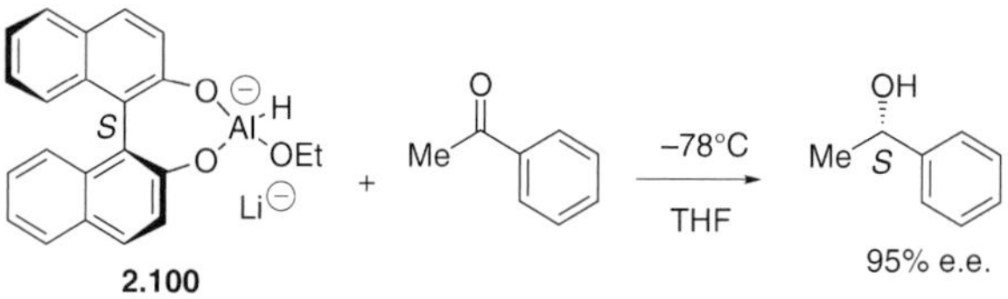

Scheme 2.50

Noyori postulated that S-**2.100** gives the S-alcohol via the transition state **2.101** where the methyl group is in the axial position (Figure 2.16).[64] In the

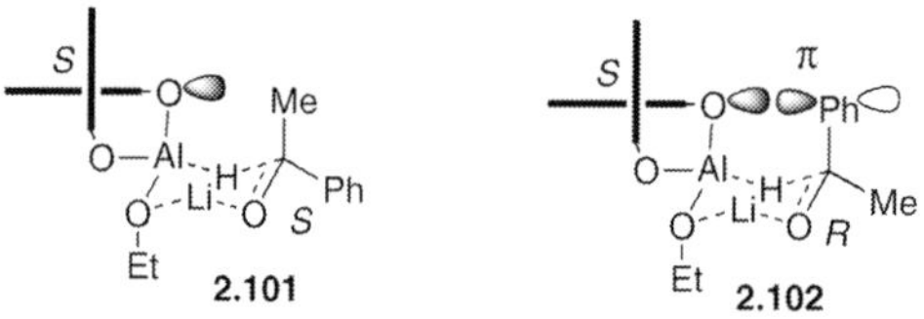

Figure 2.16

alternative transition state **2.102** there is an unfavourable electronic n–π type interaction that develops between the axially orientated binaphthyloxy oxygen and the axial phenyl group which is not present in **2.101**.

Chiral trialkylboranes have also been utilised as asymmetric reducing agents.[65] Midland *et al.* reported that the pinene derived borane **2.103** (Alpine borane) reduces acetylinic ketones such as **2.104** in reasonable e.e. (Scheme 2.51).[66] Later, Brown and Pai modified the original procedure and

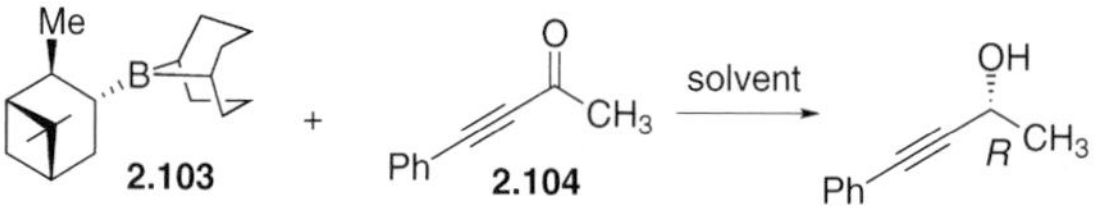

Solvent	e.e.
THF	72
neat	98

Scheme 2.51

found that the yields, reaction rate and e.e. values improved when the reductions were conducted in the absence of solvent.[67,68]

The mechanism for reduction with trialkylborane **2.103** and related compounds involves a hydride transfer from the pinene auxiliary to the ketone via a six-membered transition state (Scheme 2.52). The preferred transition state

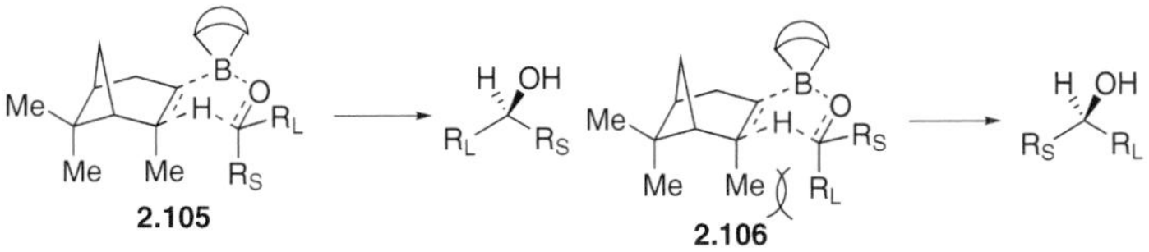

Scheme 2.52

2.105 has the R_s (small) group of the unsymmetrical ketone orientated pseudoaxially. In the higher energy transition state, **2.106**, there is a larger steric interaction which develops between the methyl group on the pinene residue and the R_L (large) group.

The final example of an enantioselective reducing agent presented is that of diisopinocampheylborane **2.107** reported by Brown *et al.*[69,70] Compound **2.107** is more reactive than Alpine borane and reacts with acetophenone to give the corresponding alcohol **2.108** in very high e.e. (Scheme 2.53).

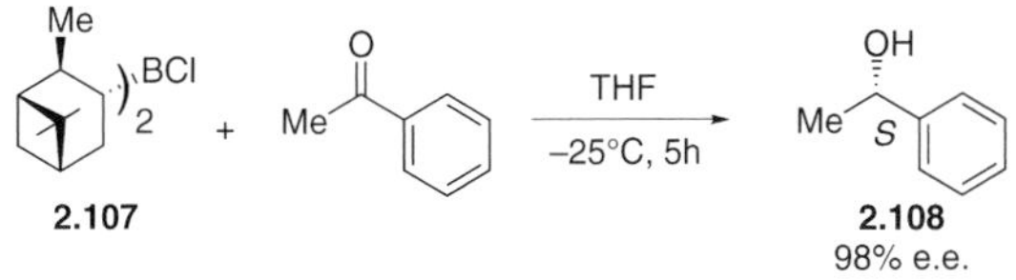

Scheme 2.53

References

1. Eliel, E.L., in *Asymmetric Synthesis*, Ed. J.D. Morrison, Academic Press, Orlando, FL, **1983**, vol. 2, p. 125.
2. Bartlett, P.A., *Tetrahedron*, **1980**, *45*, 2.
3. Cram, D.J. and Elhafz, F.A.A., *J. Am. Chem. Soc.*, **1952**, *74*, 5828.
4. Cram, D.J. and Kopecky, K.R., *J. Am. Chem. Soc.*, **1959**, *81*, 2748.
5. Cornforth, J.W., Cornforth, R.H. and Mathew, K.K., *J. Chem. Soc.*, **1959**, 112.
6. Karabatsos, G.J., *J. Am. Chem. Soc.*, **1967**, *89*, 1367.
7. Chérest, M., Felkin, H. and Prudent, N., *Tetrahedron Lett.*, **1968**, 2199.
8. Anh, N.T. and Eisenstein, O., *Nouv. J. Chem.*, **1977**, *1*, 61.

9. Bürgi, H.B., Dunitz, J.D. and Shefter, E.J., *J. Am. Chem. Soc.*, **1973**, *95*, 5065.

10. Bürgi, H.B., Dunitz, J.D., Lehn, J.M. and Wipff, G., *Tetrahedron*, **1974**, *30*, 1563.

11. Heathcock, C.H. and Flippin, L.A., *J. Am. Chem. Soc.*, **1983**, *105*, 1667.

12. Lodge, E.P. and Heathcock, C.H., *J. Am. Chem. Soc.*, **1987**, *109*, 3353.

13. Yamamoto, Y. and Maruyama, K., *J. Am. Chem. Soc.*, **1985**, *107*, 6411.

14. Nakada, M., Urano, Y., Kobayashi, S. and Ohno, M., *J. Am. Chem. Soc.*, **1988**, *110*, 4826.

15. Still, W.C. and McDonald III, J.H., *Tetrahedron Lett.*, **1980**, *21*, 1031.

16. Still, W.C. and Schneider, J.A., *Tetrahedron Lett.*, **1980**, *21*, 1035.

17. Eliel, E.L., Koskimies, J.K. and Lohri, B., *J. Am. Chem. Soc.*, **1978**, *100*, 1614.

18. Reetz, M.T., Hüllmann, M. and Seitz, T., *Angew. Chem., Int. Ed. Engl.*, **1987**, *26*, 477.

19. Keck, G.E. and Castellino, S., *J. Am. Chem. Soc.*, **1986**, *108*, 3847.

20. Fischer, J.C., Horton, D. and Weckerle, W., *Carbohydr. Res.*, **1977**, *59*, 459.

21. Overman, L.E. and McCready, R.J., *Tetrahedron Lett.*, **1982**, *23*, 2355.

22. Frye, S.V. and Eliel, E.L., *Tetrahedron Lett.*, **1986**, *27*, 3223.

23. Kahn, S.D., Henre, W.J. and Keck, G.E., *Tetrahedron Lett.*, **1897**, *28*, 279.

24. Keck, G.E. and Castellino, S., *Tetrahedron Lett.*, **1987**, *28*, 281.

25. Keck, G.E., Castellino, S. and Wiley, M.R., *J. Org. Chem.*, **1986**, *51*, 5478.

26. Pohland, A. and Sullivan, H.R., *J. Am. Chem. Soc.*, **1953**, *75*, 4458.

27. Angiolini, L., Bizzarri, P.C. and Tramontini, M., *Tetrahedron*, **1969**, *25*, 4211.

28. Andrisano, R. and Angiolini, L., *Tetrahedron*, **1970**, *26*, 5247.

29. Herold, P., *Helv. Chim. Acta*, **1988**, *71*, 354.

30. Mukaiyama, T., *Tetrahedron*, **1981**, *37*, 4111.

31. Collum, D.B., McDonald III, J.H. and Still, W.C., *J. Am. Chem. Soc.*, **1980**, *102*, 2117.

32. Collum, D.B., McDonald III, J.H. and Still, W.C., *J. Am. Chem. Soc.*, **1980**, *102*, 2118.

33. Collum, D.B., McDonald III, J.H. and Still, W.C., *J. Am. Chem. Soc.*, **1980**, *102*, 2120.

34. Nicolaou, K.C., Claremon, D.A. and Barnette, W.E., *J. Am. Chem. Soc.*, **1980**, *102*, 6611.

35. Horita, K., Nagato, S., Oikawa, Y. and Yonemitsu, O., *Tetrahedron Lett.*, **1987**, *28*, 3253.

36. Horita, K., Oikawa, Y., Nagato, S. and Yonemitsu, O., *Tetrahedron Lett.*, **1988**, *29*, 5143.

37. Horita, K., Oikawa, Y., Nagato, S. and Yonemitsu, O., *Chem. Pharm. Bull.*, **1989**, *37*, 1717.

38. Horita, K., Nagato, S., Oikawa, Y. and Yonemitsu, O., *Chem. Pharm. Bull.*, **1989**, *37*, 1705.

39. Horita, K., Noda, I., Tanaka, K., Oikawa, Y. and Yonemitsu, O., *Tetrahedron*, **1993**, *49*, 5997.

40. Evans, D.A., Kaldor, S.W., Jones, T.K., Clardy, J. and Stout, T.J., *J. Am. Chem. Soc.*, **1990**, *112*, 7001.

41. Carreira, E.M. and Bois, J.D., *J. Am. Chem. Soc.*, **1995**, *117*, 8106.

42. Evans, D.A., Barrow, J.C., Leighton, J.L., Robichaud, A.J. and Sefkow, M., *J. Am. Chem. Soc.*, **1994**, *116*, 12 111.

43. Morihira, K., Hara, R., Kawahara, S., Nishimori, T., Nakamura, N., Kusama, H. and Kuwajima, I., *J. Am. Chem. Soc.*, **1998**, *120*, 12 980.

44. Solladié, G., in *Asymmetric Synthesis*, Ed. J.D. Morrison, Academic Press, Orlando, FL, **1983**, vol. 2, p. 157.

45. Nozaki, H., Aratani, T., Toraya, T. and Noyori, R., *Tetrahedron*, **1971**, *27*, 905.

46. Mukaiyama, T., Soai, K., Sato, T., Shimizu, H. and Suzuki, K., *J. Am. Chem. Soc.*, **1979**, *101*, 1455.

47. Johnson, W.S., Frei, B. and Gopalan, A.S., *J. Org. Chem.*, **1981**, *46*, 1512.

48. Reetz, M.T., Kükenhöhner, T. and Weinig, P., *Tetrahedron Lett.*, **1986**, *27*, 5711.

49. Mazaleyrat, J.-P. and Cram, D.J., *J. Am. Chem. Soc.*, **1981**, *103*, 4585.

50. Sakuraba, H. and Ushiki, S., *J. Am. Chem. Soc.*, **1990**, *31*, 5349.

51. McGarvey, G.J. and Kimura, M., *J. Org. Chem.*, **1982**, *47*, 5420.

52. Seebach, D., Bech, A.K., Imwinkelried, R., Roggo, S. and Wonnacott, A., *Helv. Chim. Acta*, **1987**, *70*, 954.

53. Weber, B. and Seebach, D., *Angew. Chem., Int. Ed. Engl.*, **1992**, *31*, 84.

54. Weber, B. and Seebach, D., *Tetrahedron*, **1994**, *50*, 6117.
55. Tan, L., Chem, G., Tillyer, R.D., Grabowski, E.J.J. and Reider, P.J., *Angew. Chem., Int. Ed. Engl.*, **1999**, *38*, 711.
56. Narasaka, K. and Pai, F.-C., *Tetrahedron*, **1984**, *41*, 2233.
57. Chen, K.-M., Hardtmann, G.E., Prasad, K., Repic, O. and Shapiro, M.J., *Tetrahedron Lett.*, **1987**, *28*, 155.
58. Evans, D.A. and Hoveyda, A.H., *J. Org. Chem.*, **1990**, *55*, 5190.
59. Evans, D.A., Chapman, K.T. and Carreira, E.M., *J. Am. Chem. Soc.*, **1988**, *110*, 3560.
60. Carreño, M.C., Ruano, J.L.G., Martín, A.M., Pedregal, C., Rodriguez, J.H., Rubio, A., Sanchez, J. and Solladié, G., *J. Org. Chem.*, **1990**, *55*, 2120.
61. Singh, V.K., *Synthesis*, **1992**, 605.
62. Yamaguchi, S. and Mosher, H.S., *J. Org. Chem.*, **1973**, *38*, 1870.
63. Marshall, J.A. and Wang, X.-J., *J. Org. Chem.*, **1991**, *56*, 4913.
64. Noyori, R., Tomino, I., Tanimoto, M. and Nishizawa, M., *J. Am. Chem. Soc.*, **1984**, *106*, 6709.
65. Midland, M.M., Greer, S., Tramonato, A. and Zderic, S.A., *J. Am. Chem. Soc.*, **1979**, *101*, 2352.
66. Midland, M.M., McDowell, D.C., Hatch, R.L. and Tramontano, A., *J. Am. Chem. Soc.*, **1980**, *102*, 867.
67. Brown, H.C. and Pai, G.G., *J. Org. Chem.*, **1982**, *47*, 1606.
68. Brown, H.C. and Pai, G.G., *J. Org. Chem.*, **1985**, *50*, 1384.
69. Chandrasekharan, J., Ramachandran, P.V. and Brown, H.C., *J. Org. Chem.*, **1985**, *50*, 5446.
70. Brown, H. C., Chandrasekharan, J. and Ramachandran, P.V., *J. Org. Chem.*, **1986**, *51*, 3394.

3 Formation and alkylation of enolates

3.1 Introduction and background

The formation of carbon–carbon bonds via enolate anions has been used for over a hundred years. Examples of such reactions include named reactions such as the Claisen–Schmidt[1–4] condensation, the Claisen condensation (acetoacetic ester condensation)[2,5] and the Reformatsky[6] reaction (Scheme 3.1). However,

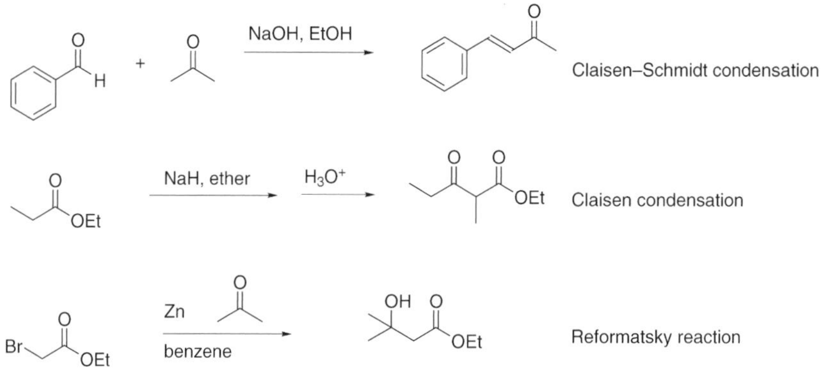

NaOH, EtOH — Claisen–Schmidt condensation

NaH, ether H_3O^+ — Claisen condensation

Zn   benzene — Reformatsky reaction

Scheme 3.1

the existence of the intermediate enolate anions in these reactions was not recognised until much later. The reactions in Scheme 3.1 are characterised by the formation of the reactive ion in the presence of the electrophile. Other early enolate reactions employed carbon acids considerably more acidic than common alcohols. These easily deprotonated carbon acids (acetoacetate, acetoacetic ester, malonic ester, etc.) owed their acidity to the presence of a β-dicarbonyl grouping, conferring added resonance stabilisation to the anion produced. The big advance in enolate chemistry of the twentieth century was the stoichiometric formation of enolates from carbon acids much less acidic than common alcohols. This was first reported in the 1950s by Hauser and co-workers[7–11] who formed the lithium enolate of ethyl acetate and t-butyl acetate by the reaction of the esters with $LiNH_2$ in liquid ammonia (Scheme 3.2).

However, it was soon found that lithium dialky- and disilylamides have greater synthetic utility than the unsubstituted amides, as a result of the

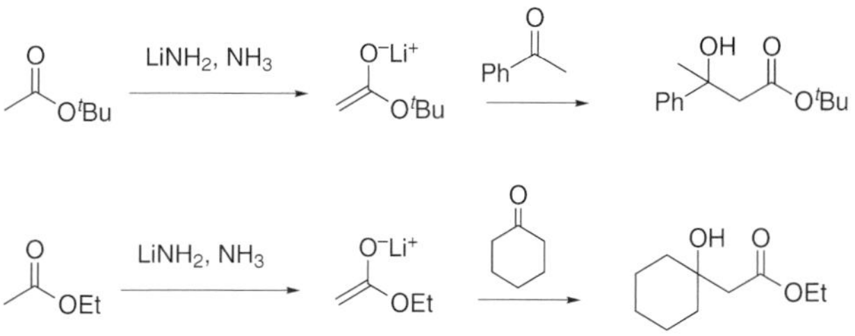

Scheme 3.2

hydrophobic alkyl groups which confer greater solubility in common organic solvents (ether, THF, etc.) and, because of their increased steric hindrance, they show reduced nucleophilicity. The most useful of these amides has proved to be lithium diisopropylamide (LDA). Other commonly used non-nucleophilic bases include the lithium, sodium and potassium salts of bis(trimethylsilyl)amine (hexamethyldisilazane), which were first introduced in the 1960s. These are somewhat less basic than LDA, but they are still sufficiently basic to deprotonate unhindered ketones or esters. Some early examples of the stoichiometric formation of enolates are shown in Scheme 3.3.[12-14]

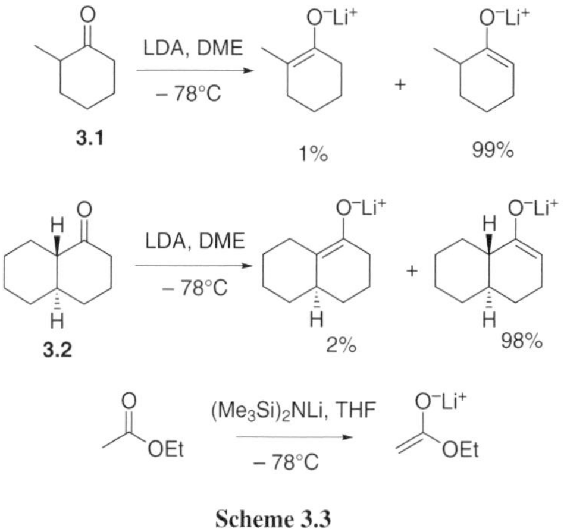

Scheme 3.3

The early history of enolate chemistry has been well reviewed and a number of excellent summaries have been published.[15,16] This chapter will focus mainly on the formation and reaction of stoichiometric acyclic enolates, with emphasis on their stereoselective formation, alkylation and reactions with electrophiles.

3.2 Alkali metal enolates

3.2.1 Consideration of regiochemistry

The enolisation of **3.1** and **3.2** in Scheme 3.3 introduces the possibility for formation of regioisomers of the enolate. The predominant enolisation on the less-hindered side of the ketone suggests that the reaction was occurring under 'kinetic' control with the proton removed from the less-hindered α-carbon. However, where the deprotonation occurs under equilibrating conditions 'thermodynamic' control results in the formation of the more stable enolate. In some cases it is possible to form either enolate selectively, as indicated in Scheme 3.4.[17] Thus enolisation with lithium N-isopropylcyclohexylamide

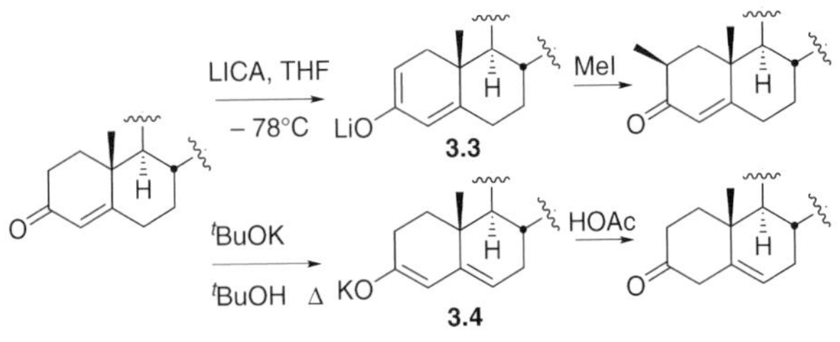

Scheme 3.4

gave the cross-conjugated dienolate **3.3** (kinetic control) whereas potassium *t*-butoxide in *t*-butyl alcohol gives the fully conjugated dienolate **3.4** (thermodynamic control).

It should be noted that alkylation of potassium and sodium enolates is problematic in that rapid proton transfer often results in the production of *di* and even *tri* substituted products (Scheme 3.5).[18]

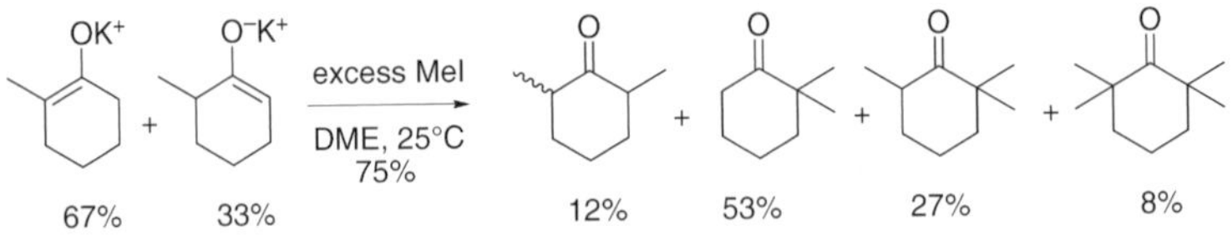

Scheme 3.5

For the less-basic lithium enolates, the rate of alkylation with relatively unhindered alkylating agents is significantly faster compared with proton transfer, making them more synthetically useful than their sodium or potassium counterparts.

3.2.2 Alkali metal enolate structures

When formulating the structures of enolates in solution a number of important issues must be considered, including the degree of aggregation and the possibility for O- and C-metallation. It has now been well documented that lithium amides and enolates exist as aggregates varying from dimeric to tetrameric in ethereal solvents (Figure 3.1).[15,16]

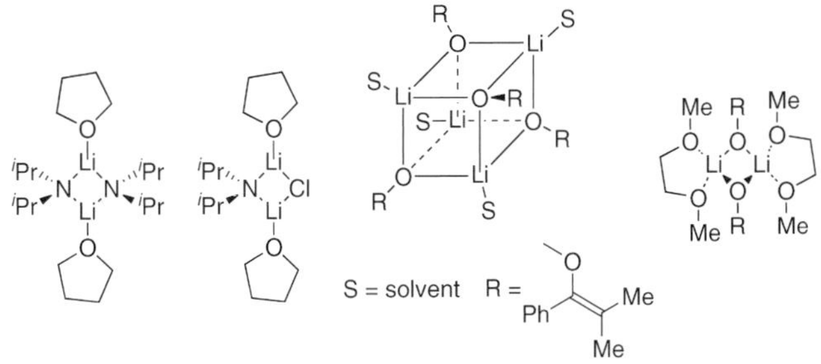

Figure 3.1

The enolates of the group I, II and III metal cations exist as the O-metallated tautomers.[19] However O-alkylation is not usually a problem with the lithium and less-electropositive metal enolates unless highly reactive alkylating agents such as dialkyl sulfates or α-chloro ethers are used in relatively polar solvents. This has traditionally been explained by reasoning that hard electrophilic reagents should attack the hard oxygen atom of the enolate, if it is accessible, whereas soft reagents should attack the soft α-carbon.[20,21] More recently it has been observed experimentally[22] and predicted by *ab initio* calculation[23] that the free enolate anions will react on oxygen irrespective of the electrophile. It thus appears that coordination of the metal cation with the oxygen reduces its reactivity to the extent that C alkylation is usually preferred in solution.

However, the reaction of enolates with the highly reactive trialkylsilyl chlorides occurs at oxygen and is often used to trap the enolates as their trialkyl silyl enol ethers.[24,25] This technique has been used to determine the structures of enolates in solution and the ratio of any isomeric forms.

3.2.3 Enolate stereochemistry

For the enolisation of cyclic ketones the question of enolate geometry does not arise, but when acyclic ketones or esters are employed it is possible to form either (Z)-O- or (E)-O-enolates. The first systematic investigation of the enolate configuration was reported by Dubois and Dubois[26] in studies which indicated that there might be a relationship between the enolate geometry and the relative configuration of the two sp^3 stereocentres formed in the aldol reaction between an enolate and an aldehyde. The first clue as to control of enolate geometry came in the important paper published by Ireland and Willard[27] in 1975, which reported that the enolate geometry of an ester enolate is dependant on the polarity of the medium, with LDA in pure THF giving the (E)-O-enolate, and LDA in THF/HMPA giving the (Z)-O-enolate (Scheme 3.6).

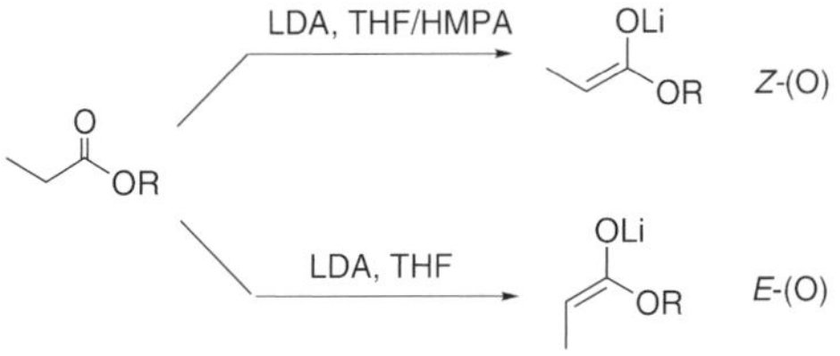

Scheme 3.6

The enolate geometry was deduced from the stereochemistry of the products formed from an ester enolate version of the Claisen rearrangement (now known commonly as the Ireland–Claisen reaction, which is discussed in Chapter 7, section 7.3.5) and assumes a chair-like transition state, analogous to that rigorously established for the Cope and Claisen rearrangements.

The stereochemistry of enolisation for a variety of carbonyl compounds is summarised in Table 3.1. Notably, it is apparent that the nature of the R group, the base and the solvent can all have an effect on the stereochemistry. The best studied of all these systems is 3-pentanone, and it has been found that treatment with sterically demanding bases under kinetic control in THF give predominantly the (E)-O-enolate. The selectivity for LDA [70:30, E-(O):Z-(O)] is lower than with the more hindered lithiumtetramethylpiperidide (LTMP) [84:16, E-(O):Z-(O)]. On the other hand, the silazide bases LHMDS and (Bn$_2$N)$_2$SiLi give more of the (Z)-O-enolate. Enolisation of esters and thioesters under kinetic control with LDA gives mainly the (E)-O-enolate, but enolisation in the presence of HMPA gives mostly the (Z)-O-enolate. Last, enolisation of amides and ketones where the R group is bulky gives mainly the (Z)-O-enolate under kinetic control with LDA.

Ireland et al.[33] proposed a cyclic six-membered transition state between the carbonyl compound and the dialkylamide base to explain the observed results, and Moreland and Dauben[35] have presented theoretical calculations that support this model. In this model it is proposed that in THF solution the lithium cation coordinates to the carbonyl oxygen and the removal of the α-proton is intramolecular, occurring via a closed chair-like transition state. Analysis of the two alternative chair transition states reveals that the transition state structure

Table 3.1 Stereochemistry of enolate formation in ethyl ketones, esters and amides

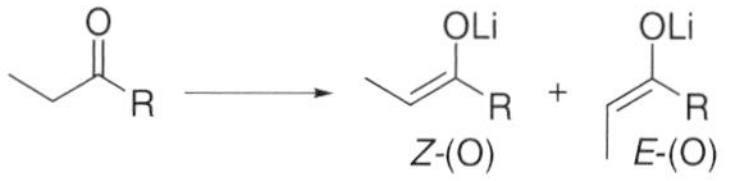

R	Base	Solvent	(E)-O-enolate	(Z)-O-enolate	Reference
Et	LDA	THF	70	30	28
Et	LTMP	THF	84	16	28
Et	LDA	THF + 23% HMPA	8	92	29
Et	LHMDS	THF	34	66	28
Et	$(Bn_2N)_2SiLi$	THF	< 3	> 97	30
OCH_3	LDA	THF	95	5	28
OCH_2H_3	LDA	THF	94	6	31
OCH_2H_3	LDA	THF + 23% HMPA	15	85	31
O^tBu	LDA	THF	95	5	32,33
S^tBu	LDA	THF	90	10	34
NEt_2	LDA	THF	< 3	> 97	16
$C(CH_3)_3$	LDA	THF	2	98	28
C_6H_5	LDA	THF	2	98	28

3.5 leading to the Z-(O) isomer is disfavoured by the nonbonding 1,3-diaxial interaction between the methyl groups of the enolate and the bulky ligands on the nitrogen. Similarly, the transition state **3.6** leading to the (E)-O-enolate is disfavoured by allylic strain (R ↔ CH$_3$) which increases as R gets larger (Scheme 3.7). This is consistent with the observation that more E-(O) is formed

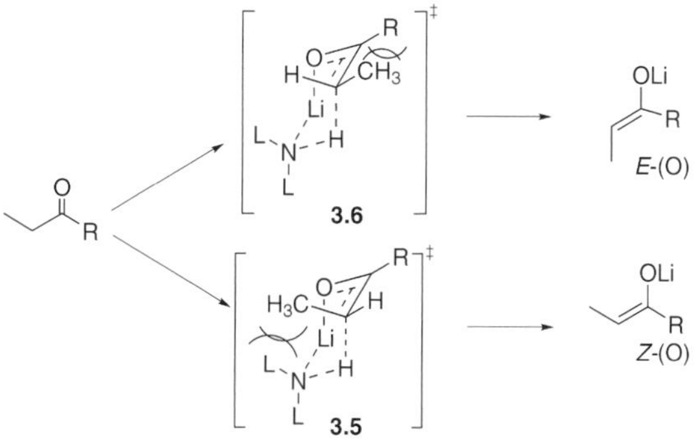

Scheme 3.7

as the ligands on the amide base become larger and more (Z)-O-enolate forms as R becomes larger.

It has been proposed[33] that when the deprotonations are carried out in the presence of HMPA, the HMPA solvates the lithium cation and thus disrupts the six-centred transition state. Now the transition state **3.7** leading to the Z-(O) isomer has no unfavourable steric interactions, whereas the transition state **3.8** leading to the (E)-O-enolate still possesses the allylic strain (R ↔ CH₃) (Scheme 3.8). Thus the formation of more (Z)-O-enolate in the presence of HMPA can be rationalised.

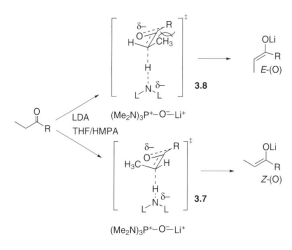

Scheme 3.8

3.2.4   Selective alkylation of enolates

Alkylation of a nonterminal enolate results in the formation of a new stereo-centre. The selective formation of (E)-O- or (Z)-O-enolates in acyclic systems is necessary for control of the stereocentre produced upon alkylation. Furthermore, facial selectivity of the attack on the enolate must be controlled to give only one new stereocentre (Scheme 3.9).

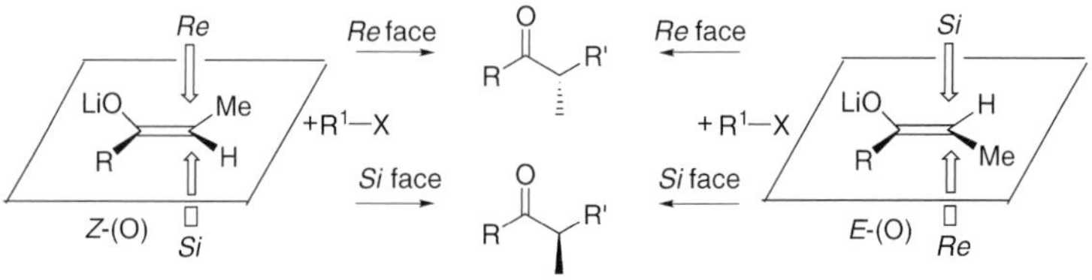

Scheme 3.9

Existing evidence indicates that *C*-alkylation of metal enolates with common electrophiles occurs by an S_N2 mechanism. During this reaction there is an interaction of the highest occupied molecular orbital (HOMO) of the enolate with the lowest unoccupied molecular orbital (LUMO) of the alkylating agent. Simple analysis would suggest that the electrophile should approach in a plane perpendicular to the enolate to give maximum orbital overlap between the developing C—C bond and the π-orbital of the carbonyl group in the transition state (Scheme 3.10).

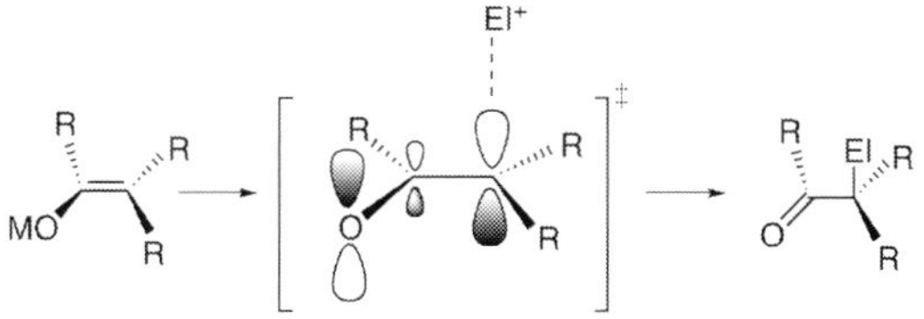

Scheme 3.10

However *ab initio* molecular orbital calculations indicate that the angle of approach is actually larger than 90°.[23] It had previously been suggested[36] that the path of the electrophile is tilted away from the right angle to the enolate plane because of repulsive interaction between the electrophile LUMO and the oxygen atom of the enolate (HOMO) (Figure 3.2). This is in accord with the

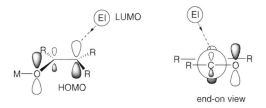

Figure 3.2

accepted Bürgi–Dunitz trajectory for attack of simple nucleophiles on carbonyl groups (See Chapter 2, section 2.1.2).

A number of highly efficient methods have been developed which can be employed for the selective formation of enolates with controlled geometry and the facial selectivity with electrophiles. These methods are thus useful in the stereoselective construction of carbon–carbon bonds and will be discussed in the following sections.

3.2.5 Enamine alkylations

Prior to the use of lithium enolates for alkylations, enamines were used in an attempt to overcome the problems of regioselectivity and polyalkylation that plagued attempts to alkylate sodium and potassium enolates of ketones and aldehydes. Methods for the synthesis of enamines, and the contribution of Stork's group in particular, have been reviewed.[37–39] However, such reactions are only partly successful, as alkylation of the enamine gives an iminium salt, which can transfer a proton to the starting enamine to give an alkylated enamine which can be further alkylated. In this way mixtures of products may be isolated after hydrolysis, and yields from enamine alkylations are thus often rather low (Scheme 3.11).

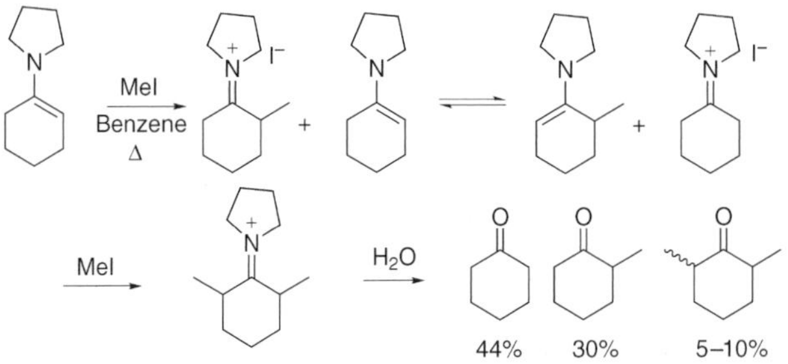

Scheme 3.11

However, it is noteworthy that the enantioselective alkylation of enamines has been carried out. Whitesell and Felman[40] reported that the cyclohexanone enamine **3.9** from (+)-*trans*-2,5-dimethylpyrrolidine underwent

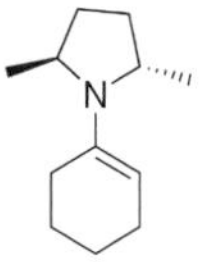

3.9

alkylation followed by hydrolysis to give 2-substituted cyclohexanones with the R-configuration in good yield and high enantiomeric excess.

3.2.6 Metallated imines and hydrazone alkylations

Although the use of enamines overcomes some of the problems associated with alkylation of potassium and sodium enolates, their use in synthesis is still somewhat limited. A far more versatile and flexible system which has been highly developed in various forms involves metallated imines and metallated hydrazones.

Metallated imines are now usually formed from imines derived from enolisable carbonyl compounds, by deprotonation with lithium dialkylamides (e.g. LDA), but deprotonation can be achieved with Grignard reagents or organolithium reagents. An example of the alkylation of a metallated imine derived from an aldehyde with use of ethylmagnesium bromide as the base is seen in Scheme 3.12.[41]

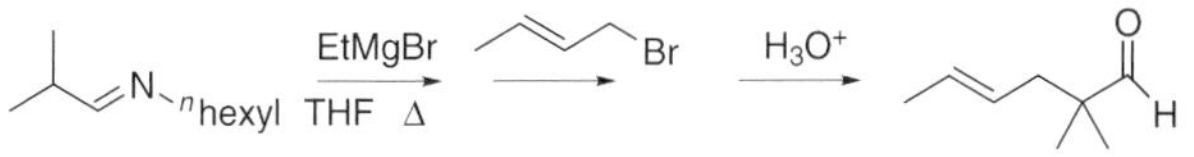

Scheme 3.12

The successful application of metallated imines can in part be attributed to the fact that they undergo proton transfer more slowly than corresponding metal enolates. Another factor leading to their versatility is that they are potent nucleophiles and undergo reaction with weakly electrophilic species such as epoxides. A list of electrophiles that achieve *C*-alkylation of metallated imines has been published by Whitesell and Whitesell.[39]

When imines are metallated they usually give *syn* metallated species, which have been rationalised by invoking a destabilising electrostatic repulsion between the metal–nitrogen bond and the imine β-carbon which lie in the same plane for the *anti* isomer[42] (Scheme 3.13).

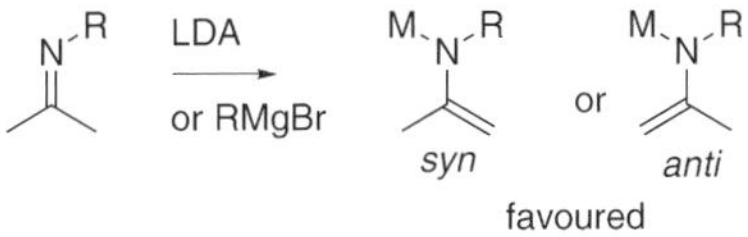

Scheme 3.13

Attempted stereoselective alkylation of metallated chiral imines has proved to give good results, provided a β-oxygen is present in the imine component. This allows formation of a five-membered ring chelate, including the metal cation, which makes the system more rigid and restricts the conformations available for alkylation. Two different transition states have been proposed for these types of reactions; first, Meyers *et al.*[43,44] proposed transition state **3.10**, and Whitesell and Whitesell[45] proposed transition state **3.11** (Scheme 3.14).

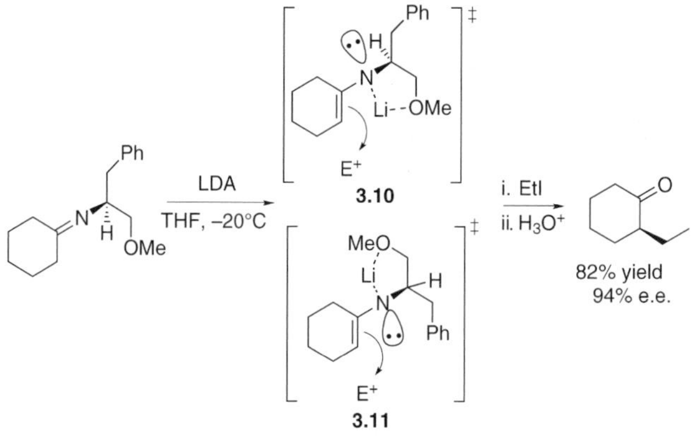

Scheme 3.14

When acyclic ketones are used it is possible to form either the (Z) or (E) lithiated imines on deprotonation. The (E)-enolates are usually formed kinetically, but in some cases they can be isomerised to the more stable (Z)-enolates by heating under reflux in THF. In the example shown in Scheme 3.15, alkylation

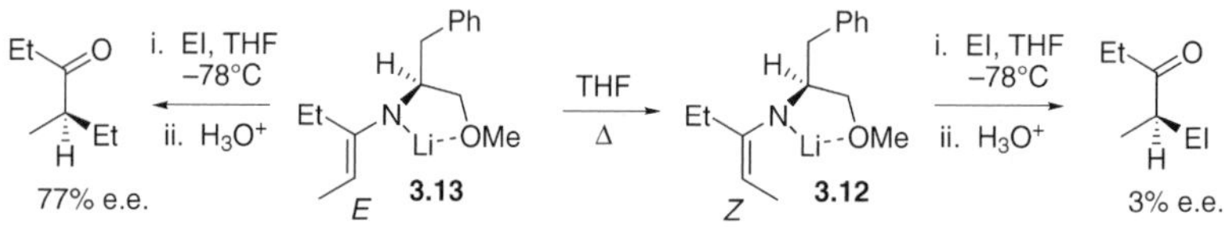

Scheme 3.15

of the kinetic enolate **3.12** gave the product with almost no selectivity; however, heating the enolate at reflux in THF converts it to thermodynamic (E)-enolate **3.13** which is alkylated with reasonable selectivity.[46]

Metallated N,N-dialkylhydrazones exhibit similar regioselectivity and stereoselectivity to metallated imines but have higher reactivity and generally give higher yields. Deprotonation of hydrazones is normally carried out with lithium dialkylamides or alkyllithium reagents and, in contrast to the corresponding imines, there is little preference for deprotonation *syn* or *anti* to the dialkyl-amino group (Scheme 3.16).[47]

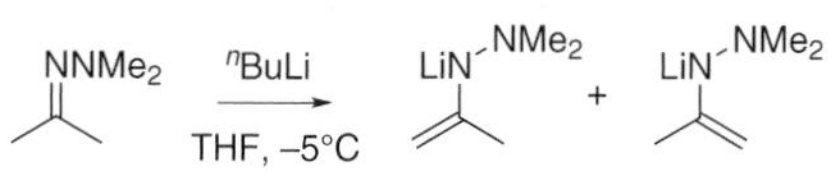

Scheme 3.16

The deprotonation of unsymmetrical hydrazones occurs with high selectivity at the least-substituted α-carbon. Since proton exchange of metallated hydrazones is very slow, C-alkylation at the less-substituted carbon can be readily achieved. Double alkylation can be carried out in a one-pot sequence starting with the dimethyl hydrazone of acetone, as seen in Scheme 3.17.[48]

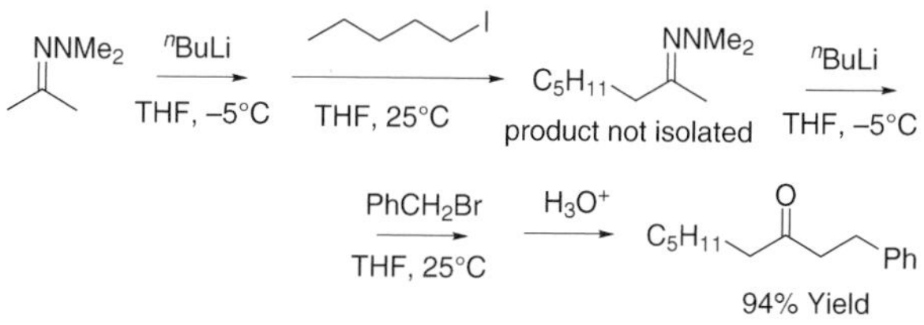

Scheme 3.17

A powerful and highly versatile technique for the enantioselective alkylation of chiral hydrazones has been introduced by Enders and co-workers. They utilised lithiated hydrazones derived from either the (R)- or (S)-1-amino-2-methoxypyrrolidine (RAMP or SAMP).[49] These reagents have been widely utilised in the total synthesis of natural products and are effective auxiliaries for the alkylation of aldehydes and ketones. A simple example[50] of the alkylation of a pentan-3-one derivative with propyl iodide to give **3.14** as a single isomer is shown in Scheme 3.18.

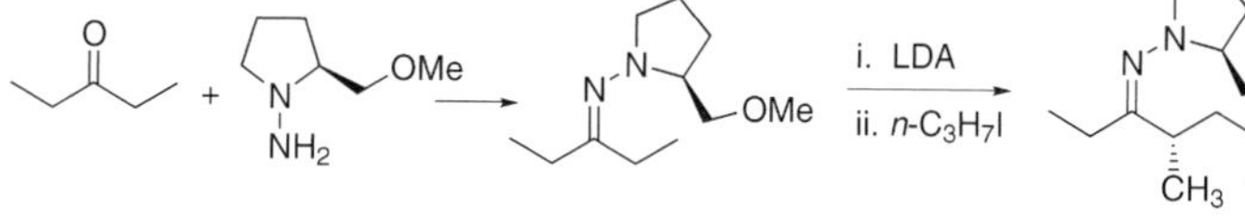

Scheme 3.18

A major advantage of these chiral hydrazones (RAMP or SAMP) is that their derivatives of aldehydes or ketones (cyclic and acyclic) all yield mainly the $(E)_{CC}$- and $(Z)_{CN}$-lithiated species on deprotonation with LDA in etheral solvents under kinetic control. These chelated lithiated hydrazones are rather conformationally rigid and exhibit high facial selectivity leading to alkylations with optical yields in excess of 90%. Modelling studies by Collum, Clardy and co-workers[51,52] indicate that the alkylations take place via a transition state where the electrophile attacks from the opposite face to an η^4-coordinate lithium cation (Scheme 3.19).

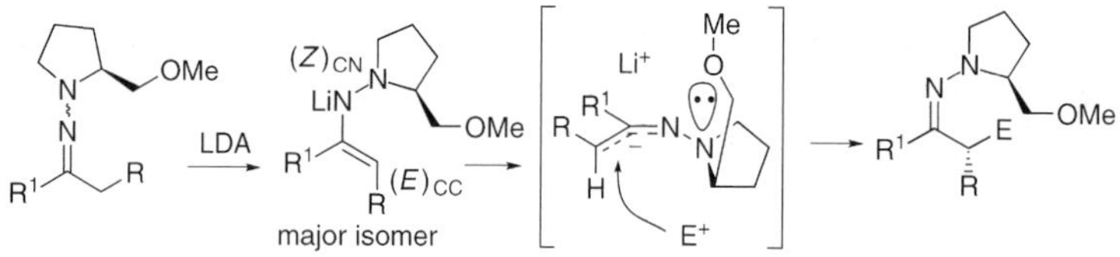

Scheme 3.19

3.2.7 Alkylation of carboxylic acid derivatives

A variety of carboxylic acid derivatives have been employed as chiral enolate precursors for asymmetric alkylation reactions. A notable example is the (S)-proline amide derivatives introduced by Evans et al.[53,54] and by Sonnet and Heath.[55] Deprotonation of these amides with LDA gives the (Z)-O-enolates, and the stereochemical outcome of the alkylation (92:8, **3.15**:**3.16**) can be explained by the assumption that the electrophile attacks from the less-hindered face of the enolate (i.e. the face opposite to the CH$_2$OLi which is coordinating to the enolate oxygen (Scheme 3.20).

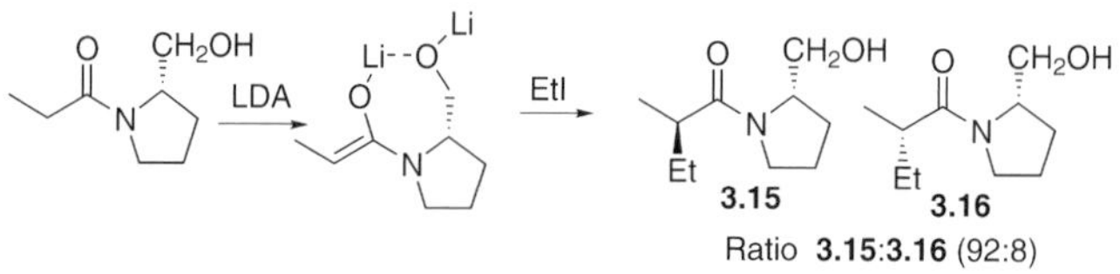

Scheme 3.20

This argument is supported by the finding that use of some protecting groups on the free alcohol results in the opposite alkylation selectivity (22:78, **3.17**:**3.18**, Scheme 3.21). In this case it appears that steric effects rather than

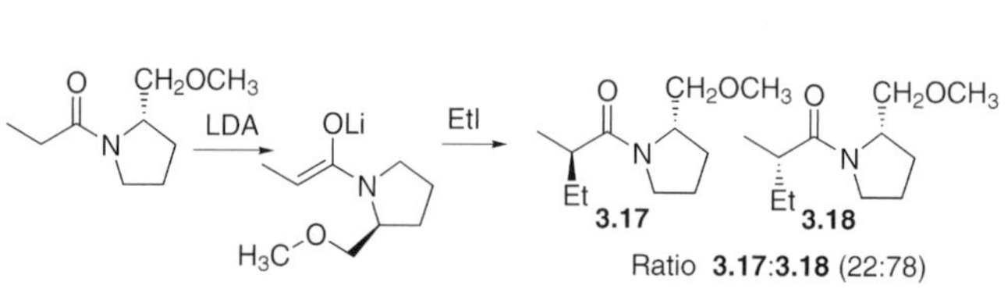

Scheme 3.21

chelation controls the low-energy rotamer of the amide bond, and thus the alkylation stereochemistry.

This procedure has proved useful in a number of total syntheses, including the total synthesis of ionomycin[54] by Evans et al., in which alkylation of a primary alkyl iodide was achieved with high stereoselectivity (Scheme 3.22).

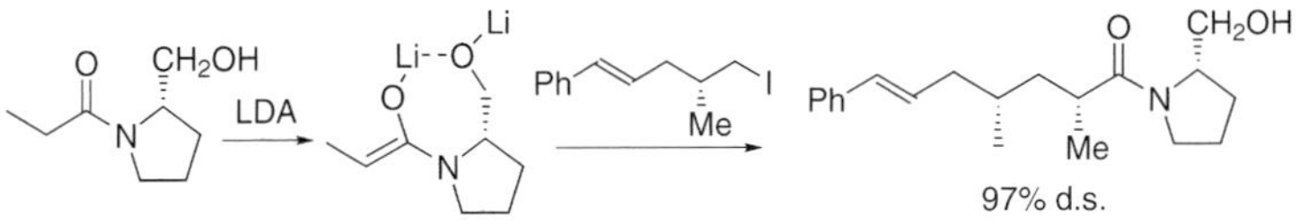

Scheme 3.22

These enolates are sufficiently reactive to open terminal epoxides, but, interestingly, the opposite facial selectivity is observed (Scheme 3.23).[56] This

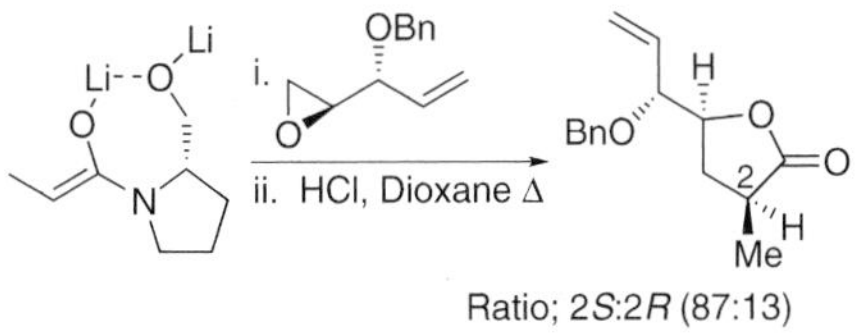

Ratio; 2S:2R (87:13)

Scheme 3.23

can be understood if one considers that in the transition state the oxygen of the epoxide coordinates to the CH_2OLi and is thus directed towards what might seem to be the more hindered face of the enolate.

The related C(2) symmetric 2,5-disubstituted pyrrolidines **3.19** have also been employed in enantioselective alkylation reactions and give exceptionally high selectivity, as shown in Scheme 3.24.[57,58] For these auxiliaries the

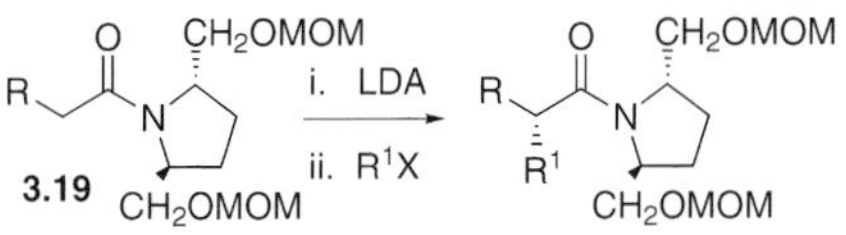

R	R¹X	Yield	d.s.
Me	BnBr	80	98
BnO	iPrOTf	76	98.5
(CH₃)₂C=N	HC≡CCH₂Br	95	99

Scheme 3.24

controlling factor is primarily the steric effect of the C(2) symmetric substituent. The sense of the chirality produced is the same as that for the protected proline amide derivatives reported by Evans (Scheme 3.21).

The metal enolates derived from the N-acylated oxazolidinones used extensively by Evans and others for aldol chemistry can be selectively alkylated with reactive electrophiles. These enolates are significantly less reactive than the proline derivatives but they are rapidly alkylated with allylic and benzylic halides and α-halo esters; their chemistry has been reviewed by Evans.[15] In these reactions the metal coordinates to the carbonyl of the oxazolidinone, locking the molecule into a conformation where alkylation occurs predominantly from the opposite face to the alkyl group on the auxiliary (Scheme 3.25).[59]

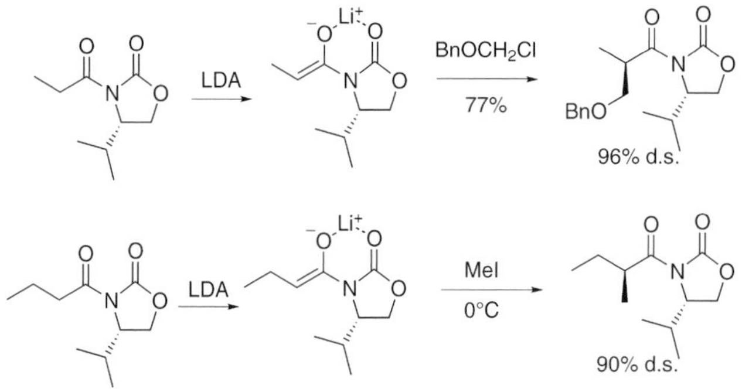

Scheme 3.25

An interesting example of how the two different configurations of the stereocentre produced in the alkylation can be obtained by control of the enolate geometry is seen in the work of Helmchen.[60,61] In this system the propionate ester of a camphor-derived auxiliary forms the (E)-O-enolate on deprotonation with lithium dicyclohexylamide in THF. Alkylation of this enolate from the less-hindered face gives predominantly the product **3.20** with one configuration of the new stereocentre. However, by carrying out the enolisation of the ester in the presence of HMPA, the (Z)-O-enolate is formed which is alkylated to give the product **3.21** with the opposite configuration of the stereocentre (Scheme 3.26).

The enolisation and alkylation of N-acylated sultam derivatives has been developed by Oppolzer and co-workers;[62] the resultant chelated (Z)-O-enolate reacts selectively on the *Re* face with a variety of electrophiles (Scheme 3.27).

The preparation of various α-amino acids[63] by using the nitrogen protected derivative **3.22** of this auxiliary has been achieved. Deprotonation with butyl lithium at −78°C, followed by treatment with a variety of alkyl iodides, gives the alkylated product in good yield. Nitrogen deprotection is achieved by acid

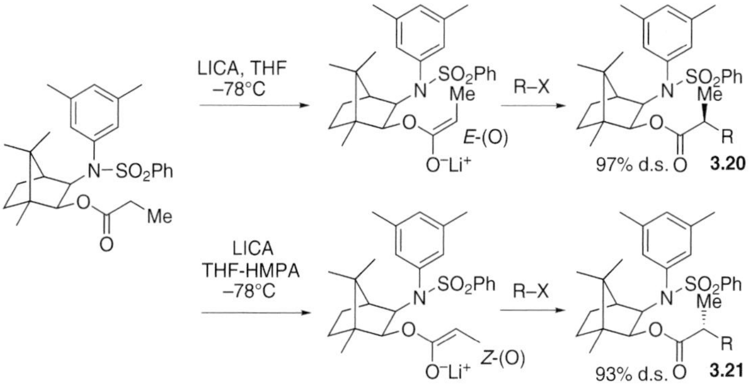

Scheme 3.26

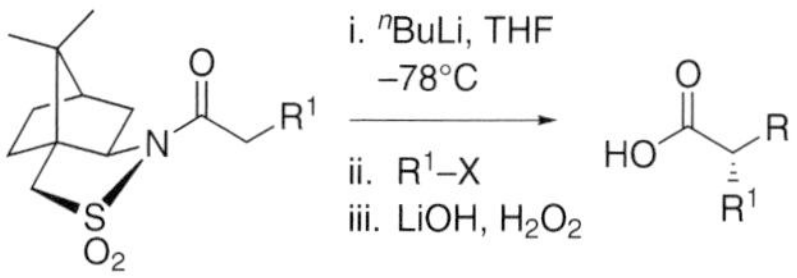

Scheme 3.27

hydrolysis and the sultam removed by means of lithium hydroxide in THF–water (Scheme 3.28).

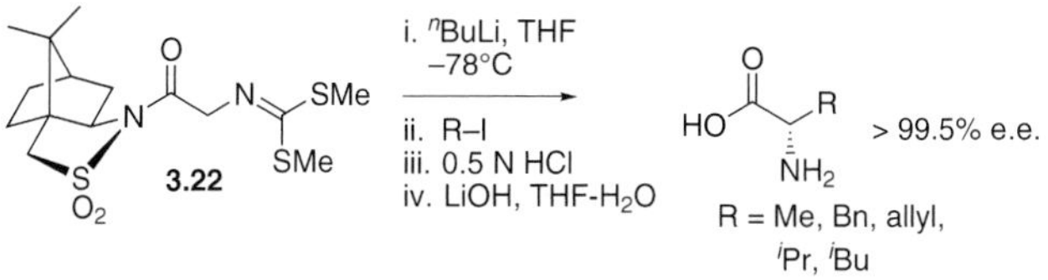

Scheme 3.28

A method for the preparation of α-amino acids via alkylation of an auxiliary-based glycine derivative has been developed by Myers *et al.*[64] In this example pseudoephidrine glycinamide **3.23** is deprotonated with two equivalents of lithium diisopropyl amide in the presence of lithium chloride. Initial deprotonation occurs at the nitrogen and hydroxyl, but upon warming to 0°C equilibration to the (Z)-O-enolate occurs and C-alkylation takes place with generally good yield and selectivity (Scheme 3.29). A major advantage of this methodology is the easy removal of the auxiliary by self-catalysed hydrolysis.

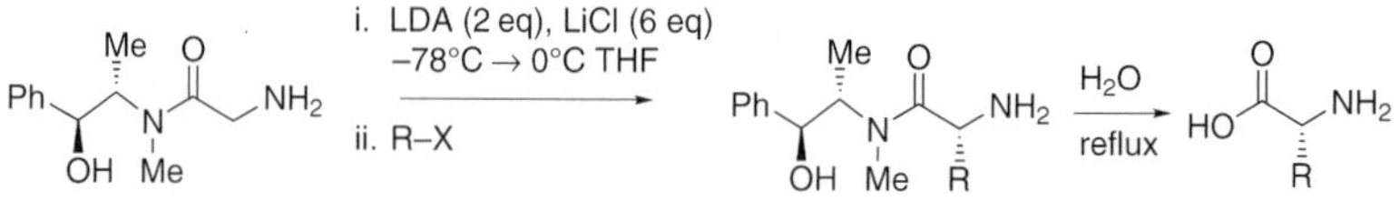

3.23

R	Yield	d.s.
Et–I	82	97
$H_2C{=}CHCH_2Br$	80	93
Bn–Br	79	91

Scheme 3.29

The same pseudoephidrine auxiliary has been used for the synthesis of α-alkylated carboxylic acid derivatives.[65] Again, deprotonation with two equivalents of base in the presence of lithium chloride gives the (*Z*)-*O*-enolate which is alkylated with high selectivity (Scheme 3.30). Furthermore, the auxiliary can

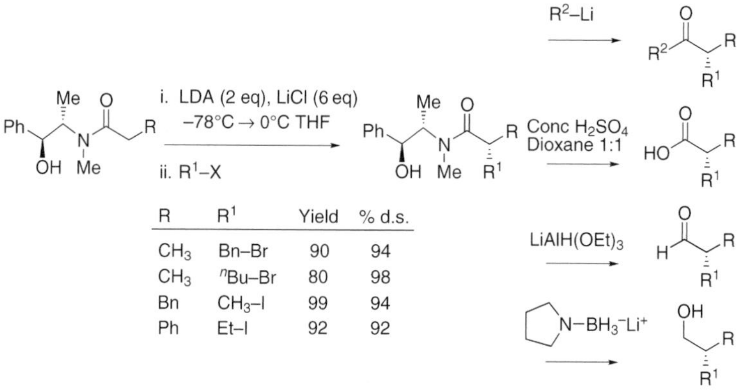

R	R^1	Yield	% d.s.
CH_3	Bn–Br	90	94
CH_3	^{n}Bu–Br	80	98
Bn	CH_3–I	99	94
Ph	Et–I	92	92

Scheme 3.30

be removed by using a number of different conditions to give a variety of products ranging from carboxylic acids to alcohols.

A versatile and often used method for the synthesis of α,α-dialkylcarboxylic acids is via alkylation of the Meyers' oxazolines. This method has been thoroughly reviewed.[66] In the example shown in Scheme 3.31 the formation of the (*S*)-acid **3.24** in 72%–78% e.e. is attributed to the rigid chelated intermediate **3.25** where the lithium cation is located on the underside of the molecule and the phenyl group shields the top face from attack. The presence of the other azaenolate **3.26** gives the opposite configuration, reducing the e.e. of the product.

The generation of carbon–carbon bonds to form a quaternary asymmetric centre has long been a challenge for the synthetic organic chemist. This has been effectively achieved in a limited number of systems by alkylation of enolate anions. For example, the generation of quaternary asymmetric centres in

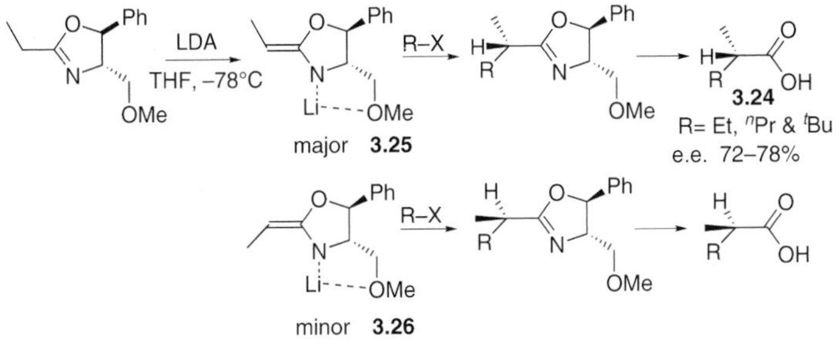

Scheme 3.31

cyanoacetic acid derivatives of the C(2) symmetric pyrrolidine derivative **3.27** has been achieved.[67] The success of these reactions is attributed to the small size of the cyano group, reducing the steric congestion at the enolate centre and the formation of the same enolate from both isomers of the monoalkylated starting material (Scheme 3.32).

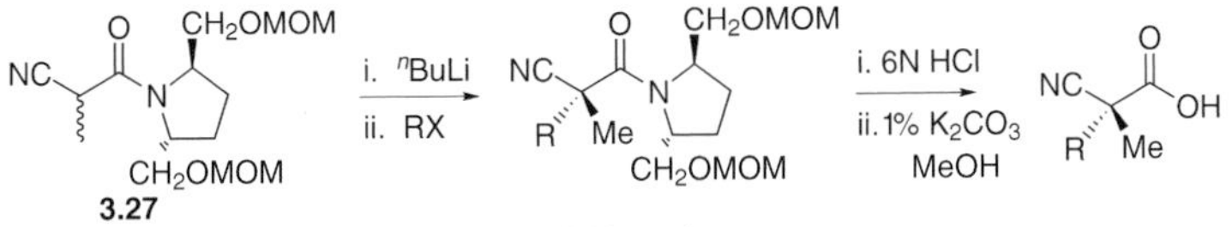

RX	Yield	% d.s.
Et–I	96	90
H₂C=CHCH₂Br	96	90
Bn–Br	96	85

Scheme 3.32

Another instance where alkylation of a dianion enolate forms a quaternary centre is seen in the example below, which utilised a 8-phenylmenthol derived chiral auxiliary,[68] which forms a half ester with 2-methylmalonic acid (Scheme 3.33).

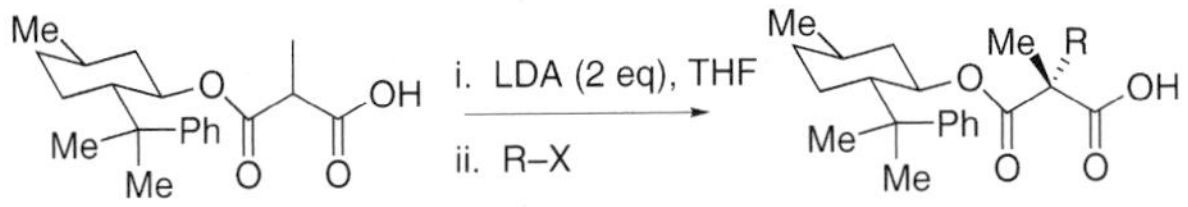

RX	Yield	% d.s.
Et–I	83	80
Pr–I	72	80
H₂C=CHCH₂–I	77	87.5
Bn–Br	72	92.3

Scheme 3.33

One of the best and certainly one of the most used methods for the generation of quaternary centres by alkylation of enolate anions is that developed by Meyers and Wanner.[69] This process involves the sequential alkylation of bicyclic lactams and has been reviewed.[70] The example shown in Scheme 3.34

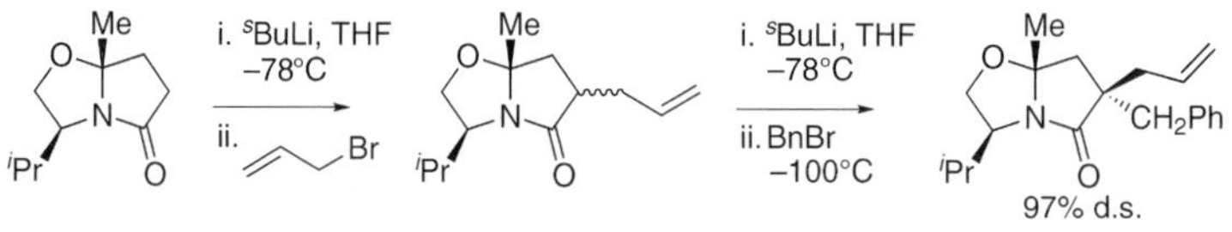

Scheme 3.34

is typical in that the first alkylation is nonselective but the critical second alkylation is highly facially selective. The selectivity of this alkylation has been rationalised[71] by the postulate that the lone pair of electrons on the convex β-face of the molecule perturb the π-system of the enolate thus favouring *endo* attack by the electrophile.

3.3   Stereoselective enolate oxidations and amination

The hydroxylation of an enolate results in conversion of a carbonyl compound into an α-hydroxy ketone and this has been carried out stereoselectively in a number of cases. One method for the selective oxidation of the enolates substituted with the oxazolidinone auxiliary, by means of an oxaziridine, has been developed by Evans and co-workers[72] and was applied to the total synthesis of calyculin (Scheme 3.35).[73] This example is significant, as the enolate is rather

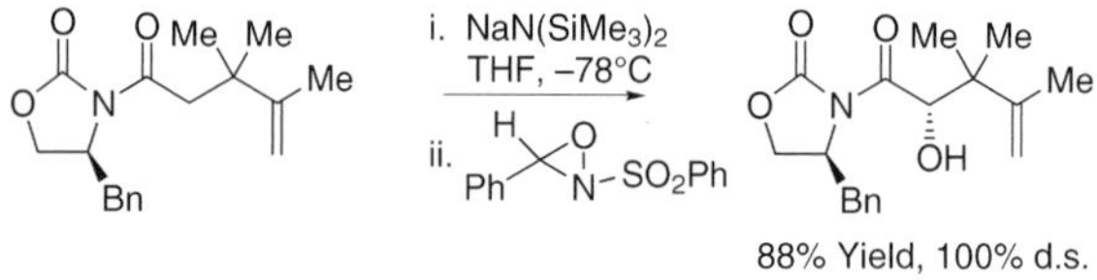

Scheme 3.35

hindered yet the reaction proceeds with high yield and only one isomer of the product is observed.

Another successful case of control from a chiral auxiliary is seen in the camphor-derived auxiliary **3.28** shown in Scheme 3.36. The potassium enolate

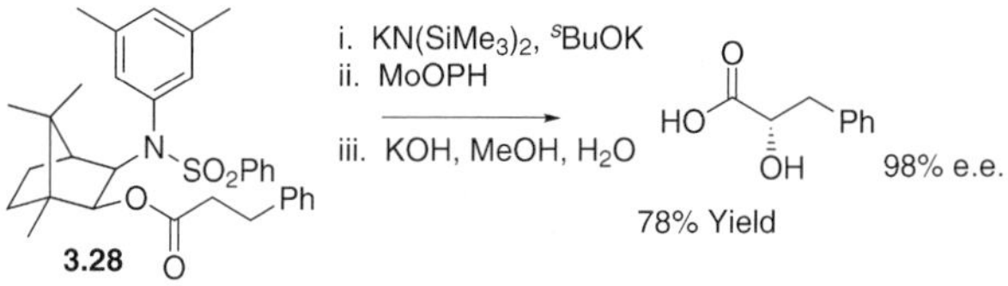

Scheme 3.36

is oxidised with $MoO_5 \cdot Py \cdot HMPT$ (MoOPH) complex.[74] In this case the presence of an excess of $K(^{sec}BuO)$ is required to obtain this high selectivity.

Similarly, it is possible to generate chiral amino acids by electrophilic amination of *N*-acyl benzyloxazolidinone enolates (Scheme 3.37). The

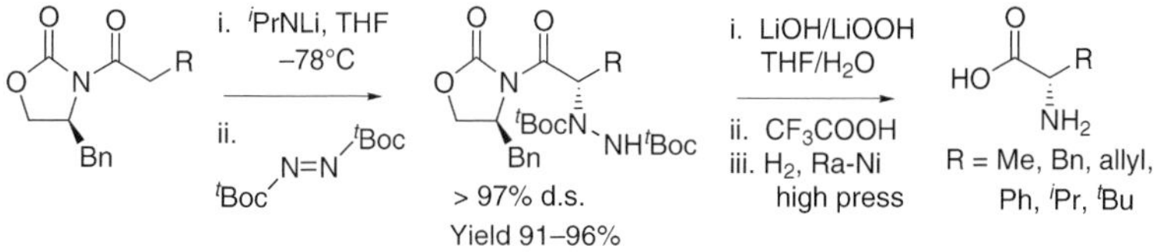

Scheme 3.37

protocols of Evans *et al.*[75] and Trimble and Vederas[76] are proving to be efficient methods, and an advantage of this approach is that the nature of the alkyl group in the amino acid target is not as critical as in the alternative alkylation approach discussed earlier (Section 3.2.7).

Evans and Britton have also demonstrated the efficient direct azide transfer to chiral enolates (Scheme 3.38).[77] In this case the use of the highly electropositive

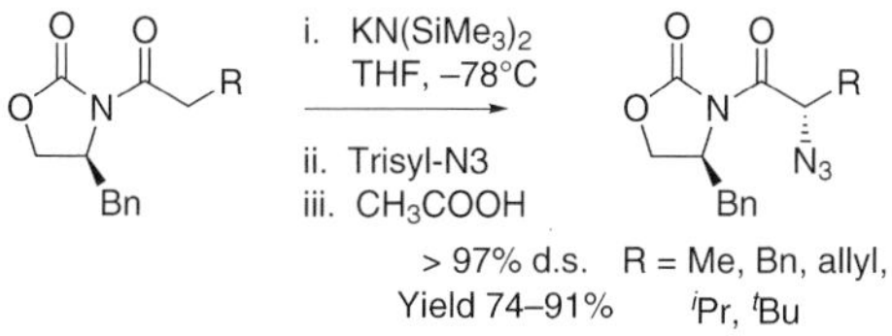

Scheme 3.38

K^+ counterion of the enolate, generated with potassium hexamethydisilazide, and the sterically demanding trisyl azide gives the best yield and selectivity. The sense of the asymmetric induction is consistent with electrophilic attack on the *Si* face of the (*Z*)-*O*-enolate.

In an interesting reaction sequence Oppolzer and Tamura[78,79] have introduced a new method for the preparation of chiral α-hydroxylaminoacids via the oxidation of enolates from *N*-acylated bornane-10,2-sultams **3.29** with 1-chloro-1-nitrosocyclohexane as a practical [HONH⁺] equivalent. The auxiliary is readily removed under normal hydrolysis conditions (Scheme 3.39).

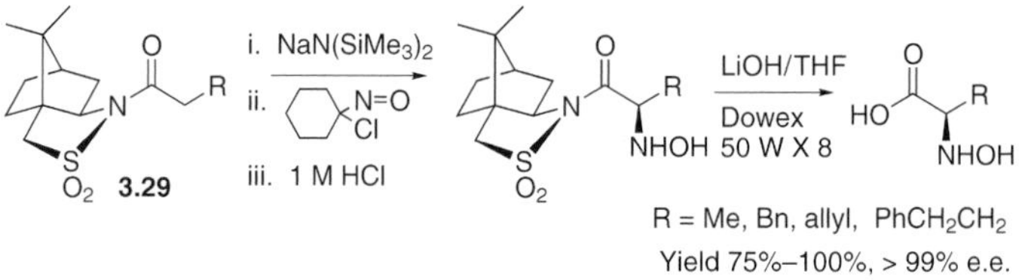

R = Me, Bn, allyl, PhCH₂CH₂
Yield 75%–100%, > 99% e.e.

Scheme 3.39

A simple method for the conversion of the hydroxylamines to the amino acids using zinc/aqueous HCl is also reported.

3.4   Formation and reaction of other enolates

3.4.1   Boron enolates

Boron enolates are now regarded as highly versatile and effective intermediates in stereoselective synthesis and are generally prepared from the carbonyl compound by treatment with a dialkylboron triflate and a tertiary amine (usually triethylamine or diisopropylethylamine) according to the Mukaiyama method.[80–82] This method has been extensively used in the development of boron aldol reactions. The boron enolates produced by the Mukaiyama method are generally believed to form under kinetic control, by deprotonation of the carbonyl compound complexed to the Lewis acid. This usually leads to the formation of the (*Z*)-*O*-enolate (Scheme 3.40).

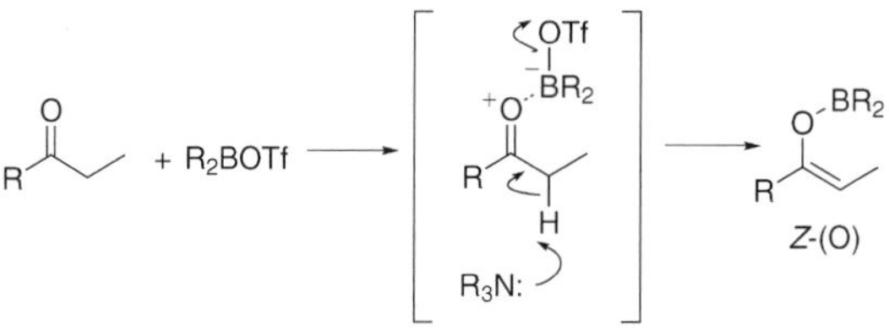

Scheme 3.40

Formation of the (E)-O-enolate was possible by modification of the Mukaiyama procedure. An early report is that of Masamune and Van Horn who found that either the (Z)-O- or (E)-O-enolate of cyclohexyl ethyl ketone could be formed by variation of the boron triflate used in the enolisation procedure (Scheme 3.41).[83]

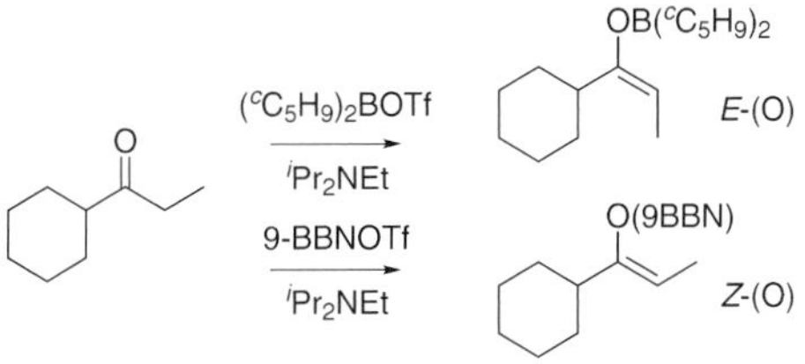

Scheme 3.41

This selectivity has been explained in terms of deprotonation of *iso*-meric Lewis acid complexes where the Lewis acid is either *cis* or *trans* relative to the acidic α-hydrogens (Scheme 3.42).

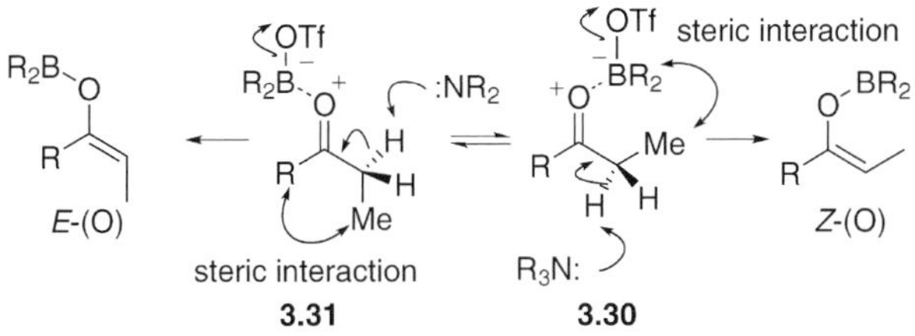

Scheme 3.42

For both complexes there is some steric interaction, either between the methyl group of the enolate and the ligands on the boron for conformer **3.30** or between the methyl group of the enolate and the R group attached to the carbonyl for conformer **3.31**. This rationalises the observation that the larger the ligands on the boron the more (E)-O-enolate is formed. It has also been found that use of the boron chloride reagents instead of the triflates favours the formation of (E)-O-enolate, and a general protocol for the formation of this enolate has been developed by Brown *et al.*[84] with use of dicyclohexylboron chloride and triethylamine. A theoretical study of the enolisations of ketones by dialkylboron chlorides and triflates has been carried out by Goodman and Paterson.[85,86]

From a practical perspective the selective formation of the (Z)-O-enolate **3.32** can generally be achieved by using the boron triflate with small ligands

(^{n}Bu or Et) and a sterically hindered base, whereas the (E)-O-enolate **3.33** can be formed by reaction of the boron chloride with large ligands (cyclohexyl) and a small base (Et$_3$N or Me$_2$NEt) (Scheme 3.43).

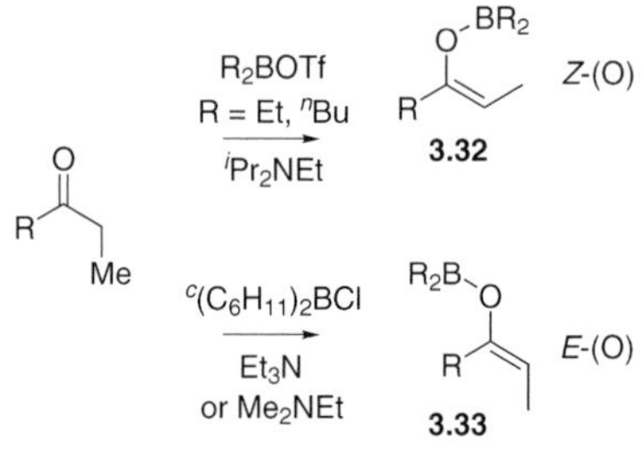

Scheme 3.43

The major limitation of boron enolate chemistry is that it is not presently possible to form boron enolates from carbonyl compounds which are significantly less acidic than ketones. Also, boron enolates are not highly reactive and therefore alkylations are generally not possible. However, these enolates are very effective in the aldol reaction with aldehydes, and this application is discussed in Chapters 4 and 5.

The boron enolates have been shown to undergo α-bromination with NBS in a stereoselective manner. These bromides have been converted into the azides with inversion of configuration by reaction with azide ion (Scheme 3.44).[87,88]

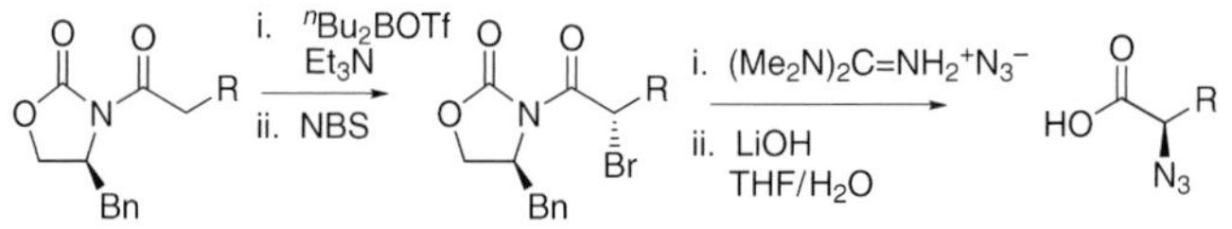

Scheme 3.44

3.4.2   *Titanium enolates*

A versatile method for the direct preparation of titanium enolates from a variety of ketones has been developed by the Evans group, in which the carbonyl compound is first complexed with TiCl$_4$ at $-78°C$ for five minutes followed by the addition of either triethylamine or diisopropylethylamine to form the enolate.[89] These enolates have proved to undergo facile alkylation with reactive electrophiles that react by S$_N$1-like paths, such as α-haloethers, acetals and orthoesters (Scheme 3.45).

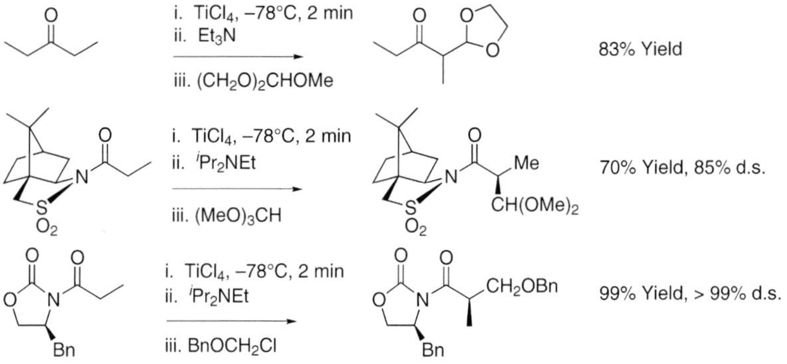

Scheme 3.45

With the oxazolidinone-based auxiliary, the titanium is postulated to coordinate to the carbonyl of the amide, and the electrophile attacks from the face opposite to the benzyl group (Scheme 3.46).

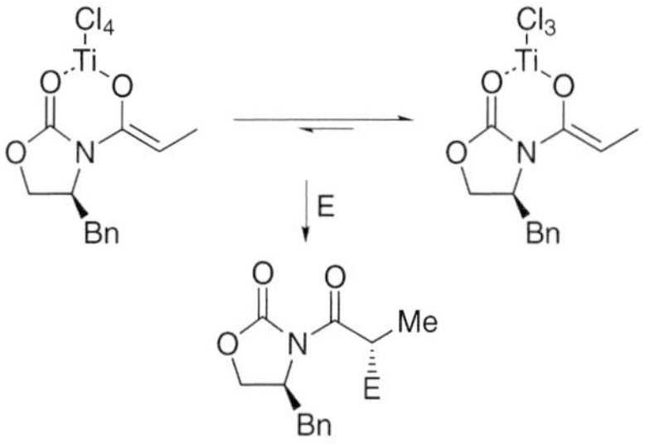

Scheme 3.46

References

1. Schmidt, J.G., *Ber. Dtsch. Chem. Ges.*, **1880**, *13*, 2341.
2. Claisen, L. and Claparède, A., *Ber. Dtsch. Chem. Ges.*, **1881**, *14*, 349.
3. Schmidt, J.G., *Ber. Dtsch. Chem. Ges.*, **1881**, *14*, 1459.
4. Claisen, L., *Justus Liebigs Ann. Chem.*, **1899**, *306*, 322.
5. Claisen, L. and Lowman, O., *Ber. Dtsch. Chem. Ges.*, **1887**, *20*, 651.
6. Reformatsky, S., *Ber. Dtsch. Chem. Ges.*, **1887**, *20*, 1210.
7. Hauser, C.R. and Puterbaugh, W.H., *J. Am. Chem. Soc.*, **1951**, *73*, 2972.
8. Hauser, C.R. and Puterbaugh, W.H., *J. Am. Chem. Soc.*, **1953**, *75*, 1068.
9. Hauser, C.R. and Lindsay, J.K., *J. Am. Chem. Soc.*, **1955**, *77*, 1050.
10. Hauser, C.R. and Lendicer, D., *J. Org. Chem.*, **1957**, *22*, 1248.
11. Dunnavant, W.R.H. and Hauser, C.R., *J. Org. Chem.*, **1960**, *25*, 503.
12. House, H.O. and Trost, B.M., *J. Org. Chem.*, **1965**, *30*, 2502.

13. House, H.O., Gall, M. and Olmstead, H.D., *J. Org. Chem.*, **1971**, *36*, 2361.
14. Rathke, M.W., *J. Am. Chem. Soc.*, **1970**, *92*, 3222.
15. Evans, D.A., in *Asymmetric Synthesis*, Ed. J.D. Morrison, **1984**, Academic Press, Orlando, FL, vol. 3, p. 1.
16. Heathcock, C.H., in *Modern Synthetic Methods*, Ed. R. Scheffold, Verlag Helvetica Chimica Acta, Basel, **1992**, vol. 6, p. 1.
17. Lee, R.A., McAndrews, C., Patel, K.M. and Reusch, W., *Tetrahedron Lett.*, **1973**, 965.
18. House, H.O. and Kramer, V., *J. Org. Chem.*, **1963**, *28*, 3362.
19. Mekelburger, H.B. and Wilcox, C.S., in *Comprehensive Organic Synthesis*, Ed. B.M. Trost, Pergamon, Oxford, **1991**, vol. 2, p. 99.
20. Pearson, R.G., *J. Chem. Educ.*, **1968**, *45*, 643.
21. Klopman, G., *J. Am. Chem. Soc.*, **1968**, *90*, 223.
22. Jones, M.E., Kass, S.R., Filley, J., Barkly, R.M. and Ellison, G.B., *J. Am. Chem. Soc.*, **1985**, *107*, 109.
23. Houk, K.N. and Paddon-Row, M.N., *J. Am. Chem. Soc.*, **1986**, *108*, 2659.
24. House, H.O., Czuba, L.J., Gall, M. and Olmstead, H.D., *J. Org. Chem.*, **1969**, *34*, 2324.
25. Dubois, J.E. and Fellmann, P., *Tetrahedron Lett.*, **1975**, 1225.
26. Dubois, J.-E. and Dubois, M., *Tetrahedron Lett.*, **1967**, 4215.
27. Ireland, R.E. and Willard, A.K., *Tetrahedron Lett.*, **1975**, 3975.
28. Heathcock, C.H., Buse, C.T., Klieschick, W.A., Pirrung, M.C., Sohn, J.E. and Lampe, J., *J. Org. Chem.*, **1980**, *45*, 1066.
29. Fataftah, Z.A., Kopka, I.E. and Rathke, M.W., *J. Am. Chem. Soc.*, **1980**, *102*, 3959.
30. Masamune, S., Elingboe, J.W. and Choy, W., *J. Am. Chem. Soc.*, **1982**, 5526.
31. Ireland, R.E., Wipf, P. and Armstrong III, J.D., *J. Org. Chem.*, **1991**, *56*, 650.
32. Ireland, R.E. and Mueller, R.E., *J. Am. Chem. Soc.*, **1972**, *94*, 5897.
33. Ireland, R.E., Mueller, R.H. and Willard, A.K., *J. Am. Chem. Soc.*, **1976**, *98*, 2868.
34. Evans, D.A. and McGee, L.R., *Tetrahedron Lett.*, **1980**, *21*, 3975.
35. Moreland, D.W. and Dauben, W.G., *J. Am. Chem. Soc.*, **1985**, *107*, 2264.
36. Agami, C., Levisalles, J. and Lo Cicero, B., *Tetrahedron*, **1979**, *35*, 961.
37. Stork, G., Brizzolara, A., Landsman, H. and Terrell, R., *J. Am. Chem. Soc.*, **1963**, *85*, 207.
38. Hickmott, P.W., *Tetrahedron*, **1982**, *38*, 3363.
39. Whitesell, J.K. and Whitesell, M.A., *Synthesis*, **1983**, 517.
40. Whitesell, J.K. and Felman, S.W., *J. Org. Chem.*, **1977**, *42*, 1663.
41. Stork, G. and Dowd, S.R., *J. Am. Chem. Soc.*, **1963**, *85*, 2178.
42. Houk, K.N., Strozier, R.W., Rondan, N.G., Fraser, R.R. and Chuaqui-Offermanns, N., *J. Am. Chem. Soc.*, **1980**, *102*, 1426.
43. Meyers, A.I., Williams, D.R. and Druelinger, M., *J. Am. Chem. Soc.*, **1976**, *98*, 3032.
44. Meyers, A.I., Williams, D.R., Erickson, G.W., White, S. and Druelinger, M., *J. Am. Chem. Soc.*, **1981**, *103*, 3081.
45. Whitesell, J.K. and Whitesell, M.A., *J. Org. Chem.*, **1977**, *42*, 377.
46. Meyers, A.I., Williams, D.R., White, S. and Erickson, G.W., *J. Am. Chem. Soc.*, **1981**, *103*, 3088.
47. Ludwig, J.W., Newcomb, M. and Bergbreiter, D.E., *J. Org. Chem.*, **1980**, *45*, 4666.
48. Yamashita, M., Matsumiya, K., Tanabe, M. and Suemitsu, R., *Bull. Chem. Soc. Jpn.*, **1985**, *58*, 407.
49. Enders, D., in *Asymmetric Synthesis*, Ed. J.D. Morrison, Academic Press, Orlando, FL, **1984**, vol. 3, p. 275.
50. Enders, D., Eichenauer, H., Baus, U., Schubert, H. and Kremer, A.M., *Tetrahedron*, **1984**, *40*, 1345.
51. Collum, D.B., Kahne, D., Gut, S.A., DePue, R.T., Mohamadi, F., Wanat, R.A., Clardy, J.C. and Van Duyne, G., *J. Am. Chem. Soc.*, **1984**, *106*, 4865.
52. Wanat, R.A. and Collum, D.B., *J. Am. Chem. Soc.*, **1985**, *107*, 2078.

53. Evans, D.A. and Takacs, J.M., *Tetrahedron Lett.*, **1980**, *21*, 4233.
54. Evans, D.E., Dow, R.L., Shih, T.L., Takacs, J.M. and Zahler, R., *J. Am. Chem. Soc.*, **1990**, *112*, 5290.
55. Sonnet, P.E. and Heath, R.R., *J. Org. Chem.*, **1980**, *45*, 3137.
56. Askin, D., Volante, R.P., Ryan, K.M., Reamer, R.A. and Shinkai, I., *Tetrahedron Lett.*, **1988**, *29*, 3137.
57. Kawenami, Y., Ito, Y., Kitagawa, T., Taniguchi, Y. and Yamaguchi, M., *Tetrahedron Lett.*, **1984**, *25*, 857.
58. Enomoto, M., Ito, Y., Katsuki, T. and Yamaguchi, M., *Tetrahedron Lett.*, **1985**, *26*, 1343.
59. Evans, D.A., Ennis, M.D. and Mathre, D.J., *J. Am. Chem. Soc.*, **1982**, *104*, 1737.
60. Schmierer, R., Grotenmeirer, G., Helmchen, G. and Selim, A., *Angew. Chem., Int. Ed. Engl.*, **1981**, *20*, 207.
61. Helmchen, G., Selim, A., Dorsch, D. and Taufer, I., *Tetrahedron Lett.*, **1983**, *24*, 3213.
62. Oppolzer, W., Moretti, R. and Thomi, S., *Tetrahedron Lett.*, **1989**, *30*, 5603.
63. Oppolzer, W., Moretti, R. and Thomi, S., *Tetrahedron Lett.*, **1989**, *30*, 6009.
64. Myers, A.G., Gleason, J.L. and Taeyoung, Y., *J. Am. Chem. Soc.*, **1995**, *117*, 8488.
65. Myers, A.G., Yang, B.R., Chen, H. and Gleason, J.L., *J. Am. Chem. Soc.*, **1994**, *116*, 9361.
66. Lutomski, K.A. and Meyers, A.I., in *Asymmetric Synthesis*, Ed. J.D. Morrison, Academic Press, Orlando, FL, **1984**, vol. 3, p. 213.
67. Hanamoto, T., Katsuki, T. and Yamaguchi, M., *Tetrahedron Lett.*, **1986**, *27*, 2463.
68. Ihara, M., Takashashi, M., Taniguchi, N., Yasui, K., Niitsuma, H. and Fukomoto, K., *J. Chem. Soc., Perkin Trans.*, **1991**, *1*, 525.
69. Meyers, A.I. and Wanner, K.T., *Tetrahedron Lett.*, **1985**, *26*, 2047.
70. Romo, D.A. and Meyers, A.I., *Tetrahedron*, **1991**, *47*, 9503.
71. Meyers, A.I. and Wallace, R.H., *J. Org. Chem.*, **1989**, *54*, 2509.
72. Evans, D.A. and Morrissey, M.M., *J. Am. Chem. Soc.*, **1985**, *107*, 4346.
73. Evans, D.A., Gage, J.R. and Leighton, J.L., *J. Am. Chem. Soc.*, **1992**, *114*, 9434.
74. Gamboni, R. and Tamm, C., *Helv. Chim. Acta*, **1986**, *69*, 615.
75. Evans, D.A., Britton, T.C., Dorow, R.L. and Dellaria, J.D., *J. Am. Chem. Soc.*, **1986**, *108*, 6395.
76. Trimble, L.A. and Vederas, J.C., *J. Am. Chem. Soc.*, **1986**, *108*, 6397.
77. Evans, D.A. and Britton, T.C., *J. Am. Chem. Soc.*, **1987**, *109*, 6881.
78. Oppolzer, W. and Tamura, O., *Tetrahedron Lett.*, **1990**, *31*, 991.
79. Oppolzer, W., Tamura, O. and Deerberg, J., *Helv. Chim. Acta*, **1992**, *75*, 1965.
80. Mukaiyama, T. and Inoue, T., *Chem. Lett.*, **1976**, 559.
81. Inoue, T., Uchimaru, T. and Mukaiyama, T., *Chem. Lett.*, **1977**, 153.
82. Inoue, T. and Mukaiyama, T., *Bull. Chem. Soc. Jpn.*, **1980**, 174.
83. Van Horn, D.E. and Masamune, S., *Tetrahedron Lett.*, **1979**, 2229.
84. Brown, H.C., Dhar, R.K., Bakshi, R.K., Pandiarajan, P.K. and Singaram, B., *J. Am. Chem. Soc.*, **1989**, *111*, 3441.
85. Goodman, J.M., *Tetrahedron Lett.*, **1992**, *33*, 7219.
86. Goodman, J.M. and Paterson, I., *Tetrahedron Lett.*, **1992**, *33*, 7223.
87. Evans, D.A., Britton, T.C., Ellman, J.A. and Dorow, R.L., *J. Am. Chem. Soc.*, **1990**, *112*, 4011.
88. Evans, D.A., Ellman, J.A. and Dorow, R.L., *Tetrahedron Lett.*, **1987**, *28*, 1123.
89. Evans, D.A., Urpí, F., Somers, T.C., Clark, J.S. and Bilodeau, M.T., *J. Am. Chem. Soc.*, **1990**, *112*, 8215.

4 Substrate-directed aldol reactions

4.1 The mechanistic basis of selectivity in aldol reactions

4.1.1 Background

Polypropionate natural products are characterised by the presence of a linear carbon chain with methylation and oxygenation on alternate carbons along the chain. The biosynthesis of these natural products is known to occur by condensation of acetate or propionate equivalents in a linear fashion building up the carbon chain.[1,2] The quest for efficient methods to construct polyketide natural products has driven the development of stereocontrolled crossed-aldol reactions between acyclic aldehydes and ketones. This biomimetic approach to the synthesis of the diverse array of structures has resulted in the total synthesis of a series of natural products with an everincreasing structural and stereochemical complexity.

The early development of aldol reactions involved the use of base which at equilibrium would only partially deprotonate the carbonyl compound, and the reactions were carried out in the presence of the electrophile under equilibrating conditions. These methods have been thoroughly reviewed[3,4] and will not be elaborated in this text. This chapter will focus on the aldol reactions of stoichiometric acyclic enolates formed by using bases much more basic than the carbonyl compounds or the combination of Lewis acids and weak bases. This 'directed' aldol reaction was first reviewed in 1982[5] and has been developed extensively since then.

4.1.2 Stereochemistry and transition state models

We discussed both the regioselective and the stereochemical aspects associated with the formation of stoichiometric enolates in the previous chapter. The aldol reaction involves the reaction of these enolates with a carbonyl compound (usually an aldehyde) to produce an alcohol and a new stereocentre. If we first consider the simple case where the enolate is not substituted at the α-carbon, then reaction with an aldehyde produces one stereocentre (i.e. two isomers, which are enantiomers if R^1 and R^2 are achiral, but diastereoisomers if R^1 or R^2 are chiral; Scheme 4.1).

These reactions are usually considered to occur via closed six-membered cyclic Zimmerman–Traxler[6] transition states, in which coordination between the aldehyde carbonyl oxygen and the enolate metal centre occurs. For the reaction of an enolate, unsubstituted at the α-carbon, attack on the *Si* face of the

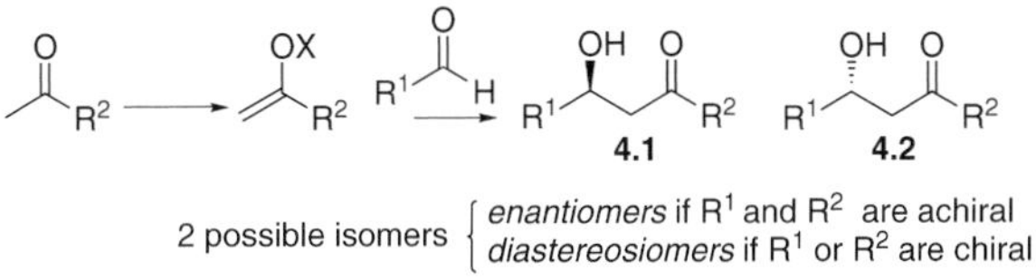

Scheme 4.1

aldehyde can occur via either transition state (ts) **I** or **IV**, and attack on the *Re* face can occur through transition states **II** or **III** (Scheme 4.2).

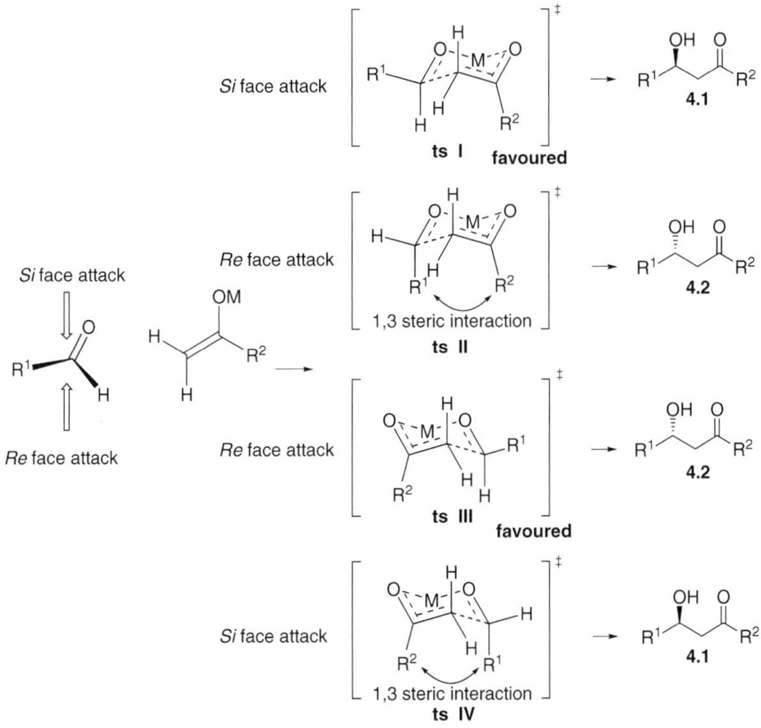

Scheme 4.2

The preferred transition states (**I** and **III**) are those in which the alkyl group of the aldehyde occupies an equatorial position. If R^1 or R^2 do not contain any further stereocentres then transition states **I** and **III** will be enantiomeric and thus equal in energy. If either R^1 or R^2 contain a further stereocentre then this is no longer the case and selective formation of product **4.1** over **4.2**

is possible. High levels of selectivity are most often obtained by using chiral
enolates. However, in some cases the effect of a chiral aldehyde can be significant and in cases where both enolate and aldehyde are chiral the effect of
the aldehyde should be considered. This is the area of double stereodifferentiating reactions, and is considered later in the chapter. Alternatively, if there are
chiral ligands coordinating to the metal, this again makes the transition states
diastereomeric and has been used to achieve *enantioselective* reactions (here
4.1 and **4.2** are enantiomers and this type of reaction is discussed in Chapter 5).
Chair transitions states have been discussed above, but often *boat* transition
states are implicated in the reaction of unsubstituted enolates, and these will be
addressed later.

The reaction of enolates which are substituted at the α-carbon is somewhat
more complicated in that now the bond formations give two new stereocentres
(i.e. four possible isomers, **4.3–4.6**). There are two enantiomeric pairs (**4.3** and
4.4) and (**4.5** and **4.6**) if R^1, R^2 and R^3 are achiral. If any one of R^1, R^2 and R^3
are chiral then they are all diastereomers (Scheme 4.3).

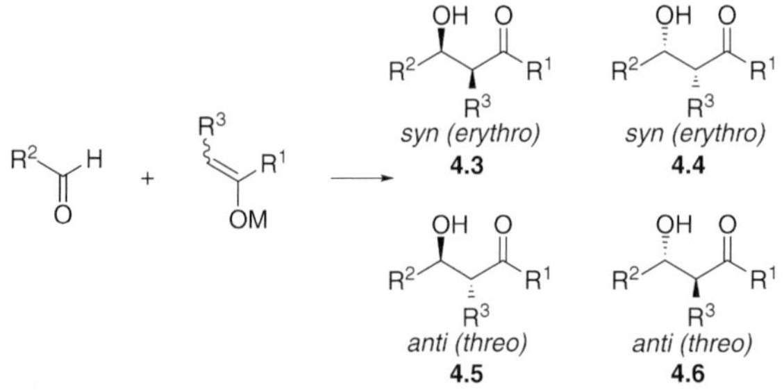

Scheme 4.3

If we consider the reaction between a (Z)-O-enolate and an aldehyde
(Scheme 4.4), which occurs through a closed six-membered cyclic Zimmerman–
Traxler[6] transition state, then reaction of the *Si* face of the aldehyde with the *Re'*
face of the enolate is possible via transition state **V** giving the *syn* product **4.3**.
Alternatively, attack on the *Re* face of the aldehyde by the *Re'* face of the enolate
by transition state **VI** gives the *anti* aldol **4.6**. As is the case for the unsubstituted enolates, the preferred transition state is **V** in which the alkyl group of the
aldehyde is equatorially orientated and 1,3 steric interaction (cf. transition state
VI) is avoided. This accounts for the observation that (Z)-O-enolates usually
give *syn* aldol products. It should not be forgotten that attack on the *Re* face
of the aldehyde by the *Si'* face of the enolate via a transition state which is
enantiomeric to transition state **V** is also possible and this would give the aldol
product **4.4** (compare transitions states **I** and **III** in Scheme 4.2).

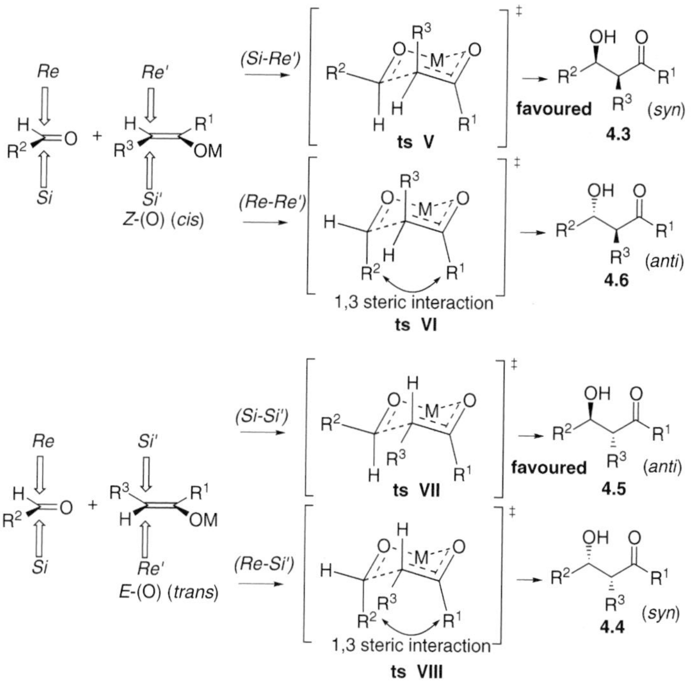

Scheme 4.4

Similarly, the observed preference for (E)-O-enolates to give *anti* aldol products is rationalised by a preference for transition state **VII** over **VIII** on the grounds that **VIII** is destabilised by a steric interaction between R^1 and R^2. It is thus apparent that control of the stereoselection in an aldol reaction first requires selective formation of the enolate. Once formed, the enolates react with greater or lesser stereoselectivity, under the influence of a number of factors, including the nature of R^1, R^2 and the metal. Thus when selecting an α-substituted enolate for a synthetic procedure, it must be remembered that different enolates (with lithium, boron, titanium, tin, etc.) can have quite different selectivities. The reaction and synthetic utility of these different enolates will be considered in the following discussion.

4.1.3 *Thermodynamic* vs *kinetic control*

Implicit in the discussion above, where isomeric product ratios are rationalised by considering the relative energies of competing transition states, is that the reactions are occurring under *kinetic* control. Although this is true in the major-ity of cases, it is possible for aldol products to undergo *syn–anti* equilibration either by deprotonation or more commonly by reverse aldolisation. Group I and II enolates undergo aldol reaction quite rapidly at $-78°C$ but the $\Delta G°$ of reaction is not normally very negative. Thus reverse aldol reactions can occur under relatively mild conditions and have at least some influence on the aldol stereochemistry. The most important factor is the basicity of the enolate. The more basic the enolate the less likely it is to undergo reverse aldolisation. Thus the enolates of esters and amides show little tendency toward reversal whereas ketone aldolates often undergo reversal under mild conditions; it should, however, be remembered that if a reaction is 80:1 *syn* selective then the reverse reaction must occur 80 times for the *syn* aldol to convert to the *anti* aldol. This is seen in Scheme 4.5 where the aldolate **4.7** from anisaldehyde and

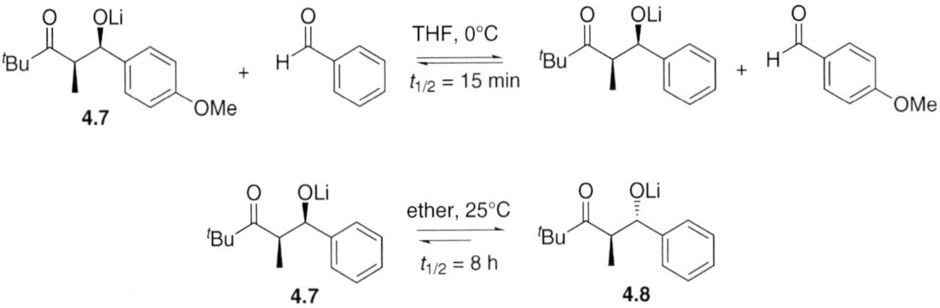

Scheme 4.5

ethyl-*t*-butyl ketone exchanges with benzaldehyde with a half-life ($t_{1/2}$) of 15 min. Yet equilibration from the *syn* isomer to the *anti* isomer **4.8** has a half-life of 8 h at 25°C. This long half-life has been attributed[7] to the high *syn* selectivity of the (Z)-O-enolate formed in the reverse aldol reaction.

4.2   Reactions of lithium enolates

The use of preformed enolates in stereoselective aldol reactions goes back to the work of Dubois and Fellmann[8] who addressed the implication of enolate geometry on the *syn* to *anti* ratio of the aldol products. They observed that the lithium (Z)-O-enolate **4.9** of diethyl ketone gave a *syn* to *anti* ratio of 88:12 on reaction with pivaldehyde whereas the (E)-O-enolate gave a *syn* to *anti* ratio of almost 1:1 (Scheme 4.6).

    That the (Z)-O-enolates show high *syn* selectivity has been shown to be generally true provided the group attached to the carbonyl end of the enolate (R in Scheme 4.7) is sterically large. This has been examined by Heathcock *et al.*[7,9,10] and is compared with the selectivity of the (E)-O-enolates in Scheme 4.7. In contrast, the enolates of these simple (E)-O-enolates show a general trend to relatively low levels of selectivity. Why simple (E)-O-enolates should be

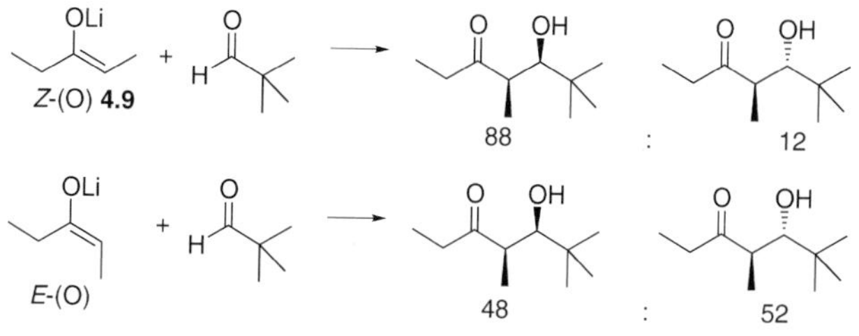

Scheme 4.6

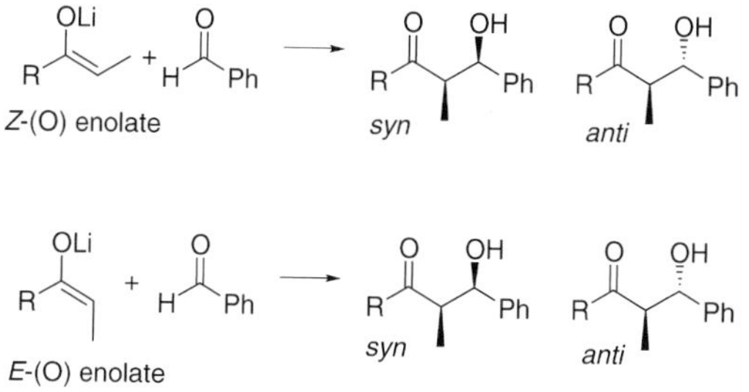

R	% syn	% anti
H	50	50
Et	90	10
^{i}Pr	90	10
^{t}Bu	98.7	1.3

R	% syn	% anti
H	60	40
Et	40	60
^{i}Pr	50	50

Scheme 4.7

significantly less selective than their Z-(O) counterparts is not explained by simple analysis of the putative transition states (**V** to **VIII**; Scheme 4.4), and alternative *boat* transition states have been postulated.[11]

This high *syn* selectivity observed for (Z)-O-enolates with large groups attached led to the development of the reagent **4.10** for selective aldol reactions.[12] This reagent, because of the bulky α-dimethyltrimethylsiloxy groups, enolises with LDA to give the (Z)-O-enolate, which reacts with a variety of aldehydes to give *syn* aldol products. The synthetic utility of this reagent is seen in the removal of the achiral auxiliary to make it an effective *syn* selective propanal synthon (Scheme 4.8).

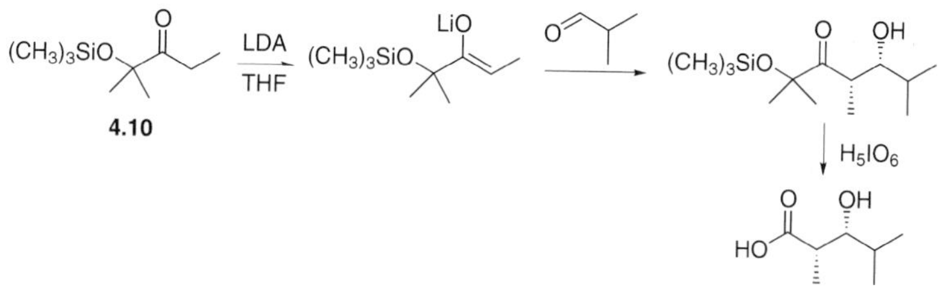

Scheme 4.8

A similar *syn* selective propanal synthon **4.11** has been reported more recently by Mori *et al.*[13] (Scheme 4.9).

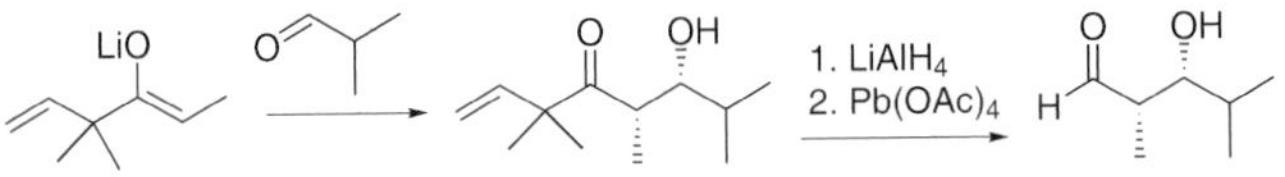

4.11

Scheme 4.9

In the previous two examples the relative stereochemistry (*syn* or *anti*) was controlled but there was no other stereocontrol. It is possible to control further the course of the reaction of these achiral enolates if they are allowed to react with chiral aldehydes. A significant example of stereocontrol by the α-stereocentre of a chiral aldehyde reacting with this lithium enolate is seen in Heathcock *et al.*'s synthesis[14] of the C(1)–C(7) segment of erythronolide A. Here the lithium enolate from ketone **4.10** gives the aldol products **4.12** and **4.13** in the ratio 6:1 (Scheme 4.10). In this example, the aldehyde shows Felkin

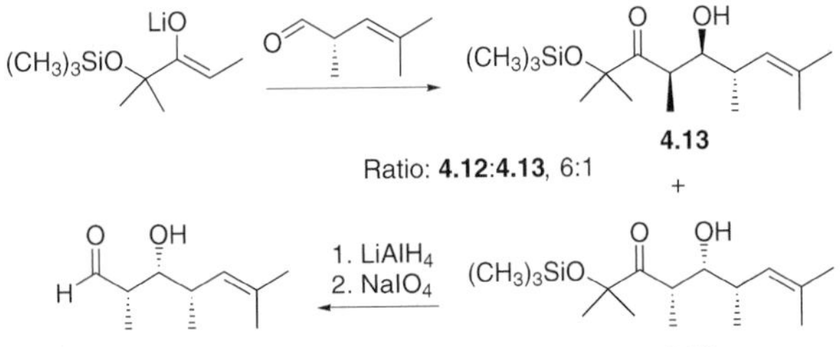

Ratio: **4.12:4.13**, 6:1

Scheme 4.10

(see Chapter 2) selectivity, but this is often not the case for the reaction of (*Z*)-*O*-enolates with aldehydes (see Section 4.3.2).

A chiral version of the bulky *Z*-(O) lithium enolate **4.10** was developed by Heathcock *et al.*,[15,16] and its reaction with a series of aldehydes is shown in Scheme 4.11. Here the chiral enolate controls the course of the reaction giving *syn–anti* isomer as the major product. A more recent paper[17] has reported an optimised experimental procedure which gives considerably higher selectivities (the results are included in Scheme 4.11).

The good-to-excellent selectivity observed for this reaction has been rationalised by a transition state in which the lithium coordinates to the three oxygen

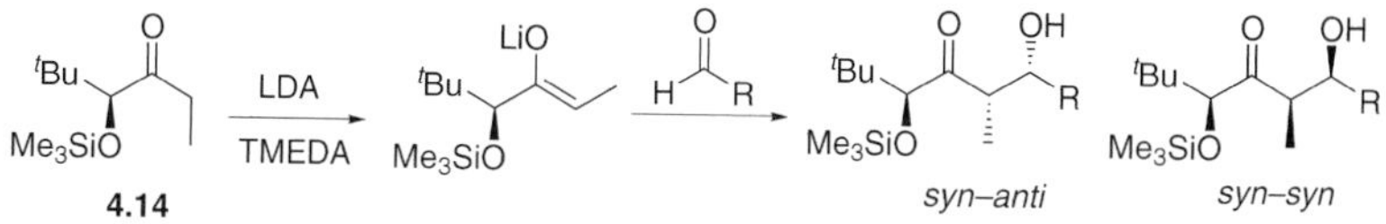

Scheme 4.11

R	% syn–anti	% syn–syn
Ph	> 95	< 5
^tBu	> 95	< 5
^iPr	> 95	< 5
$PhCH_2$	87	13
Ph_2CH	> 90	< 10

atoms (Scheme 4.12). In this transition state the small hydrogen (not the large *t*-butyl) points towards the middle of the transition state.

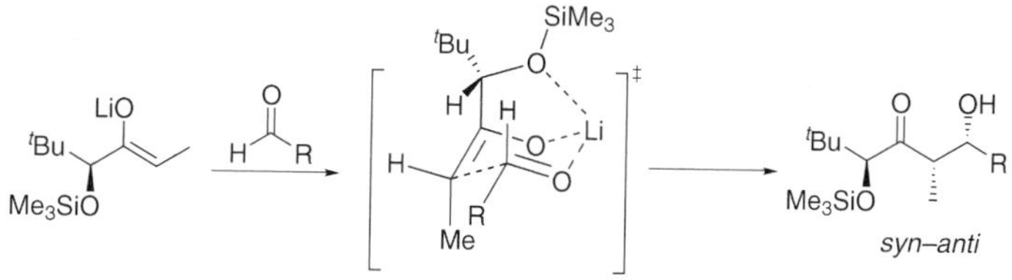

Scheme 4.12

Whereas in general it has been noted that lithium (*E*)-*O*-enolates show low selectivity, the magnesium (*E*)-*O*-enolate of ketone **4.14** prepared from the ketone with *N*-bromomagnesium-2,2,6,6-tetramethylpiperidine (BrMgTMP) in THF shows high selectivity, giving *anti* aldols in excess of 92% diastereoselectivity (Scheme 4.13).

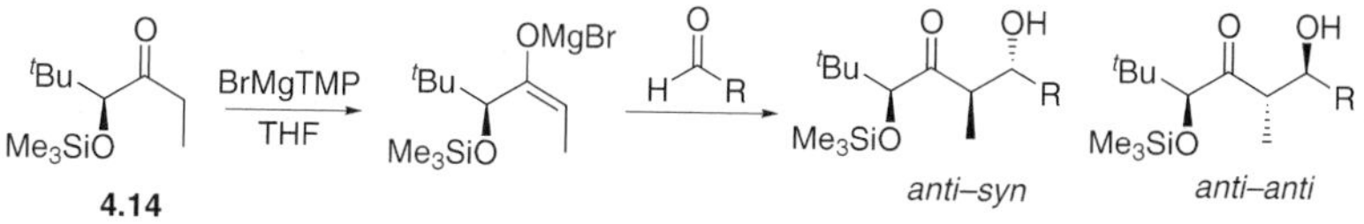

R	% anti–syn	% anti–anti
Ph	95	5
^tBu	95	5
^iPr	92	8

Scheme 4.13

Another auxiliary-based chiral enolate is the sultam **4.15** developed by Oppolzer for lithium and boron metals, which shows reasonable levels of selectivity (compare the percentage of **4.16** with that of **4.17**), as indicated in Scheme 4.14.[18] Both this enolate and the lithium enolate of **4.14** show the

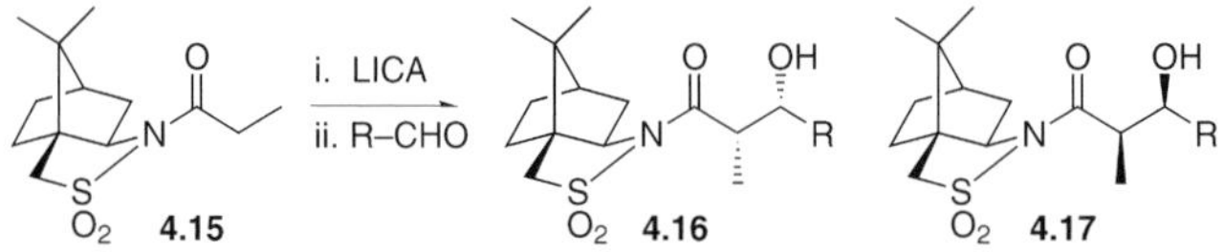

LICA = lithium	R	% **4.16**	% **4.17**	% other
N-isoprpycyclohexylamide	Ph	85	8	7
	^{t}Bu	88	7	5
	^{i}Pr	86	8	6

Scheme 4.14

opposite facial selectivity to that of their analogous boron enolates. These boron enolates are discussed in the next section.

Although esters are known to show highly selective formation of the (E)-O-enolates, in general this does not translate into good *anti* to *syn* selectivity in the aldol reaction.[11] However, notable exceptions are the enolates derived from 2,6-disubstituted phenyl propionates **4.18** which show exceptional *anti* selectivity,[10,19] offering considerable potential for use in synthesis (Scheme 4.15).

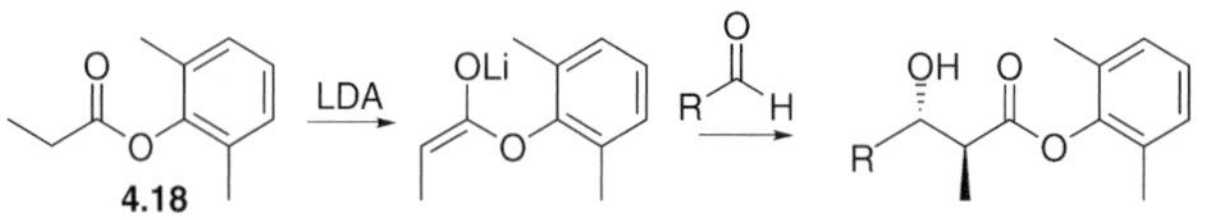

R	anti:syn
Ph	88:12
$^nC_5H_{11}$	86:14
^iPr	> 98:2
^tBu	> 98:2

Scheme 4.15

4.3   Reactions of boron enolates

4.3.1   Background

The reactions of boron enolates[20–22] generally show higher selectivity than the corresponding lithium enolate, which in part can be attributed to the shorter B—O bond compared with the Li—O bond, resulting in 'tighter' and more organised transition states. Of all the enolates used in synthesis, boron enolates are the best characterised and understood in their reaction selectivity. They exist in solution as homogenous monomers uncomplicated in terms of aggregation (their lithium counterparts are very much complicated by dimeric, tetrameric, etc., forms in solution). Importantly, the enolate geometry of boron enolates is faithfully translated into aldol stereochemistry as predicted by the closed six-membered ring transition state models (Scheme 4.4). Thus (Z)-O-enolates give *syn* aldol products whereas (E)-O-enolates give *anti* aldol products. Selective formation of the correct enolate geometry is therefore critically important. Fortunately, in the case of direct enolate formation it is usually possible to selectively prepare either enolate by enolisation of simple ethyl ketones. A combination of small ligands on boron (ethyl, *n*-butyl), a good leaving group on boron (triflate), and a bulky amine base (diisopropyl ethylamine) usually gives the (Z)-O-enolate. Conversely, the presence of bulky ligands on the boron (cyclohexyl), a poor leaving group (chloride), and a small base (triethylamine) gives the (E)-O-enolate (Scheme 4.16; see also Chapter 3, Section 3.2.1).

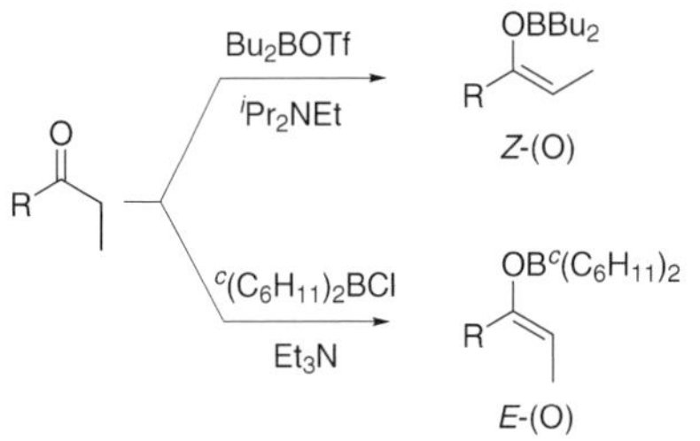

Scheme 4.16

4.3.2   Influence of chiral aldehydes

The reaction of an achiral boron enolate with a chiral aldehyde relies on the control by the aldehyde for any stereoselectivity obtained. In general, reaction of an enolate with an α-methyl aldehyde gives four possible products (Scheme 4.17).

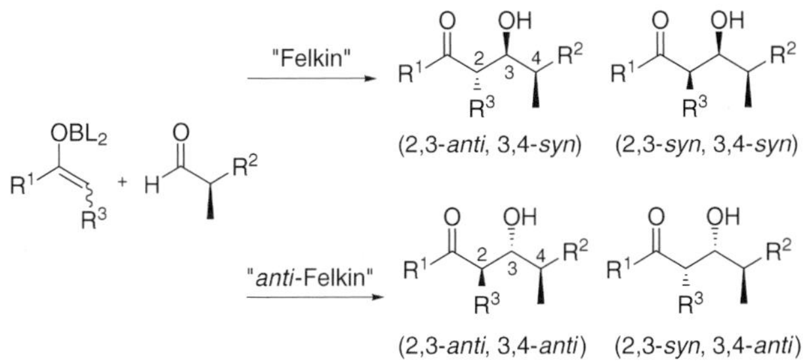

Scheme 4.17

The selectivity of the aldehyde toward attack by simple nucleophiles can be predicted by the Felkin–Anh model (Chapter 2, Section 2.1.2). Thus two of the products (one *syn* and one *anti*) are those predicted to predominate by the model, and the other two are opposite to that predicted.

We will first consider the reaction of boron enolates which are unsubstituted at the α-position (R^3=H in Scheme 4.17). Although these reactions are simple in that only one new stereocentre is formed, the influence of both the α and the β substituents on the aldehyde must be considered. It appears that the influence of an α-chiral aldehyde alone is negligible in the few cases reported without a β-stereocentre. For example, in the reaction of **4.19** with the chiral aldehyde **4.20** an equal amount of the two products is reported to form (Scheme 4.18).[20]

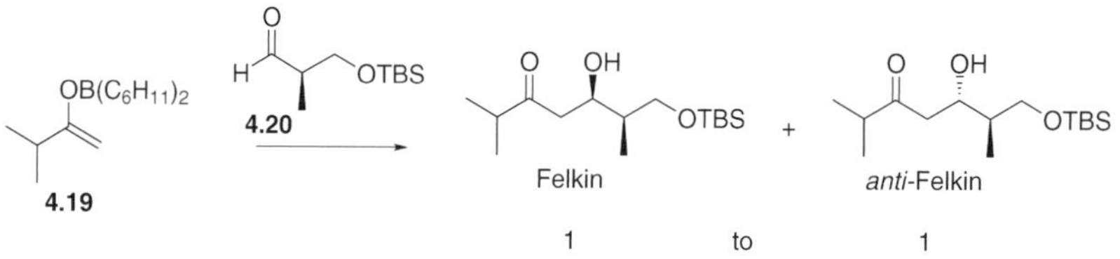

Scheme 4.18

Contrastingly, the β-stereocentre can effectively control the course of the reaction and has accordingly been examined in much greater detail. For the reaction of the *para*-methoxybenzyl (PMB) protected aldehydes **4.21** and **4.22** below, modest levels of stereocontrol are obtained regardless of the α-methyl configuration (Scheme 4.19).[23]

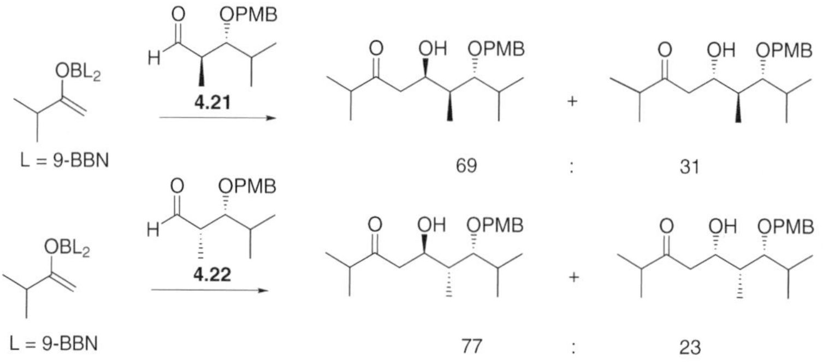

Scheme 4.19

When considering the facial selectivity with an aldehyde, the nature of the substituent at the β positions must be considered as even the protecting group at this position is known to have a significant effect, as is seen in Scheme 4.20.[24] The dominant product when R is triethylsilyl (TES) is the *anti*-Felkin product

R = MOM (84% yield, **4.23**:**4.24** = 62:38)

R = TES (74% yield, **4.23**:**4.24** = < 5: > 95)

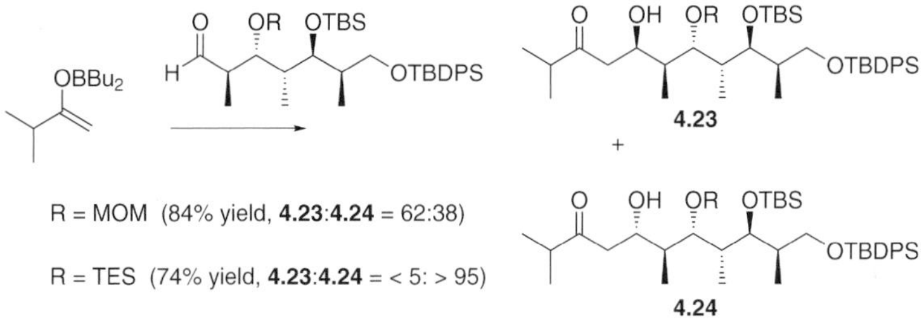

Scheme 4.20

4.24 and this reaction is postulated to proceed through a boat transition state to avoid an unfavourable steric interaction of the β-alkoxy protecting group. The effectiveness of the β-oxygen in controlling the selectivity of an aldol reaction has been noted in the reaction of α,β-unsaturated aldehydes (Scheme 4.21).[25]

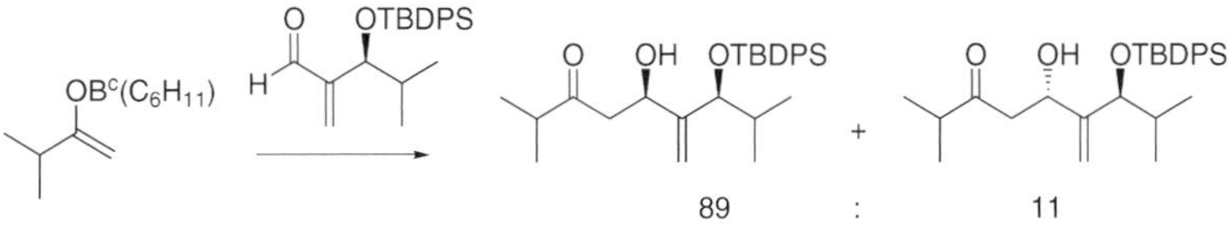

89   :   11

Scheme 4.21

The reactions of α-substituted achiral boron enolates with chiral aldehydes show significantly different levels of selectivity from those observed for their unsubstituted counterparts. For the (*E*)-*O*-enolates it appears that the α-stereocentre of that aldehyde has a much greater influence than the β-stereocentre. Reaction of the *E*-(O) dicyclohexylboron enolate of 2-methylpentan-3-one gives predominantly the Felkin product on reaction with two chiral aldehydes which differ at the β-stereocentre (Scheme 4.22).[23]

The major product from the reactions is that predicted by the Felkin model, which is usually the case for *E*-(O) boron enolates. However, the tendency is that the *Z*-(O) boron enolates favour production of the *anti*-Felkin product when reacting with chiral aldehydes. Two examples are shown in Scheme 4.23. This tendency has been rationalised[26,27] by considering the unfavourable *syn*-pentane interactions in the chair transition state of the (*Z*)-*O*-enolate leading to the Felkin product, which are not present in the transition state leading to the

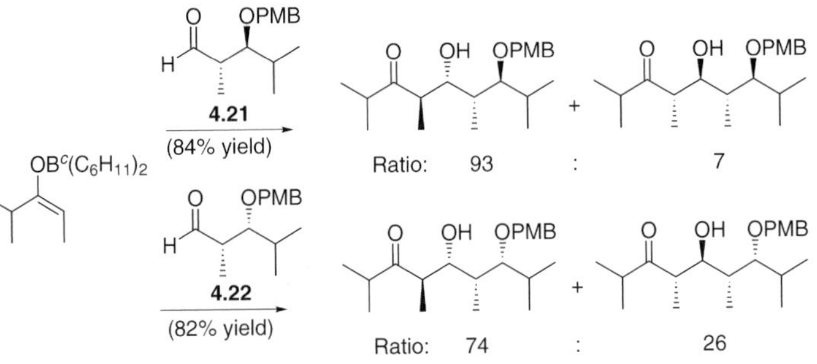

Scheme 4.22

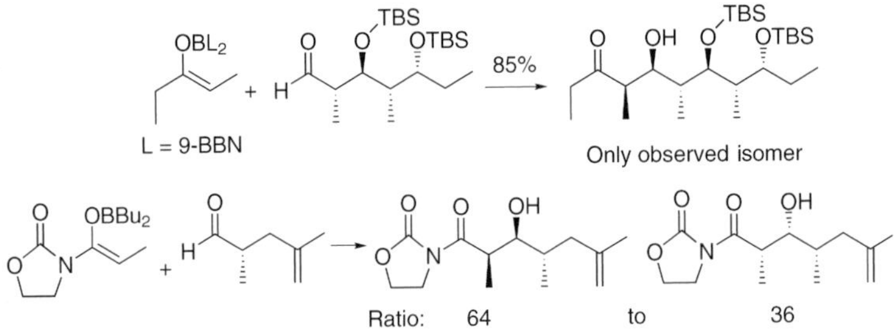

Scheme 4.23

anti-Felkin product (Scheme 4.24). For the (E)-O-enolate these unfavourable *syn* pentane interactions are present in the transition state leading to the *anti*-Felkin product, increasing the Felkin selectivity of aldehydes in their reaction with (E)-O-enolates (see Section 4.3.5 for further discussion).

However, the effect of the β-stereocentre has been shown to be significant when present in an α,β-unsaturated aldehyde as seen in Scheme 4.25, from the study by Paterson *et al.*[25]

4.3.3 Reactions of chiral enolates

While chiral aldehydes can sometimes be effective in controlling the stereochemical course of aldol reactions, the use of chiral enolates of various types has proved to be much more generally useful in the synthesis of natural products. The chirality of the ketone can be incorporated in various ways; it may be part of the carbon framework to be included into the target molecule, it may be part of the carbon framework that is to be removed at a later stage in the synthesis or it may be an auxiliary which may be removed (and at least in theory recycled). The other possibility that the chirality is part of the ligands on the boron of the enolate will be discussed in the next chapter.

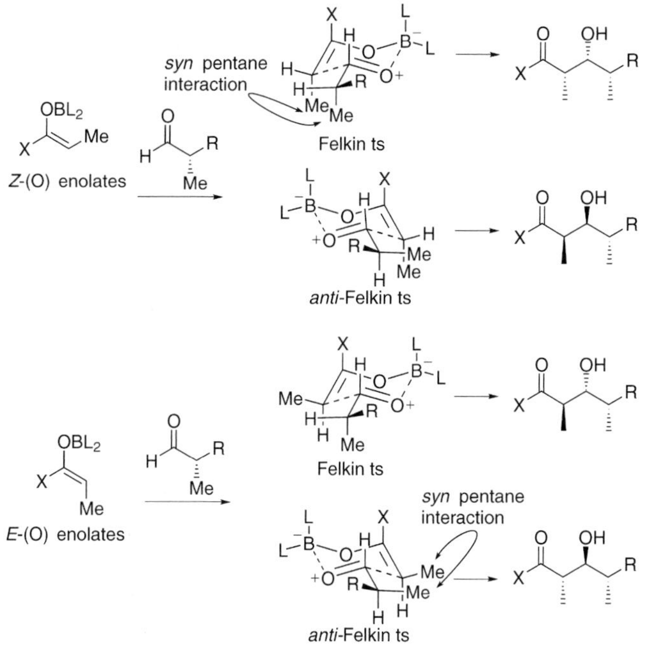

Scheme 4.24

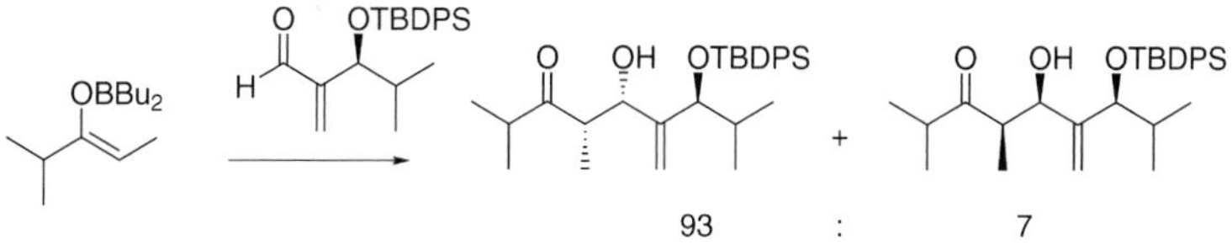

Scheme 4.25

The ethylketone **4.25** developed by Masamune and co-workers[28,29] is an early example of an effective chiral boron enolate precursor. For this ketone the dibutyl boron enolate shows high selectivity on reaction with a variety of aldehydes giving predominantly the product **4.26** over **4.27** (Scheme 4.26). The selectivity of this reaction has been explained by suggesting it proceeds through the chair transition state in which the carbon oxygen bond on the α-carbon is in an *anti* conformation relative to the enolate oxygen because of dipole–dipole repulsion, and the hydrogen not the cyclohexyl group points into the crowded part of the transition state (Scheme 4.27).

This chiral enolate of **4.25** (and *ent*-**4.25**) has been shown to overcome effectively the intrinsic facial preference of an α-chiral aldehyde in double stereodifferentiating reactions with the predominant formation of either **4.28** or **4.29** by use of the appropriate enantiomer of the ketone (Scheme 4.28).

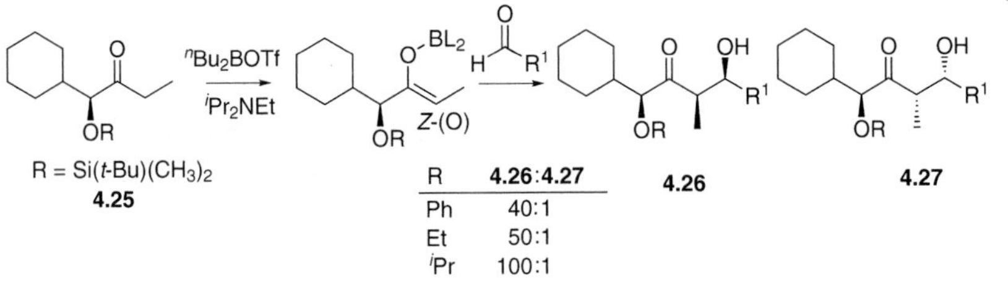

Scheme 4.26

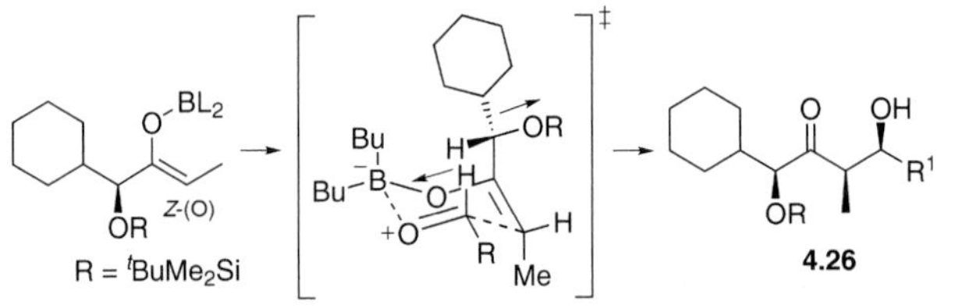

Scheme 4.27

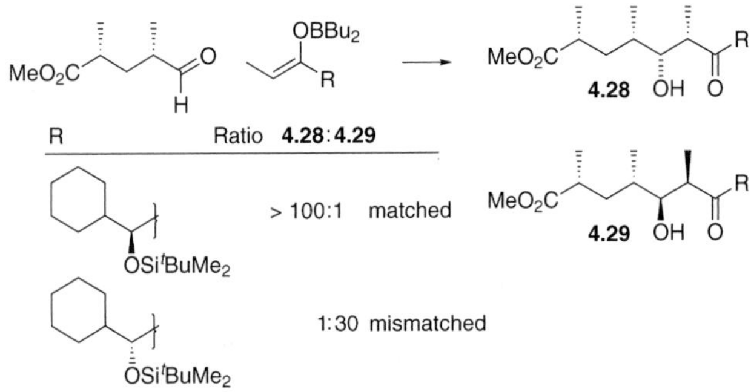

Scheme 4.28

The related chiral auxiliary developed by Heathcock *et al.* has also been reacted as its boron enolate to give *syn* aldol products with high levels of selectivity (Scheme 4.29).[17] The sense of selectivity for this reaction is opposite to that reported for the lithium enolate and can be rationalised by a transition state analogous to that shown in Scheme 4.27.

A new standard in both facial and *syn* selectivity was obtained with the chiral auxiliary based *N*-acyl-2-oxazolidinones **4.30** and **4.31** originally introduced and extensively developed by the Evans group.[11]

While alkyl esters and amides fail to enolise with boron triflate reagents under normal conditions, these imides are converted under the standard (Mukaiyama) conditions into the corresponding boron enolates which undergo

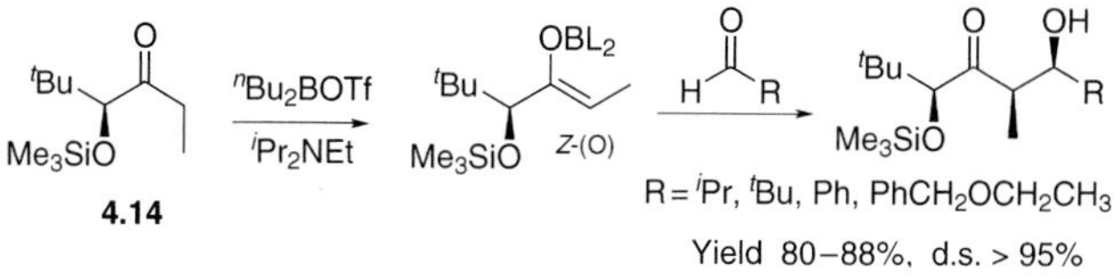

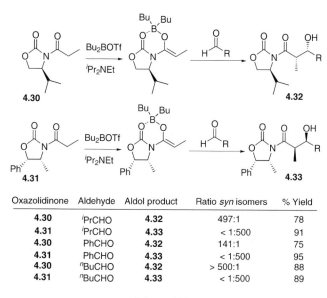

Scheme 4.29

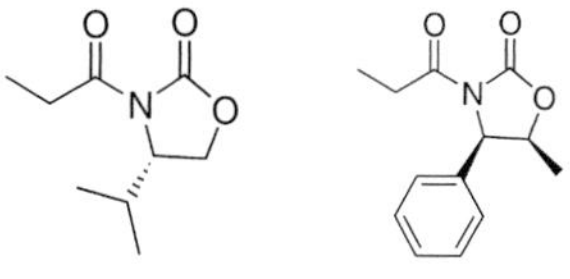

reaction with a variety of aldehydes with near perfect stereoselection. The ratio of *syn* to *anti* isomers in all cases was greater than or equal to 100:1 and the major *syn* isomer was **4.32** or **4.33** with the selectivity as indicated in Scheme 4.30.[30]

Oxazolidinone	Aldehyde	Aldol product	Ratio *syn* isomers	% Yield
4.30	*PrCHO	**4.32**	497:1	78
4.31	*PrCHO	**4.33**	< 1:500	91
4.30	PhCHO	**4.32**	141:1	75
4.31	PhCHO	**4.33**	< 1:500	95
4.30	*BuCHO	**4.32**	> 500:1	88
4.31	*BuCHO	**4.33**	< 1:500	89

Scheme 4.30

The high level of facial selectivity has been attributed to the preferred conformation of the N—C bond as indicated in Scheme 4.31 where the dipoles of the enolate oxygen and the carbonyl group of the auxiliary are opposed.

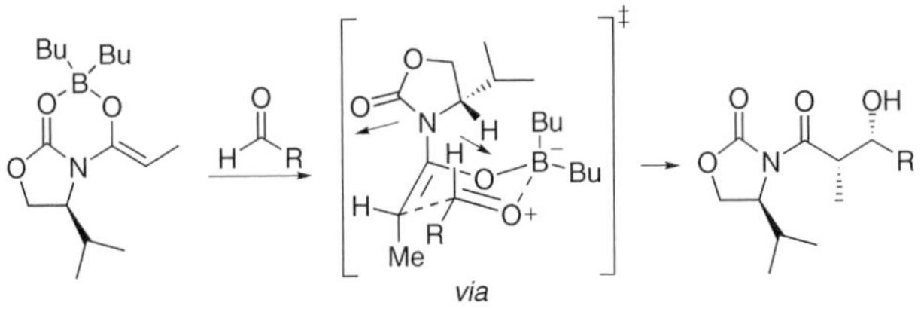

Scheme 4.31

For this rotamer of the N—C bond, the favoured transition state projects the small hydrogen (as opposed to the isopropyl) towards the sterically demanding centre of the transition state.

The power of this enolate to control the stereoselectivity of aldol reactions is considerable as seen in its reaction with a chiral aldehyde in a double stereo-differentiating reaction.[31] It is interesting to note that in this reaction the aldehyde shows *anti*-Felkin selectivity with this (Z)-O-enolate giving **4.34** in a matched reaction, but its preference is completely swamped by that of the chiral enolate in the mismatched case giving almost exclusively the product **4.35** (Scheme 4.32).

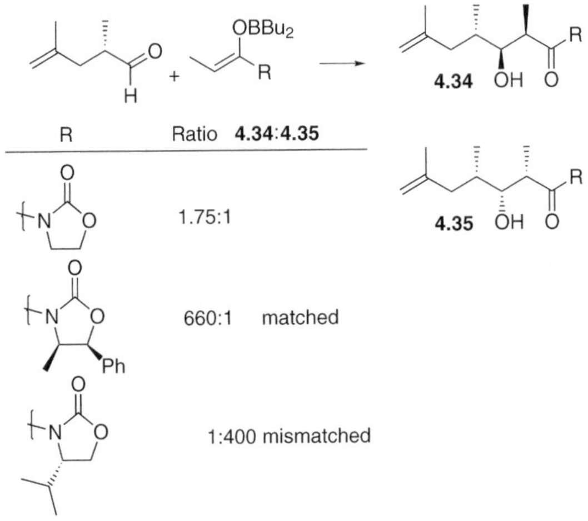

Scheme 4.32

As efficient as they are, these reagents have been somewhat superseded by related reagents **4.36** and **4.37** which are derived from L- and D-phenylalanine, both of which are available commercially at reasonable cost. A further advantage of the use of all these chiral auxiliaries is that they can be recovered and reused after removal from the reaction product.

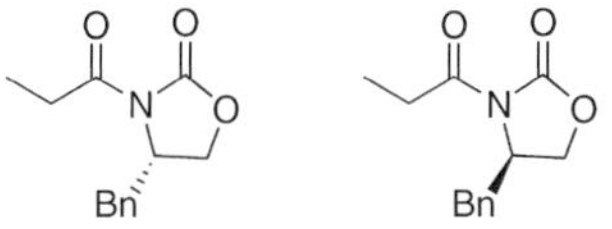

4.36 **4.37**

Another auxiliary-based system is that developed by Oppolzer and co-workers,[18,32] a bornane sultam that is prepared in a few steps from camphor. We have already seen that the lithium enolate of this species is selective in its reaction with aldehydes (Scheme 4.14), but its selectivity as the boron enolate is *reversed* to that observed with the lithium enolate (Scheme 4.33).

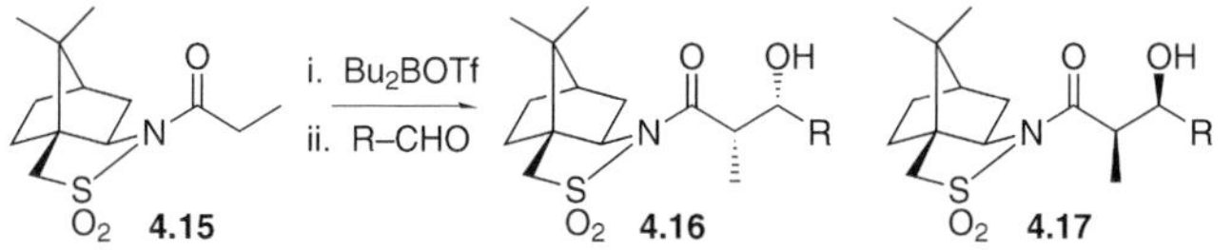

R	% **4.16**	% **4.17**
Ph	< 5	> 95
¹Pr	< 1	> 99

Scheme 4.33

The difference in selectivity between the lithium and boron enolates is explained by considering that the lithium enolate can coordinate to the sultam oxygen in the transition state, which is not possible for the boron enolate. The sultams are highly crystalline and can usually be purified to perfect (> 99%) diastereomeric purity. The versatility of boron enolates of Oppolzer's sultams is seen in the novel desymmetrisation of the aldehyde in the total synthesis of denticulatin.[33,34] In this reaction the *meso* dialdehyde **4.38** reacts (selectively at the prochiral left carbon, as drawn in Scheme 4.34) with the diethylboron

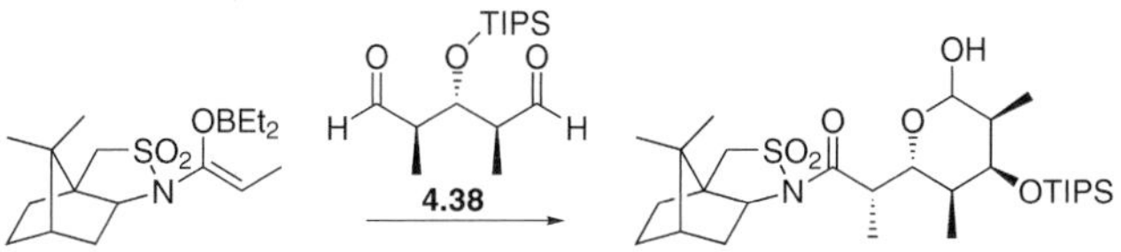

4.38

95% Yield, > 95% d.s.

Scheme 4.34

enolate of the sultam to give the lactol in high yield. The diastereoselectivity is greater than 95% and no double addition products are seen. The lack of double addition products can be attributed to the spontaneous formation of the lactol as the aldehyde reacts.

There are a number of examples of the highly selective reactions of E-(O) boron enolates derived from chiral ketones, with aldehydes giving *anti* aldols. Two prominent cases are the reactions of the dipropionate equivalents developed by the groups of Evans and Paterson.[35,36]

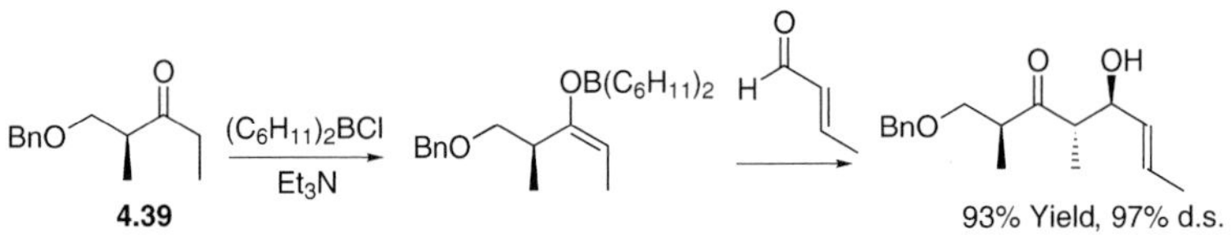

Scheme 4.35

The Paterson chiral ketone **4.39** (Scheme 4.35) undergoes highly selective *anti–anti* aldol reactions with a wide variety of aldehydes and has been used in the total synthesis of a number of natural products including swinholide A,[37] denticulatin A and B,[38,39] oleandolide[40] and muamvatin.[41] This ketone is synthesised in three steps from the commercially available methyl (S)-3-hydroxy-2-methylpropionate. The enantiomer of the ketone is similarly available from the enantiomeric ester. This reaction has been studied by means of molecular modelling which found a favourable chair transition state in which the benzyloxymethylene unit of the ketone occupies the more sterically demanding position toward the ligand on boron.[42] The preferred rotamer of the α-enolate bond has the hydrogen and enolate methyl eclipsed to minimise the A-1,3 strain. The transition state leading to the other *anti* aldol product appears to be disfavoured by electrostatic repulsions of the lone pairs on the benzyl ether substituent oxygen and the enolate oxygens (Scheme 4.36).

The dipropionate equivalent designed by Evans *et al.* are the ketones **4.40** and **4.41** which both give highly selective *anti* aldols.[35] This reaction is interesting as the methyl group situated between the two carbonyl groups is the controlling factor in the reaction, and the auxiliary stereocentre has little effect. Interestingly, the stereocentre between the two carbonyls is totally configurationally stable and it is the more accessible hydrogens which are removed upon enolisation (Scheme 4.37).

All three reagents **4.39–4.41** are effective for the stereoselective construction of polypropionate chains. The aldol products derived from **4.39–4.41** can be manipulated in a biomimetic fashion to give a dipropionate extension of the carbon chain and are readily converted to the aldehyde to allow further extension of the carbon chain (Scheme 4.38).

By the use of other metals these dipropionate equivalents can be directed to form (Z)-O-enolates, thereby leading to the two other *syn* aldol products. These reactions will be discussed in Section 4.4.3.

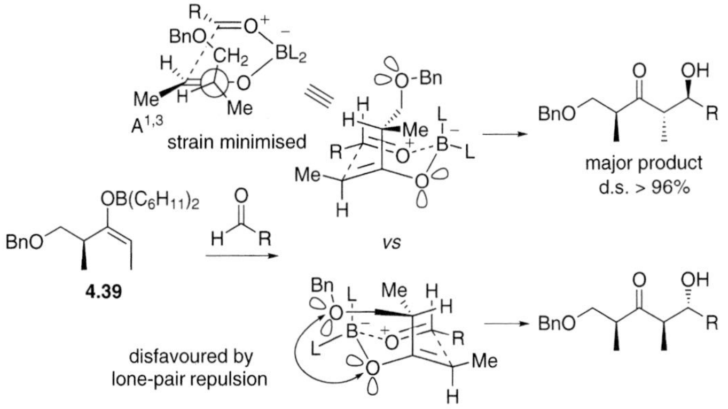

Scheme 4.36

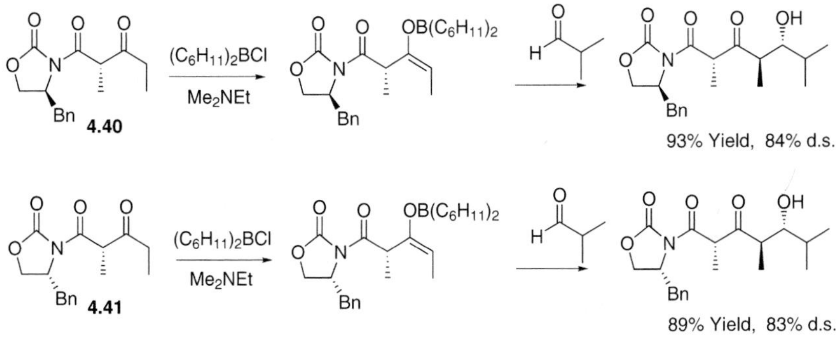

Scheme 4.37

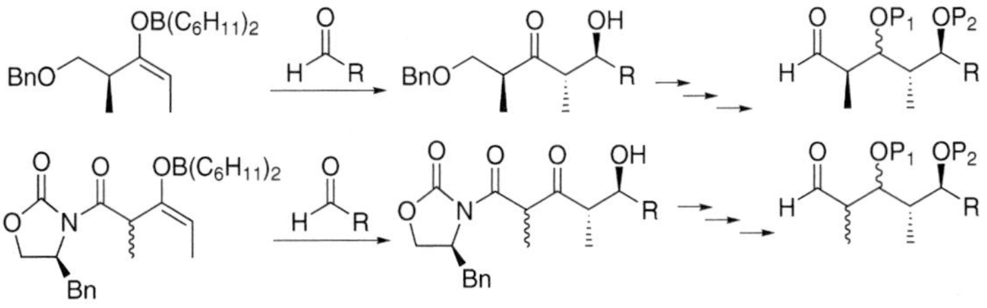

Scheme 4.38

Two recently developed chiral mono propionate equivalents are the lactate-derived reagents **4.42** and **4.43**.[43–45] These reagents differ in that they are protected either as the benzyl ether **4.42** or as the benzoate ester **4.43**. They are available in either enantiomeric form from the inexpensive (*S*)-ethyl and (*R*)-isobutyl lactate esters.

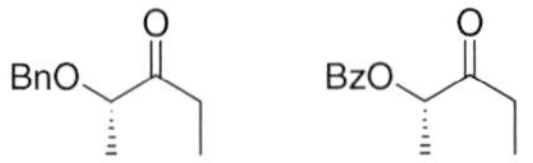

These two reagents are complementary in that they are enolised under very similar conditions to form (Z)-O- or (E)-O-enolates, respectively, which undergo highly selective reagent-controlled aldol reactions, and two examples are shown in Scheme 4.39. The aldol products formed (Scheme 4.39) can then be

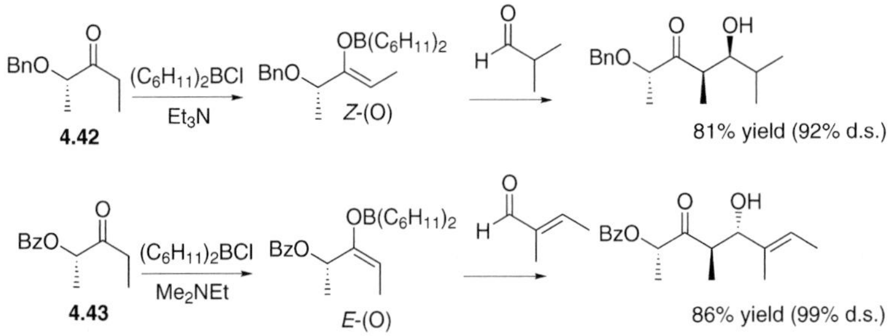

Scheme 4.39

manipulated to generate the ethyl ketone or the aldehyde for subsequent reaction as either enolate or aldehyde, as shown in Scheme 4.40.

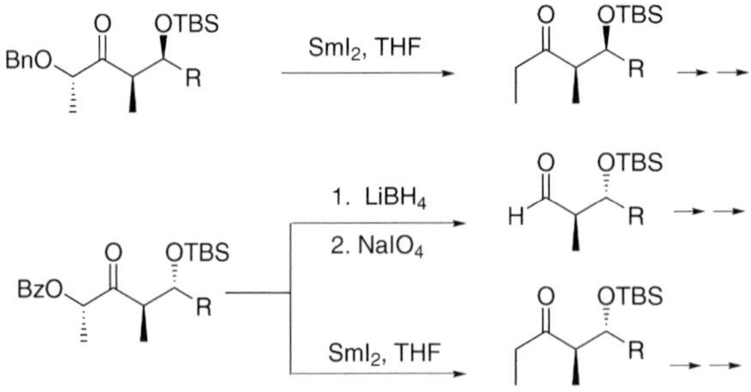

Scheme 4.40

4.3.4 Reaction of boron enolates through open transition states

The generation of *anti* aldol products from (Z)-O-enolates is possible by diverting the reaction from a cyclic transition state.[46] This is usually accomplished

by precomplexation of the aldehyde with a bulky Lewis acid, thus forcing
the reaction to occur through an open transition state. When the dibutylboron
enolate of the imide **4.30** is added at $-78°C$ to isobutyraldehyde which has
been precomplexed with Et_2AlCl, a single *anti* isomer is formed (95% d.s.)
and the minor product is a *syn* stereoisomer.[47] The *syn* isomer that is produced is the opposite to that produced in the normal Z-(O) *syn* aldol reaction
(Scheme 4.41).

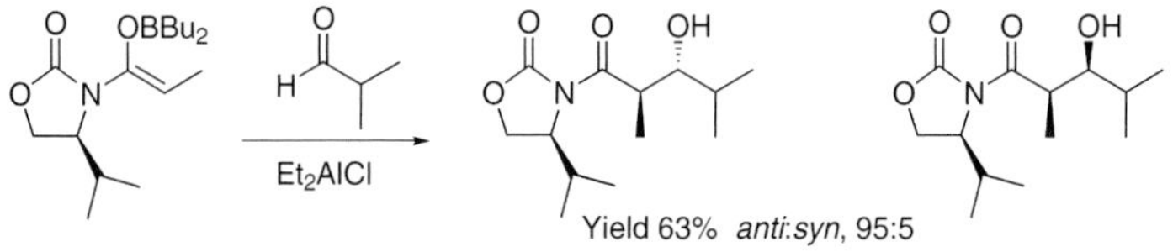

Yield 63% *anti:syn*, 95:5

Scheme 4.41

When less sterically demanding Lewis acids are used it is possible to form
the 'non-Evans' *syn* aldol product as the major product of the reaction.[47] This is
achieved by reaction of the dibutylboron enolate of the imide **4.30** in a modified
procedure where the Lewis acid $TiCl_4$ is added to the enolate in one portion,
followed by dropwise addition of the isobutyraldehyde giving the two aldol
products in the ratio of 6 to 94 (Scheme 4.42).

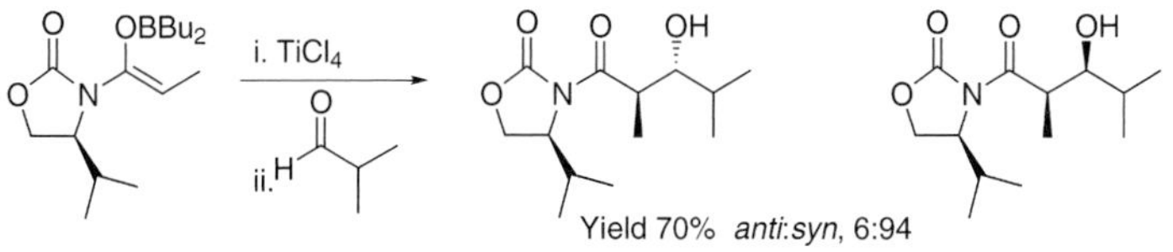

Yield 70% *anti:syn*, 6:94

Scheme 4.42

The preference for the formation of the 'non-Evans' *syn* aldol product, shown
when a small Lewis acid is employed, is rationalised by the preferred transition
state **IX** being the one where the *gauche* interactions of the forming bond are
minimised. However, when the Lewis acid is sterically demanding the Me↔LA
interactions become significant, raising the energy of transition state **IX** compared with transition state **X**, leading to preferential formation of the *anti* product
via transition state **X** (Scheme 4.43).

When considering the use of these types of aldol reactions for synthesis
with sensitive aldehydes it should be remembered that the complexation of the
aldehydes with the Lewis acids is liable to lead to undesired side-reactions.

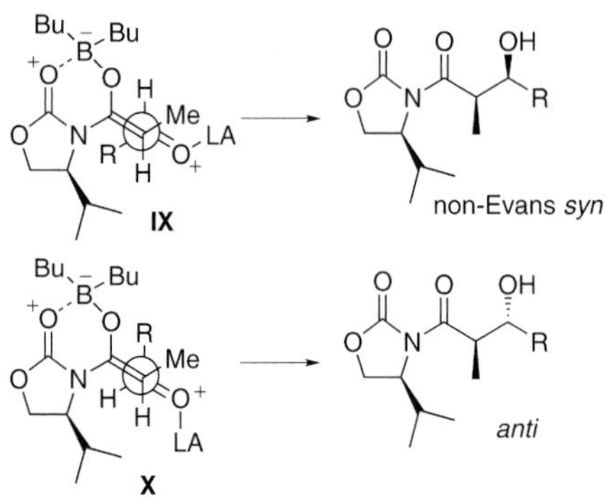

Scheme 4.43

4.3.5   Reactions of boron enolates in synthesis

The true power of boron enolates is seen in their application to the total synthesis of natural products. There are many majestic examples in the literature and a number have recently been reviewed.[20] Here we will highlight a late aldol coupling reaction in the recent total synthesis of altohyrtin C (spongistatin 2) by the Evans group.[48–50] The reaction between the dicyclohexyboron (E)-O-enolate **4.44** with the aldehyde **4.45** results in a remarkably stereoselective aldol reaction with 90% d.s. in favour of the desired product and with 70% yield (Scheme 4.44).

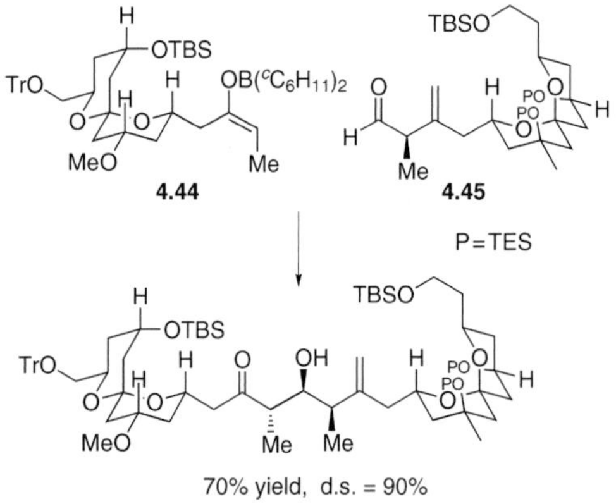

70% yield, d.s. = 90%

Scheme 4.44

This is a classic example of double asymmetric synthesis where both the aldehyde and the ketone are chiral and in principle can influence the course of the reaction. In this case, model studies of this reaction indicated that the stereoselectivity was derived primarily from the aldehyde. This was seen from the reaction of an achiral dicyclohexyboron (E)-O-enolate with a model aldehyde (Scheme 4.45). This selectivity can at least in part be attributed to the increased

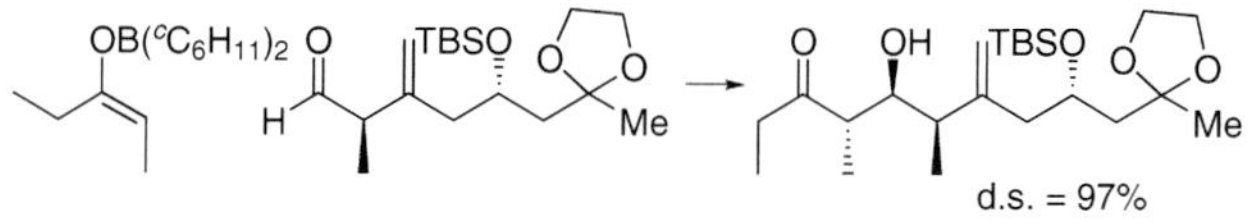

Scheme 4.45

Felkin selectivity observed for (E)-O-enolates compared with their unsubstituted counterparts (compare Schemes 4.19 and 4.22). This enhanced Felkin selectivity can be attributed to the developing *syn* pentane interaction for the *anti*-Felkin product (see Scheme 4.24).

4.4   Reactions of titanium, zirconium and tin enolates

4.4.1   Titanium(IV) enolates

Titanium is probably the best studied of all the transition metals used for aldol reactions and in many cases leads to a superior yield and selectivity compared with the corresponding lithium enolates. The Lewis acidity of the titanium metal centre can be manipulated by variation of the ligands (alkoxy, amino, chloro, cyclopentadienyl, etc.) to optimise the selectivity in appropriate cases.[51] This extends to the use of chiral ligands on the titanium for enantioselective reactions, and these are discussed in Chapter 5, Section 5.2.3. Until recently, most titanium enolates were formed by metal exchange of lithium enolates. Because these transmetallations generally proceed with retention of configuration, the stereochemistry obtained in the original deprotonation determines the configuration of the final enolate. For example the Z-(O) lithium enolate of ketone **4.25** exchanges with $ClTi(O^iPr)_3$ to give the Z-(O) titanium enolate which reacts[52] with very similar selectivity to the corresponding boron enolate (Scheme 4.46).

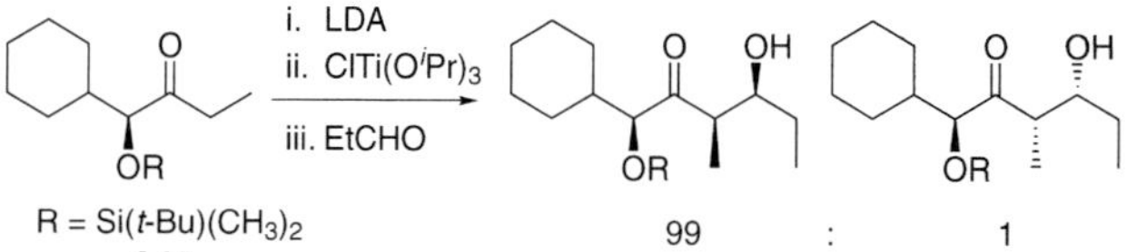

Scheme 4.46

A somewhat unusual method for the selective production of *anti* aldol products from an *E*-(O) titanium enolate has been developed by Heathcock *et al.*[17] The magnesium enolate of the chiral auxiliary **4.10** is formed as previously described (Scheme 4.13) and the magnesium is exchanged for titanium (with retention of the enolate configuration), which results in the formation of the opposite *anti* aldol product to that formed from the magnesium enolate (Scheme 4.47).

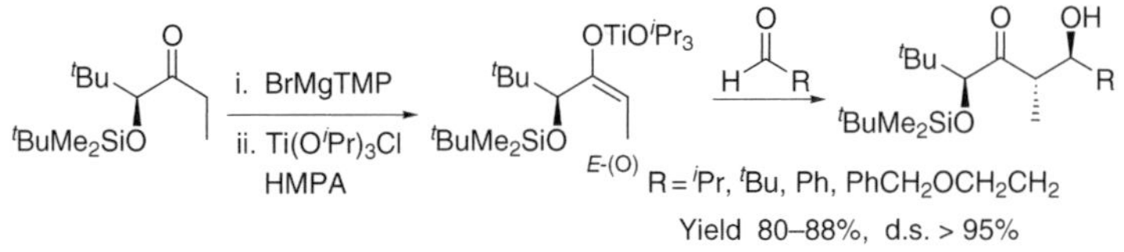

Scheme 4.47

This can be rationalised by the reaction proceeding through the chair transition state in which the titanium does not coordinate to the trimethylsilyl protected oxygen and the dipoles are opposed. On the other hand, the magnesium enolate goes through the transition state in which the magnesium coordinates to the trimethylsilyl protected oxygen of the ketone (Scheme 4.48).

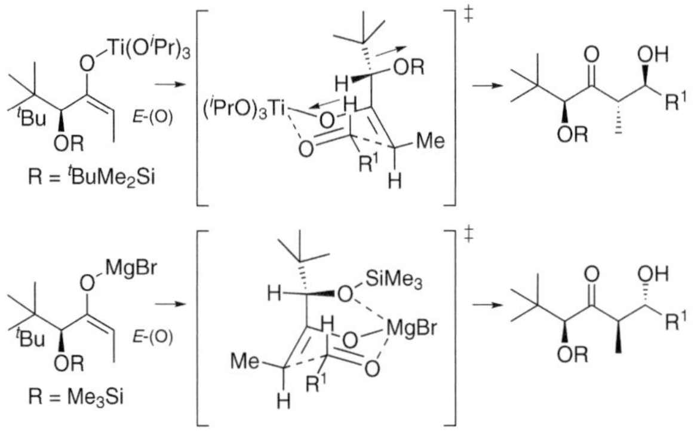

Scheme 4.48

At this point in our discussions it is worthy of note that this auxiliary developed by Heathcock can be used to produce all four possible diastereomeric products by appropriate choice of metal and of enolate geometry.[17] These conditions are summarised in Scheme 4.49.

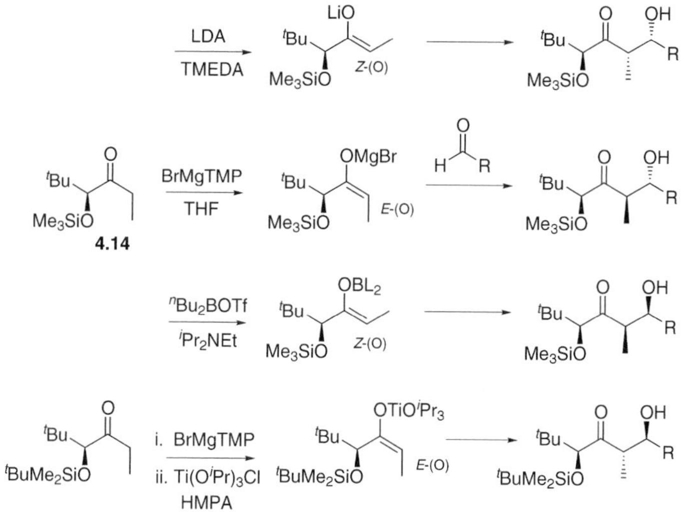

Scheme 4.49

A simple and highly effective method has been developed for the direct generation of trichlorotitanium enolates from acyclic ketones by the Evans group.[53,54] This method (as discussed in Chapter 3, Section 3.4.2) involves precomplexation of the ketone with TiCl₄ followed by addition of a hindered amine base (usually diisopropylamine) to form the enolate. The precomplexation overcomes the problem of the normal irreversible binding of the TiCl₄ and the amine base. This procedure has been carried out with a variety of ketones, some examples of which are illustrated in Scheme 4.50.

The examples in Scheme 4.50 show a general trend for high levels of substrate control for reaction of the titanium enolate.[54] Notably, the selectivity for these TiCl₄-derived enolates show the same sense as the corresponding boron enolates, suggesting chelation to the titanium is not occurring in any of the transition states. This trend is exemplified by the reaction of titanium enolate of **4.46** giving a 96:4 ratio with an excellent yield of 96%, and the 9-BBN enolate[55] shows similar selectivity (92:8), yet the yield is somewhat lower. The lithium enolate, on the other hand, favours the other *syn* isomer, presumably via a transition state where the β-oxygen coordinates to the lithium (Scheme 4.51).[56]

The use of the directly generated trichlorotitanium enolate has been shown to be a powerful method for the controlled stereochemical coupling of complex fragments in the total synthesis of natural products. In Evans *et al.*'s[57,58] total synthesis of the macrolide antibiotic rutamycin, the complex ketone and aldehyde fragments are joined by using a TiCl₄/^{i}Pr₂NEt enolisation aldol protocol, giving the product in 83% yield and with a stereoselectivity of 97:3 in favour of the necessary product for the synthesis (Scheme 4.52).

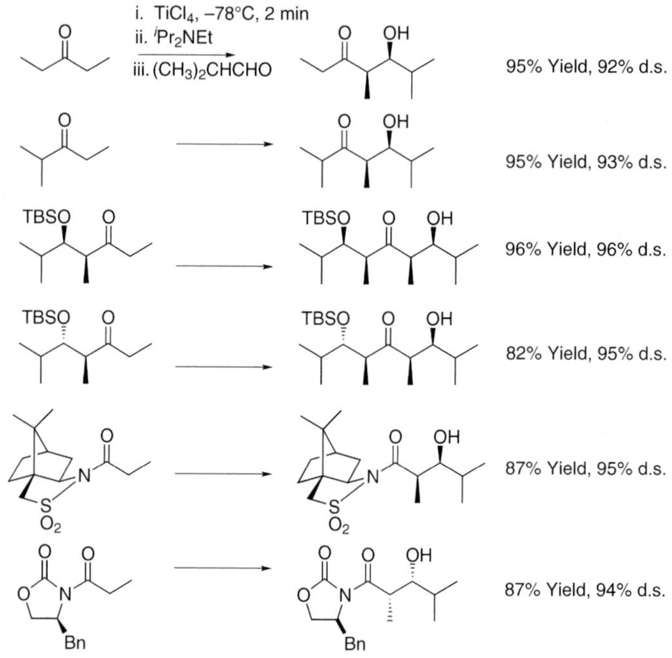

i. TiCl₄, −78°C, 2 min
ii. ᶦPr₂NEt
iii. (CH₃)₂CHCHO

95% Yield, 92% d.s.

95% Yield, 93% d.s.

96% Yield, 96% d.s.

82% Yield, 95% d.s.

87% Yield, 95% d.s.

87% Yield, 94% d.s.

Scheme 4.50

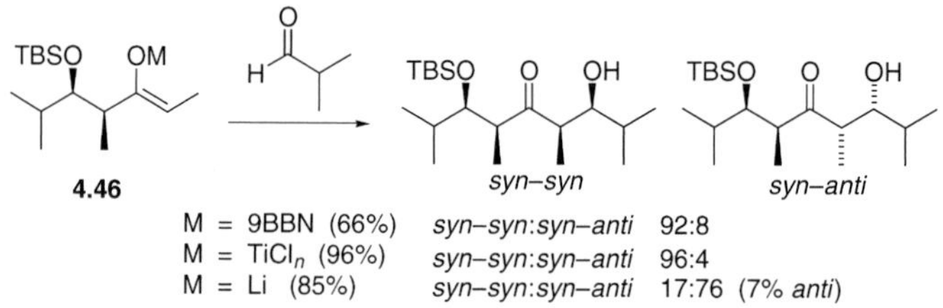

4.46

syn–syn

syn–anti

M = 9BBN (66%)   syn–syn:syn–anti   92:8
M = TiCl_n (96%)   syn–syn:syn–anti   96:4
M = Li   (85%)   syn–syn:syn–anti   17:76 (7% anti)

Scheme 4.51

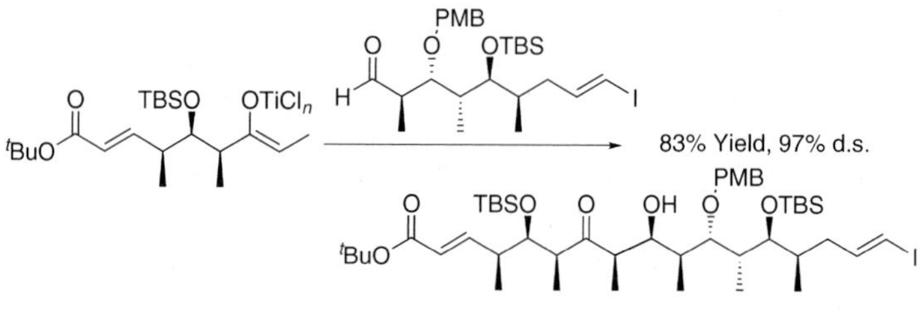

83% Yield, 97% d.s.

Scheme 4.52

This is an example of double stereodifferentiation where both the enolate and aldehyde have a facial preference. In this case, the product obtained is the Felkin product with respect to the aldehyde. As noted earlier, most aldehydes show *anti*-Felkin selectivity on reaction with (Z)-O-enolates, so one would suspect that this reaction is mismatched. However, by studying the reactions of a series of model compounds with differing stereochemistry, the reaction has been shown to be partially matched with the β-stereocentre.[59]

This quite detailed model study[59] (Scheme 4.53) determined the product ratio for the reaction of titanium enolate **4.47** with all the possible isomers of

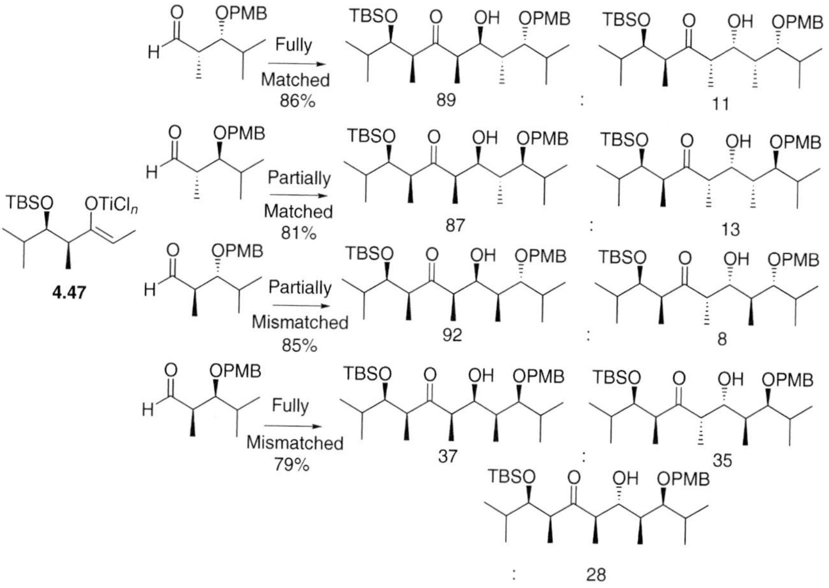

Scheme 4.53

a model α-methyl β-PMBO aldehyde. This showed the relative effect of the α and β substituents on reaction selectivity. In the partially mismatched case third from the top, the major product is the Felkin product (which is inherently disfavoured for *syn* aldols) leading to the conclusion that for this aldehyde the β-oxygen centre is the major controlling factor in the reaction.

Another interesting case of a double stereodifferentiating titanium-mediated aldol reaction is seen in the Paterson[38,39] synthesis of denticulatin. Here the reaction of a titanium enolate **4.48** with aldehyde **4.49** (80% e.e.) gives only two products which have the same stereochemistry at the newly formed stereo-centres. More of the minor product, arising from the minor enantiomer of the aldehyde, was produced than would have been the case if the rate of the reaction with both enantiomers of the aldehyde was the same. (A threefold excess of the aldehyde was used and reaction of racemic aldehyde gave a 31:69 ratio of products.) Thus the ketone reacts only on one face of the enolate and in that respect is totally selective, but the rate of reaction with the aldehyde with which it is matched (*ent*-**4.49**) is somewhat faster (Scheme 4.54).

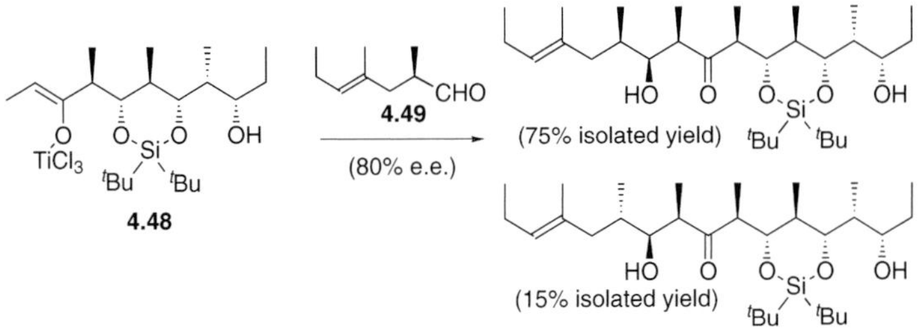

Scheme 4.54

4.4.2   *Zirconium enolates*

Like their titanium counterparts, zirconium enolates are characterised by high *syn* selectivity and good yields. These enolates are usually prepared by trans-metallation of the lithium enolate with the zirconium halide and are unusual in that they often give *syn* aldol products *irrespective* of their enolate geometry. This is seen in the reaction of a variety of enolates, generated from the lithium enolates by transmetallation with Cp_2ZrCl_2, as shown in Scheme 4.55.[60,61]

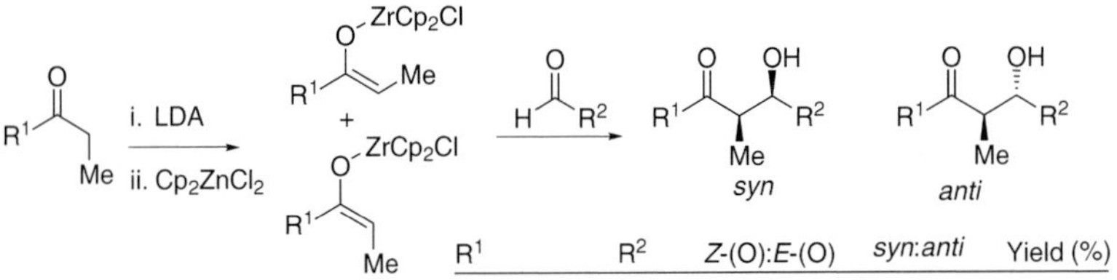

R^1	R^2	Z-(O):E-(O)	syn:anti	Yield (%)
^tBuS	Ph	10:90	93:7	70
MeO	Ph	5:95	87:13	80
^tBuO	Ph	5:95	72:28	n.r.
Ph	Ph	> 98:2	90:10	62
Et	Ph	92:8	67:33	86
Et	Ph	14:86	88:12	70
$^iPr_2N$	Ph	81:19	> 98:2	87
$^iPr_2N$	^nPr	81:19	98:2	80–90
$^iPr_2N$	^iPr	81:19	97:3	80–90
$\overset{}{N-}$ (pyrrolidine)	Ph	> 95:5	95:5	80

Scheme 4.55 n.r. = not reported.

This unusual selectivity has been studied by Evans and McGee,[60] who showed that the reaction was occurring under kinetic control. They explained the selectivity by postulating that the (Z)-O-enolates react predominantly through chair transition states, whereas the (E)-O-enolates react through boat transition states (Scheme 4.56).

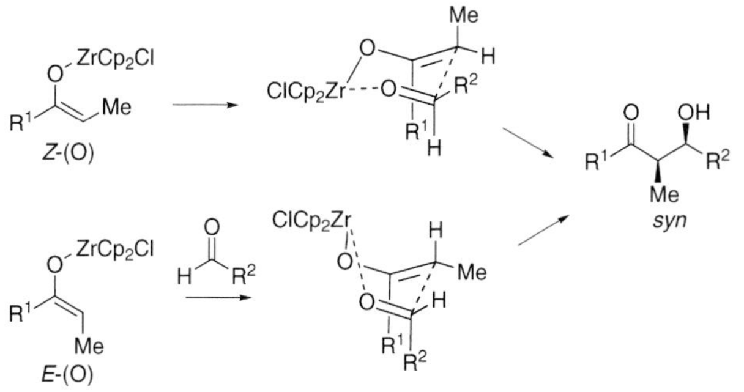

Scheme 4.56

Zirconium enolates have been employed successfully in the synthesis of natural products. For example in Deslongchamps *et al.*'s formal total synthesis of erythromycin A,[62] a highly complex spiroacetal-containing aldehyde undergoes an aldol reaction with the enolate formed from Cp_2ZrCl_2 and methyl acetate in a highly stereoselective manner. The reaction is totally *syn* selective, with the minor product being the *anti*-Felkin product (Scheme 4.57).

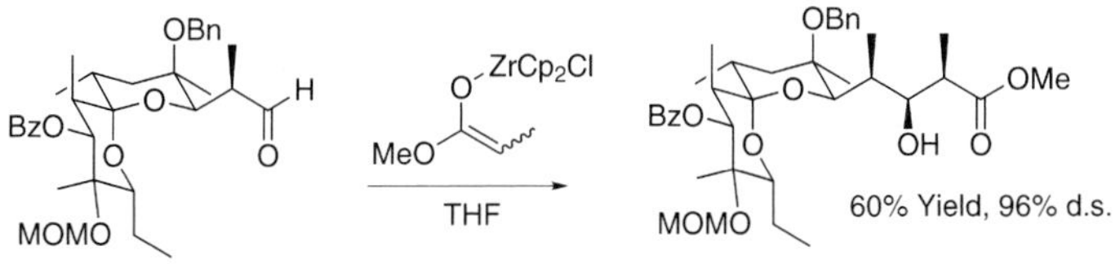

Scheme 4.57

4.4.3   Tin(II) enolates

Tin(II) enolates are readily prepared by treatment of carbonyl compounds with tin(II) triflate in the presence of an amine base.[63] These enolates are more reactive than the corresponding boron enolate and often give high levels of selectivity in aldol reactions. An interesting example is the stereoselective tin(II) aldol with a thiocarbonyl chiral auxiliary **4.50**. This auxiliary shows high selectivity in both the acetate aldol reaction and the propionate aldol reaction (Scheme 4.58).[64]

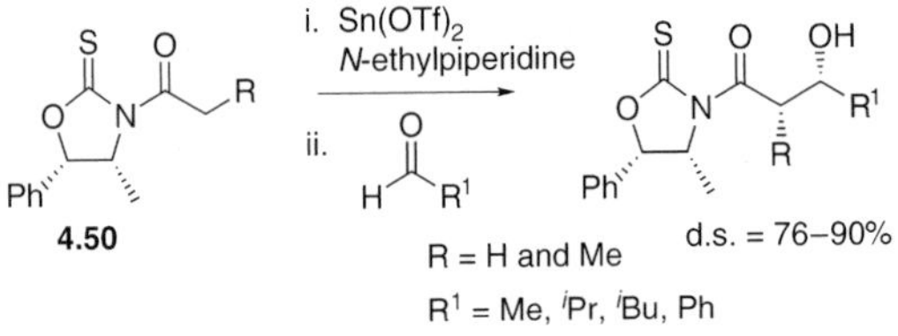

Scheme 4.58

This is significant as the Evans oxazolidinone auxiliary shows poor selectivity for acetate aldol reactions.

The different behaviour of various metal enolates is seen in the reaction of the enolates derived from the Evans dipropionate equivalent **4.40**. Here the tin(II) enolate gives the *syn* aldol product with the methyl *anti* relative to the pre-existing methyl.[53] On the other hand, reaction of the titanium(IV) enolate gives the *syn* aldol product with the methyl *syn* relative to the pre-existing methyl in the enolate. A third possible aldol product can be formed by reaction of the (*E*)-dicyclohexylboron enolate, which gives the *anti* aldol product with synthetically useful levels of control.[35] No conditions have yet been reported for the production of the remaining *anti* aldol product (Scheme 4.59).

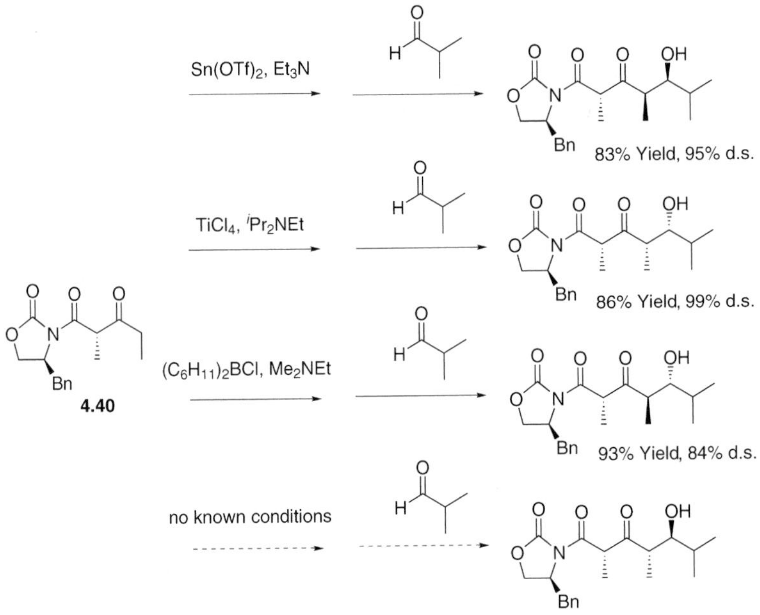

Scheme 4.59

The difference in selectivity between the Z-(O) tin(II) and the Z-(O) titanium(IV) enolates has been rationalised by invoking a transition state in which the carbonyl of the enolate coordinates to the titanium(IV) but not to the tin(II) (Scheme 4.60).

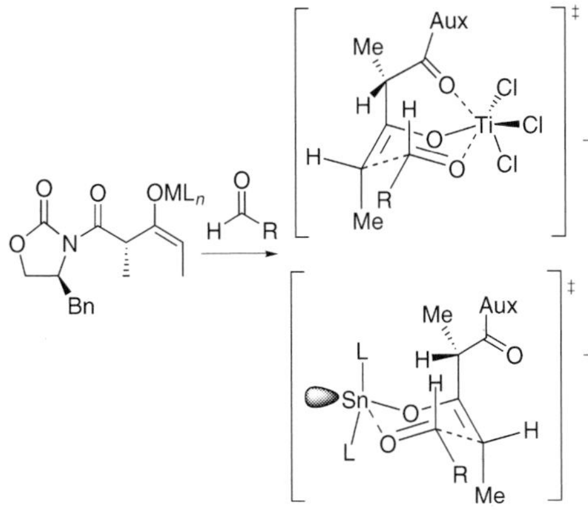

Scheme 4.60

In contrast, the reaction of the Paterson dipropionate ketone **4.30** is enolised with tin(II) triflate and undergoes a highly selective *syn* aldol reaction to give the *syn–syn* aldol as the major product.[65] This is the same as the major isomer from the titanium(IV)-mediated aldol which gives significantly lower selectivity (Scheme 4.61).

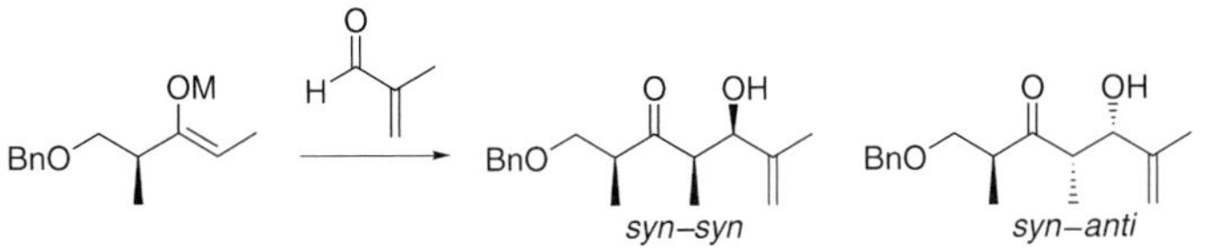

	syn–syn		*syn–anti*
M = TiCl$_n$	62	:	38
M = SnOTf (92%)	93	:	7 < 1% anti

Scheme 4.61

Paterson proposes that chelation of the benzyl ether oxygen to the tin leads to the high selectivity observed for this reaction (Scheme 4.62).[65] Either the tin enolate is behaving differently in the two systems or there must be some other explanation.

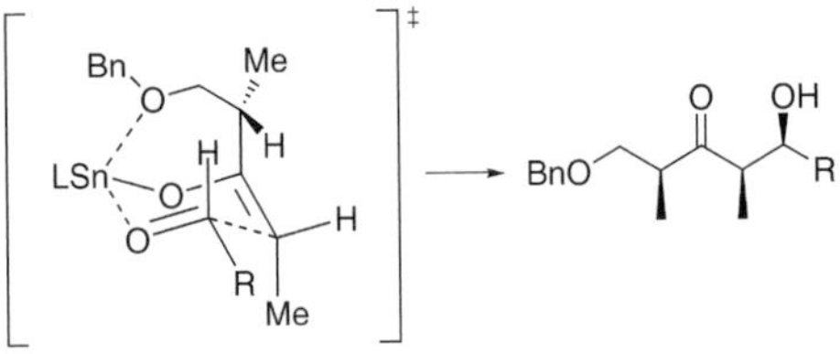

Scheme 4.62

4.5   Mukaiyama aldol reactions

The Mukaiyama Lewis-acid-catalysed reaction of silyl enol ethers with aldehydes proceeds through an open transition state where the geometry of the
enol ether is not necessarily reflected in the stereochemistry of the product. It
is interesting to note that in these reactions the β-oxygen stereocentre of the
aldehyde can have a strong directing influence, as seen in Scheme 4.63.[23,66]

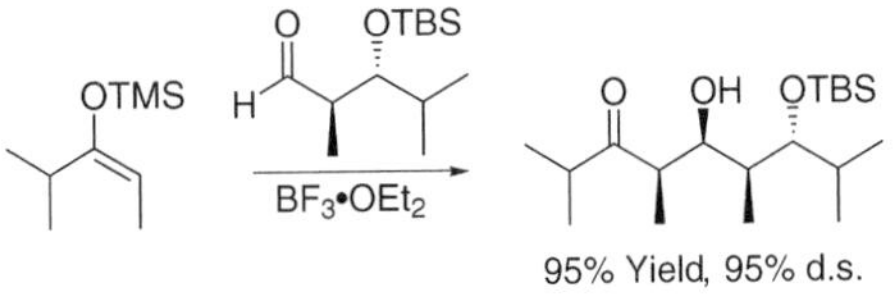

95% Yield, 95% d.s.

Scheme 4.63

In this example the (E)-O-enolate gives only the *syn* aldol products. However,
reports of efficient control by the use of chiral ketones are available. For example, the auxiliary developed by Oppolzer and Starkmann[67] can be used efficiently
to direct the outcome of reactions with achiral aldehydes (Scheme 4.64).

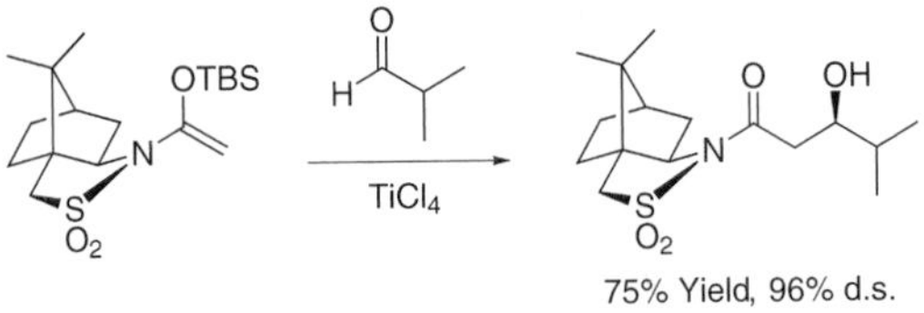

75% Yield, 96% d.s.

Scheme 4.64

The Mukaiyama aldol can often be relied on in synthesis when Felkin
control from an α-methyl ketone is desired. An example of this is in the

complex fragment coupling in the Paterson total synthesis of swinholide A.[37] This coupling occurs with the remarkably high selectivity (97% d.s.) with a yield of 91% (Scheme 4.65).

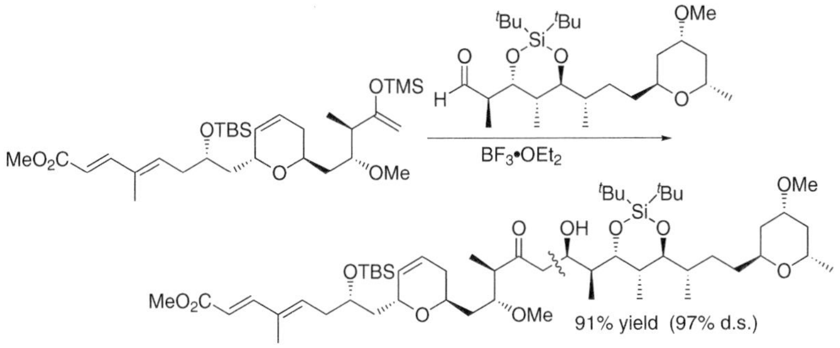

Scheme 4.65

4.6   Conclusions

In this chapter we have seen how by choice of the metal and of the enolisation conditions the selectivity of aldol reactions can be controlled to give the desired product in a large range of chemical processes. The power of the strategy resides in the selective formation of *two* stereocentres while creating a carbon to carbon bond. The combination of controlled formation of stereocentres and the building of carbon framework makes the aldol reaction a process to be considered for any synthetic strategy.

We have seen examples of where this control of the stereocentres formed in the aldol reaction comes from the aldehyde, where it is part of the ketone and where it is part of an auxiliary that is attached to the ketone and later removed. In the next chapter we consider the case where the chirality comes from a chiral reagent that is used to enolise the ketone.

References

1. Rawlings, B.J., *Natural Product Reports*, **1997**, *14*, 523.
2. Garson, M.J., *Chem. Rev.*, **1993**, *93*, 1699.
3. Nielson, A.T. and Houlihan, W.J., in *Organic Reactions*, Ed. A.C. Cope, John Wiley, New York, **1968**, vol. 16, p. 1.
4. Heathcock, C.H., in *Comprehensive Organic Synthesis*, Ed. B.M. Trost, Pergamon, Oxford, **1991**, vol. 2, p. 133.
5. Mukaiyama, T., in *Organic Reactions*, Ed. W.G. Dauben, John Wiley, New York, **1982**, vol. 28, p. 204.

6. Zimmerman, H.E. and Traxler, M.D., *J. Am. Chem. Soc.*, **1957**, *79*, 1920.
7. Heathcock, C.H., Buse, C.T., Klieschick, W.A., Pirrung, M.C., Sohn, J.E. and Lampe, J., *J. Org. Chem.*, **1980**, *45*, 1066.
8. Dubois, J.E. and Fellmann, P., *Tetrahedron Lett.*, **1975**, 1225.
9. Heathcock, C.H. and Pirrung, M.C., *J. Org. Chem.*, **1980**, *45*, 1727.
10. Heathcock, C.H., Pirrung, M.C., Montgomery, S.H. and Lampe, J., *Tetrahedron*, **1981**, *37*, 4087.
11. Evans, D.A., Nelson, J.V. and Taber, T.R., in *Topics in Stereochemistry*, Eds. L.A. Norman, E.L. Eliel and S.H. Wilen, John Wiley, New York, **1982**, vol. 13, p. 1.
12. Buse, C.T. and Heathcock, C.H., *J. Am. Chem. Soc.*, **1977**, *99*, 2337.
13. Mori, I., Ishihara, K. and Heathcock, C.H., *J. Org. Chem.*, **1990**, *55*, 1114.
14. Heathcock, C.H., Young, S.D., Hagen, J.P., Pilli, R.A. and Badertschester, U., *J. Org. Chem.*, **1985**, *50*, 2095.
15. Heathcock, C.H., Pirrung, M.C., Buse, C.T., Hagan, J.P., Young, S.D. and Sohn, J.E., *J. Am. Chem. Soc.*, **1979**, *101*, 7077.
16. Heathcock, C.H., Pirrung, M.C., Lampe, J., Buse, C.T. and Young, S.D., *J. Org. Chem.*, **1981**, *46*, 2290.
17. Van Draanen, N.A., Arsenyadis, S., Crimmins, M.T. and Heathcock, C.H., *J. Org. Chem.*, **1991**, *56*, 2499.
18. Oppolzer, W., *Pure Appl. Chem.*, **1988**, *60*, 39.
19. Pirrung, M.C. and Heathcock, C.H., *J. Org. Chem.*, **1980**, *45*, 1274.
20. Cowden, C.J. and Paterson, I., in *Organic Reactions*, Ed. L.A. Paquette, John Wiley, New York, **1997**, vol. 51, p. 1.
21. Kim, B.M., Williams, S.F. and Masamune, S., in *Comprehensive Organic Synthesis*, Ed. B.M. Trost, Pergamon, Oxford, **1991**, vol. 2, p. 239.
22. Franklin, A.S. and Paterson, I., *Cont. Org. Synthesis*, **1994**, 317.
23. Evans, D.A., Dart, M.J., Duffy, J.L., Yang, M.G. and Livingston, A.B., *J. Am. Chem. Soc.*, **1995**, *117*, 6619.
24. Gustin, D.J., VanNieuwenhze, M.S. and Roush, W.R., *Tetrahedron Lett.*, **1995**, *36*, 3443.
25. Paterson, I., Bower, S. and Tillyer, R.D., *Tetrahedron Lett.*, **1993**, *34*, 4393.
26. Roush, W.R., *J. Org. Chem.*, **1991**, *56*, 4151.
27. Gennari, C., Vieth, S., Comotti, A., Vulpetti, A., Goodman, J.M. and Paterson, I., *Tetrahedron*, **1992**, *48*, 4439.
28. Masamune, S., Choy, W., Kerdesky, F.A.J. and Imperiali, B., *J. Am. Chem. Soc.*, **1981**, *103*, 1566.
29. Masamune, S., Hirama, M., Mori, S., Ali, S.A. and Garvey, D.S., *J. Am. Chem. Soc.*, **1981**, *103*, 1568.
30. Evans, D.A., Bartroli, J. and Shih, T.L., *J. Am. Chem. Soc.*, **1981**, *103*, 2127.
31. Evans, D.A. and Bartroli, J., *Tetrahedron Lett.*, **1982**, *23*, 807.
32. Oppolzer, W., Blagg, J., Rodriguez, I. and Walther, E., *J. Am. Chem. Soc.*, **1990**, *112*, 2767.
33. Debrabander, J. and Oppolzer, W., *Tetrahedron*, **1997**, *53*, 9169.
34. Oppolzer, W., Debrabander, J., Walther, E. and Bernardinelli, G., *Tetrahedron Lett.*, **1995**, *36*, 4413.
35. Evans, D.A., Ng, H.P., Clark, J.S. and Rieger, D.L., *Tetrahedron*, **1992**, *48*, 2127.
36. Paterson, I., Goodman, J.M. and Isaka, M., *Tetrahedron Lett.*, **1989**, *30*, 7121.
37. Paterson, I., Ward, R.A., Smith, J.D., Cumming, J.G. and Yeung, K.S., *Tetrahedron*, **1995**, *51*, 9437.
38. Paterson, I. and Perkins, M.V., *Tetrahedron Lett.*, **1992**, *33*, 801.
39. Paterson, I. and Perkins, M.V., *Tetrahedron*, **1996**, *52*, 1811.
40. Paterson, I., Norcross, R.D., Ward, R.A., Romea, P. and Lister, M.A., *J. Am. Chem. Soc.*, **1994**, *116*, 11287.
41. Paterson, I. and Perkins, M.V., *J. Am. Chem. Soc.*, **1993**, *115*, 1608.
42. Bernardi, A., Gennari, C., Goodman, J.M. and Paterson, I., *Tetrahedron Asym.*, **1995**, *6*, 2613.

43. Paterson, I. and Wallace, D.J., *Tetrahedron Lett.*, **1994**, *35*, 9477.
44. Paterson, I., Wallace, D.J. and Velazquez, S.M., *Tetrahedron Lett.*, **1994**, *35*, 9083.
45. Paterson, I. and Wallace, D.J., *Tetrahedron Lett.*, **1994**, *35*, 9087.
46. Danda, H., Hansen, M.M. and Heathcock, C.H., *J. Org. Chem.*, **1990**, *55*, 173.
47. Walker, M.A. and Heathcock, C.H., *J. Org. Chem.*, **1991**, *56*, 5747.
48. Evans, D.A., Coleman, P.J. and Dias, L.C., *Angew. Chem., Int. Ed. Engl.*, **1998**, *36*, 2738.
49. Evans, D.A., Trotter, B.W., Cote, B. and Coleman, P.J., *Angew. Chem., Int. Ed. Engl.*, **1998**, *36*, 2741.
50. Evans, D.A., Trotter, B.W., Cote, B., Coleman, P.J., Dias, L.C. and Tyler, A.N., *Angew. Chem., Int. Ed. Engl.*, **1998**, *36*, 2744.
51. Paterson, I., in *Comprehensive Organic Synthesis*, Ed. B.M. Trost, Pergamon, Oxford, **1991**, vol. 2, p. 301.
52. Siegel, C. and Thornton, E.R., *Tetrahedron Lett.*, **1986**, *27*, 457.
53. Evans, D.A., Clark, J.S., Metternich, R., Novak, V.J. and Sheppard, G.S., *J. Am. Chem. Soc.*, **1990**, *112*, 866.
54. Evans, D.A., Riegler, D.L., Bilodeau, M.T. and Urpi, F., *J. Am. Chem. Soc.*, **1991**, *113*, 1047.
55. Duffy, J.L., Yoon, T.P. and Evans, D.A., *Tetrahedron Lett.*, **1995**, *36*, 9245.
56. McCarthy, P.A. and Kageyama, M., *J. Org. Chem.*, **1987**, *52*, 4681.
57. Evans, D.A. and Ng, H., *Tetrahedron Lett.*, **1993**, *34*, 2229.
58. Evans, D.A., Ng, H.P. and Rieger, D.L., *J. Am. Chem. Soc.*, **1993**, *115*, 11446.
59. Evans, D.A., Dart, M.J. and Duffy, J.L., *J. Am. Chem. Soc.*, **1995**, *117*, 9073.
60. Evans, D.A. and McGee, L.R., *Tetrahedron Lett.*, **1980**, *21*, 3975.
61. Yamamoto, Y. and Muruyama, K., *Tetrahedron Lett.*, **1980**, *21*, 4607.
62. Bernet, B., Bishop, P.M., Caron, M., Kawamato, T., Roy, B.L., Ruest, L., Sauvé, G., Soucy, P. and Deslongchamps, P., *Can. J. Chem.*, **1985**, *63*, 2810.
63. Mukaiyama, T. and Kobayashi, S., *Org. React.*, **1994**, *46*, 1.
64. Nagao, Y., Yamada, S., Kumagai, T., Ochiai, M. and Fujita, E., *J. Chem. Soc., Chem. Commun.*, **1985**, 1418.
65. Paterson, I. and Tillyer, R.D., *Tetrahedron Lett.*, **1992**, *33*, 4233.
66. Evans, D.A., Yang, M.G., Dart, M.J., Duffy, J.L. and Kim, A.S., *J. Am. Chem. Soc.*, **1995**, *117*, 9598.
67. Oppolzer, W. and Starkmann, C., *Tetrahedron Lett.*, **1992**, *33*, 2439.

5 Reagent-controlled aldol reactions

5.1 Introduction

Absolute stereochemical control of a reaction, where two achiral reagents combine to give an excess of one enantiomer of the product, requires the use of some external source of chirality. This can be in the form of a chiral reagent, a chiral catalyst or in some cases a chiral solvent. The boron enolate aldol reaction lends itself to control by the use of chiral reagents in the form of chiral ligands attached to the boron Lewis acid used in the formation of the enolate. It is these chiral ligands attached to the boron which control the enantioselectivity of the reaction.

In Scheme 5.1 the chiral borolane developed by Reetz et al.[1] is shown controlling the absolute configuration of the *anti* aldol between the thioester and

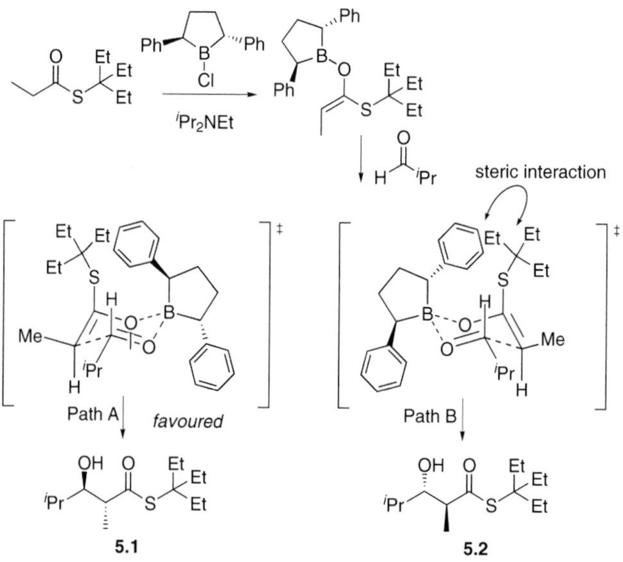

Scheme 5.1

isobutyraldehyde giving product **5.1** over its enantiomer **5.2** with an enantiomeric excess (e.e.) of 99%. That means less than 0.5% of the reaction proceeds by pathway B. This can be attributed to the unfavourable steric interaction between the phenyl attached to the borolane and the bulky group attached to the sulfur.

The chiral ligands employed on boron include those derived from natural and purely synthetic sources (the chirality invariably comes from nature, even if it is a chiral resolving reagent of some kind). Common examples are the (−)-diisopinocampheylboron chloride, (−)-Ipc$_2$BCl **5.3** a commercially available crystalline solid (the enantiomer is also commercially available), the corresponding boron triflate, **5.4**, the menthone-derived [(−)-(menth)CH$_2$]$_2$BCl **5.5** and [(−)-(menth)CH$_2$]$_2$BBr **5.6**, the diaborolidine **5.7** and the borolones **5.8** and **5.9**.

(−)-Ipc$_2$BX, X = Cl **5.3**
or OTf **5.4**

[(−)-(Menth)CH$_2$]$_2$BX, X = Cl **5.5**
or Br **5.6**

5.7

5.8

5.9

5.2 Natural-product-derived ligands

5.2.1 Isopinocampheyl ligands on boron

(−)-Diisopinocampheylboron triflate, (−)-Ipc$_2$BOTf **5.4** can be prepared by treatment of the corresponding (−)-diisopinocampheylborane with triflic acid. Treatment of ethyl ketones with this triflate in conjunction with hindered bases results in the formation of (Z)-O-enolates which undergo aldol reactions with unhindered aldehydes to give enantiomerically enriched *syn* aldol products (Scheme 5.2). For example the aldol reaction of the enolate derived from diethyl ketone and (−)-Ipc$_2$BOTf gives good yields and e.e. values of 91% and 86%, respectively, on reaction with methacrolein and crotonaldehyde, respectively (Scheme 5.2).[2,3]

An extension of this reaction has involved a *syn* selective reduction of the intermediate aldolate, giving enantiomerically enriched 1,3-*syn* diols in moderate yield (Scheme 5.3).[4]

It is interesting to note that the analogous aldol reaction with a methyl ketone and exactly the same reagent scheme gives aldol adducts which are of the opposite facial selectivity to that observed for the ethyl ketone reaction. The e.e. is

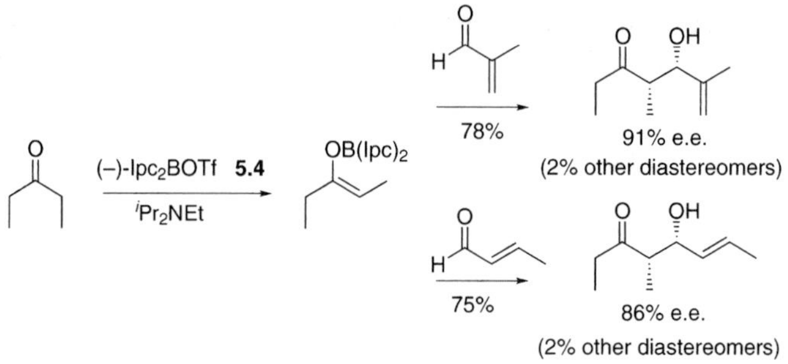

Scheme 5.2

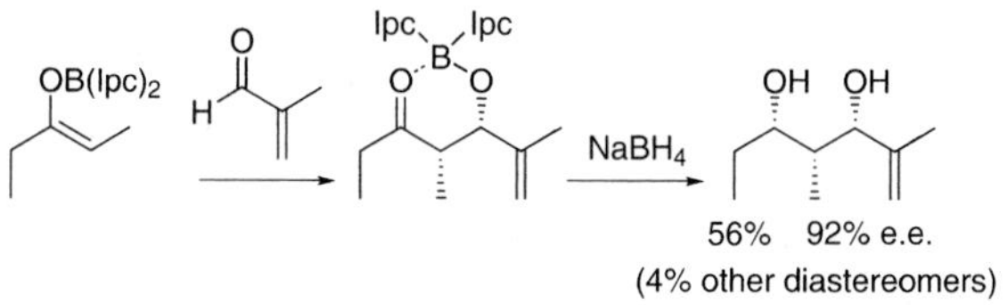

Scheme 5.3

also somewhat lower, 73% e.e. compared with the 91% above (Scheme 5.4). The difference in facial selectivity has been explained by the ethyl ketone reaction

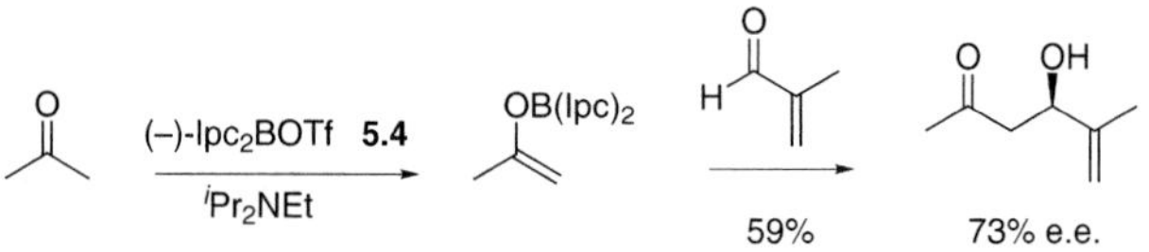

Scheme 5.4

proceeding through a chair[2] transition state whereas the methyl ketone reaction proceeds through a twist boat.[5]

While the reported yield[5] for this reaction is rather low, an improved procedure[6,7] resulting in higher yields puts the possible use of this reaction for natural product synthesis in better stead. The use of Ipc ligands on boron has been extended to the control of stereochemical outcomes in the reaction of enolates from chiral ketones. The chiral ligand will now either enforce the facial preference of the ketone in a matched sense, or work against it in a mismatched sense. The dibutyl boron enolate of the chiral methyl ketone **5.10** has considerable facial bias in its reaction with aldehydes. The use of $(-)$-Ipc$_2$BOTf to form the enolates results in a matched reaction where the selectivity increases to 95:5 from 84:16 (**5.12**:**5.11**; Scheme 5.5). The yield obtained in the Ipc-mediated

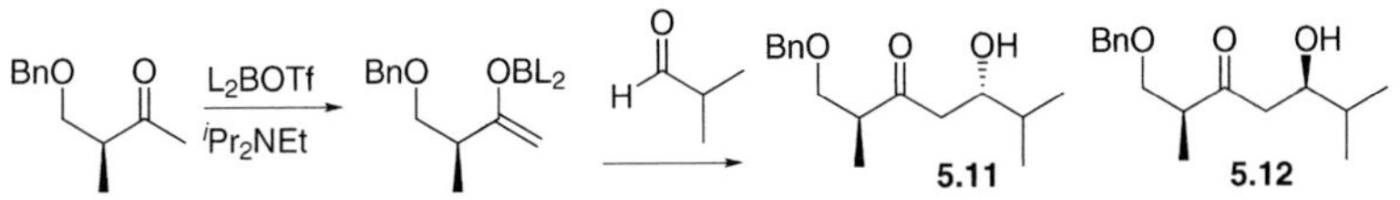

L	Yield (%)	**5.12:5.11**
^{n}Bu	45	84:16
(−)-Ipc	91	95:5

Scheme 5.5

reaction has been reported[8] to be significantly higher when ether is used as the solvent as opposed to the earlier reported ether/DCM mixtures.[9]

For the reaction of **5.10** there is no report of the yields in the mismatched sense to test if the facial preference of the ketone could be overcome by the chiral ligands on boron. While the dibutyl boron enolate of a related ketone **5.13** shows little selectivity in its aldol reactions with simple aldehydes, either *syn* isomer **5.14** or **5.15** can be formed with high selectivity by the choice of the appropriate enantiomer of Ipc$_2$BOTf (Scheme 5.6).[10]

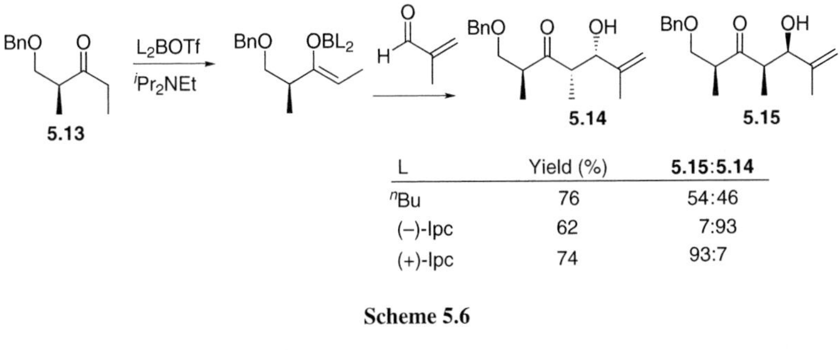

L	Yield (%)	**5.15:5.14**
^{n}Bu	76	54:46
(−)-Ipc	62	7:93
(+)-Ipc	74	93:7

Scheme 5.6

An interesting example of kinetic resolution in enolate chemistry is shown in Scheme 5.7.[11] Here a racemic mixture of ketones is enolised with (+)-Ipc$_2$BOTf to give a diastereomeric mixture of enolates. Although both enolates are formed, in one diastereoisomer the inherent facial preference of the ketone is matched with the facial preference of the (+)-Ipc ligands, making a fast-reacting enolate. The other enantiomer of the ketone produces a mismatched enolate which reacts more slowly. On reaction with half an equivalent of an achiral aldehyde, only the faster-reacting enolate reacts significantly, and it reacts to give predominantly

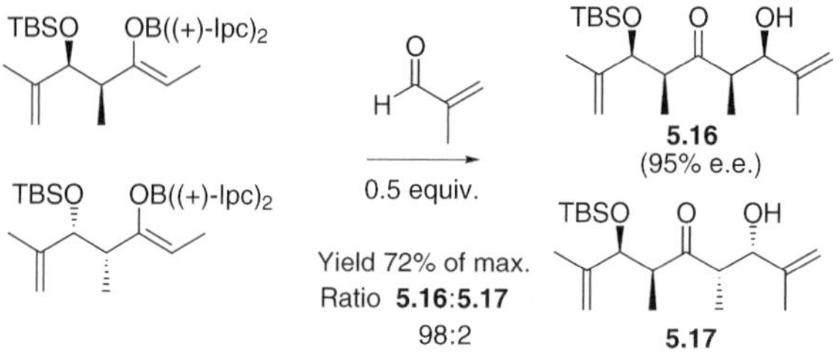

Scheme 5.7

isomer **5.16** over isomer **5.17**, in the ratio 98:2. Product **5.16** is produced with an e.e. of 95%. The recovered ketone is enriched in the slow-reacting enantiomer.

5.2.2   Menthone-derived ligands on boron

A further natural-product-derived ligand may be synthesised from menthone by methylenation. Subsequent hydroboration with chloroborane or bromoborane gives the chloride or bromide Lewis acid (Scheme 5.8).

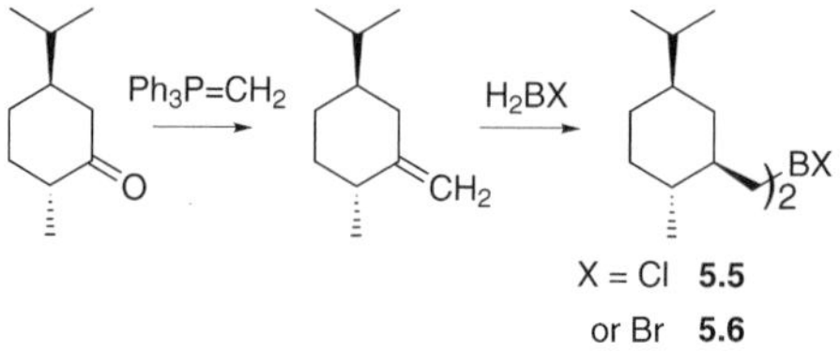

Scheme 5.8

The chloroborane reagent **5.5** has been used for the reaction of methyl ketones, giving enantioselectivities[12] comparable to those obtained with Ipc$_2$BCl.[2] (Note the selectivity of the Ipc$_2$BOTf, Scheme 5.4.) In neither case is a good yield or enantiomeric excesses obtained (Scheme 5.9). However,

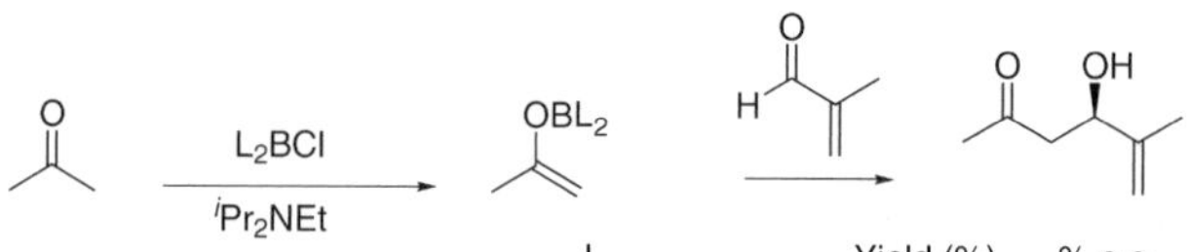

L	Yield (%)	% e.e.
(–)-Ipc	67	62
(–)-MenthCH$_2$	65	58

Scheme 5.9

when ethyl ketones are treated with the chloroborane reagents **5.3** and **5.5**, (*E*)-*O*-enolates are formed which react via chair transition states giving *anti* aldol products. The enantioselectivity obtained with $[(-)-(\text{menth})\text{CH}_2]_2\text{BCl}^{12}$ is far superior to that obtained for $(-)-\text{Ipc}_2\text{BCl}$ (Scheme 5.10).[2]

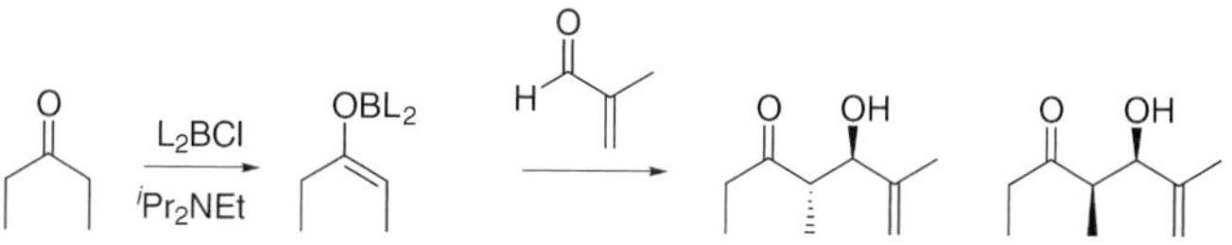

L	Yield (%)	*anti*:*syn*	% e.e. (*anti* isomer)
(–)-Ipc	80	80:20	< 20
(–)-MenthCH$_2$	62	93:7	75

Scheme 5.10

While the chloroborane reagent **5.5** does not enolise thioesters, the analogous bromoborane reagent **5.6** has been found to enolise hindered thioesters effectively and the resulting enolates react with aldehydes giving products of high enantiomeric excess (Scheme 5.11).[13] This reagent **5.6** has been used in

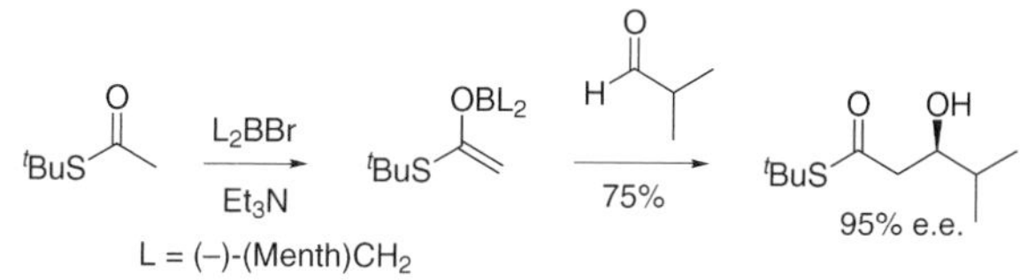

$L = (-)-(\text{Menth})\text{CH}_2$

Scheme 5.11

reactions of thioesters with chiral aldehydes in double stereodifferentiating reactions. It has been shown that the reagent control can overturn the facial bias of a protected glyceraldehyde (compare ratio of **5.18** and **5.19** in Scheme 5.12).[14,15] The dramatic effect of the ligands in this case is especially significant given that the inherent facial selectivity of this aldehyde is probably quite high, as seen from the reaction of a related achiral enolate analogue with a similar aldehyde (compare ratio of **5.20** and **5.21** in Scheme 5.13).[16]

This ligand has also been used in the *anti* aldol reaction of a variety of α-substituted thioesters both in enantioselective and in double stereodifferentiating aldol reactions. Reaction of the propionate thioester analogue with the chiral

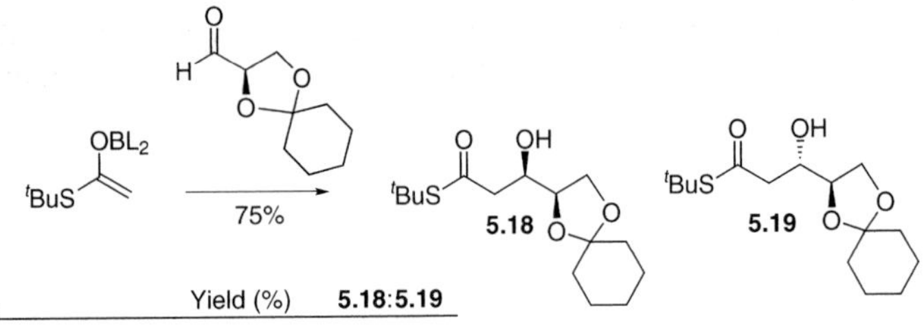

L	Yield (%)	5.18:5.19
(–)-(Menth)CH₂	72	97:3
(+)-(Menth)CH₂	75	4:96

Scheme 5.12

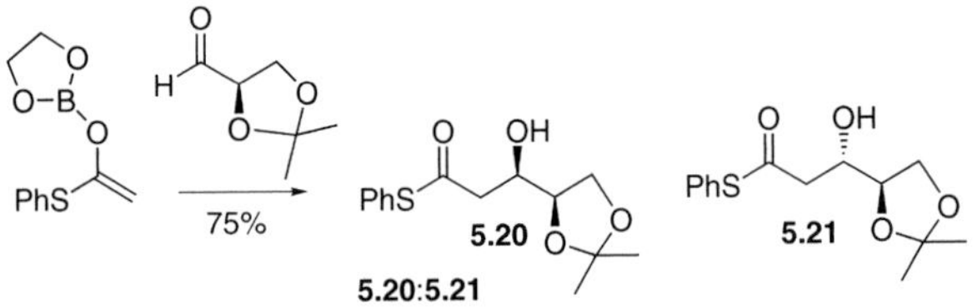

5.20:5.21
20:80

Scheme 5.13

protected glyceraldehyde has been carried out and almost total stereocontrol was obtained. By changing the chirality of the ligand, either *anti* aldol isomer can be formed with greater than 95% diastereoselectivity (compare the ratio of **5.22** and **5.23** in Scheme 5.14).[14] Unfortunately the yield for these reactions is only modest.

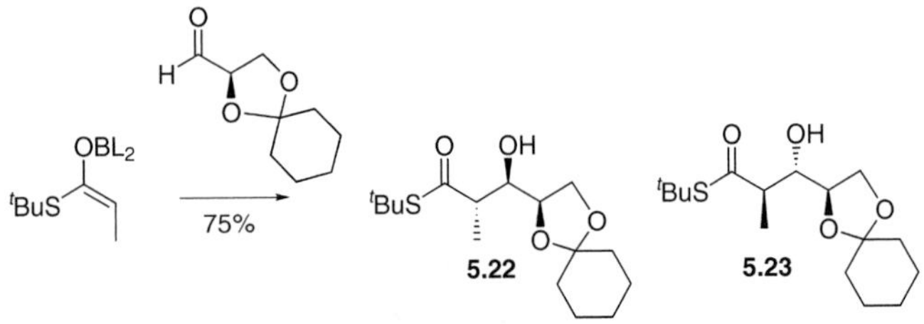

L	Yield (%)	5.23:5.22
(+)-(Menth)CH₂	50	99:1
(–)-(Menth)CH₂	45	5:95

Scheme 5.14

Other α-substituted thioesters have been used, including protected alcohols and halogens, which both form the (*E*)-*O*-enolates and give *anti* aldol products in good enantiomeric excess (Scheme 5.15).[17]

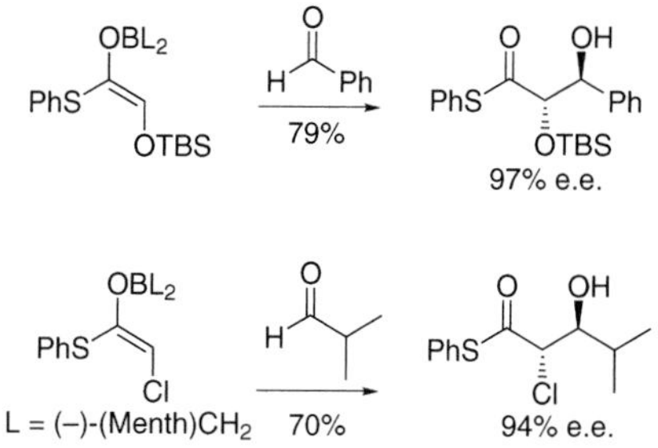

Scheme 5.15

5.2.3   D-*Glucose-derived ligands on titanium enolates*

A quite different set of conditions has been used to prepare titanium enolates which bear a D-glucose-derived chiral ligand for the aldol reactions of *t*-butyl acetate. The chiral titanium reagent **5.24** prepared[18] from cyclopentadienyl-titanium trichloride with two equivalents of (1,2;5,6)-di-*O*-isopropylidene-α-D-glucofuranose can be used to transmetallate the lithium enolate of *t*-butyl acetate and then be reacted with a variety of aldehydes to give products exhibiting high enantiomeric excess (Scheme 5.16).[19]

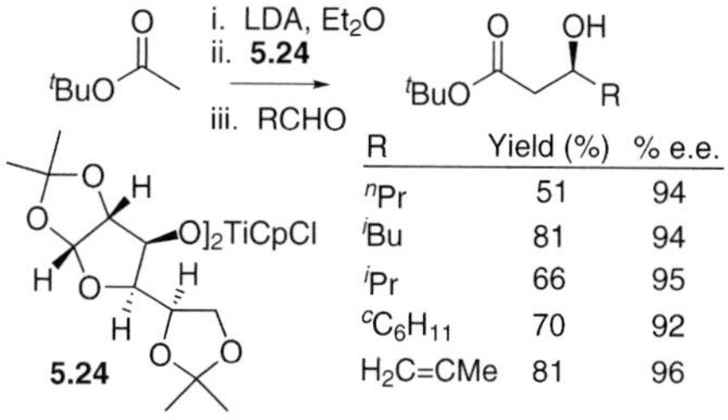

R	Yield (%)	% e.e.
^nPr	51	94
^iBu	81	94
^iPr	66	95
$^cC_6H_{11}$	70	92
$H_2C=CMe$	81	96

Scheme 5.16

This same chiral titanium reagent **5.24** has been used in a similar fashion to carry out *syn* selective aldol reactions on an *N*-protected amino acid producing an *N*-protected *syn*-β-hydroxy-α-amino acid in high enantiomeric excess (Scheme 5.17).

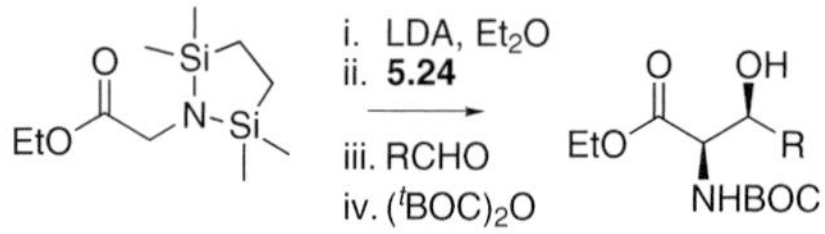

R	Yield (%)	syn:anti	% e.e.
^nPr	66	> 98:2	98
^tBu	43	> 96:4	96
$H_2C=CMe$	57	99:1	98

Scheme 5.17

5.3   Synthetic C(2) symmetric ligands

Two structurally related synthetic chiral boron reagents having a C(2) axis of symmetry, **5.8**[20,21] and **5.9**,[1,22,23] have been developed by the groups of Masamune and Reetz, respectively.   Being prepared by resolution, these reagents are thus available in both enantiomeric forms. They show very similar yield and selectivity for the aldol reaction of acetate thioesters, as seen in Scheme 5.18. Both reagents enolise the corresponding propionate thioesters to

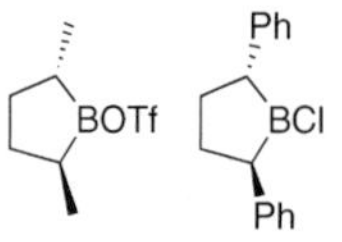

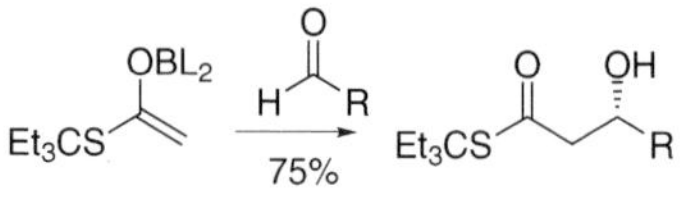

Reagent	R	Yield (%)	% e.e.
5.8	^iPr	81	86.9
5.8	C_6H_{11}	95	85.6
5.9	^iPr	72	92
5.9	C_6H_{11}	87	95

Scheme 5.18

the (*E*)-*O*-enolates which give the *anti* aldol products in good yield and enantioselectivity, as is seen in Scheme 5.19.

The reagent **5.8** has been effective in overcoming the facial bias of chiral aldehydes, as can be seen in the reaction of the acetate thioester enolate with a chiral aldehyde, as reported by Masamune and Duplantier's synthetic studies

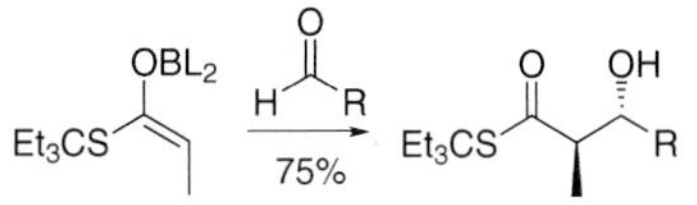

Reagent	R	Yield (%)	% e.e.
5.8	^iPr	85	95.4
5.8	C_6H_{11}	82	93.1
5.9	^iPr	82	99
5.9	C_6H_{11}	58	98.7

Scheme 5.19

towards pimaricin.[24] In this synthesis it was possible to produce either alcohol configuration, **5.25** or **5.26**, by use of the appropriate enantiomer of the chiral auxiliary, overwhelming the influence of the aldehyde (Scheme 5.20).

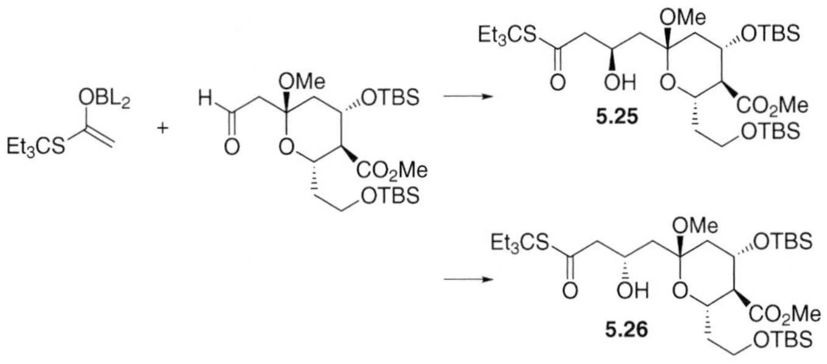

Reagent	
5.8	5.25:5.26, 91:9
ent-5.8	5.25:5.26, 11:89

Scheme 5.20

While the above reagents **5.8** and **5.9** have been useful for the enolisation of thioesters, the only reagent that has been successfully employed for the direct enolisation and enantioselective reaction of normal esters is the diazaborolidine reagent **5.27** developed by Cory and Lee.[25,26] This reagent enolises *t*-butyl propionate to the (*E*)-*O*-enolate, which reacts with benzaldehyde in good yield and enantioselectivity. However, enolisation of the benzyl esters under the same conditions gives mostly the (*Z*)-*O*-enolate, leading to the *syn* aldol again in good yield and enantioselectivity. Changing the enolisation conditions to triethylamine in a toluene/hexane mixture gives mostly the (*E*)-*O*-enolate and thus the *anti* aldol product (Scheme 5.21).

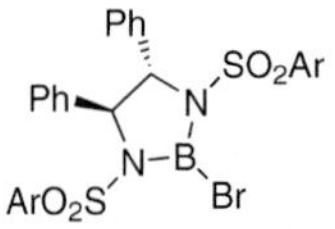

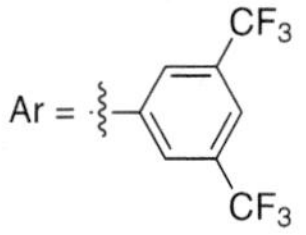

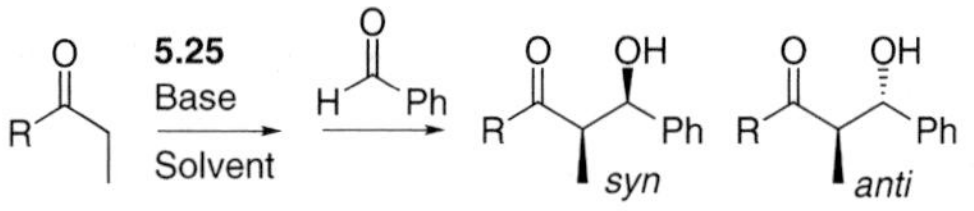

R	Base/Solvent	syn:anti	Yield (%)	% e.e. (major product)
^{t}BuO	^{i}Pr$_2$NEt/CH$_2$Cl$_2$	4:96	89	94
BnO	^{i}Pr$_2$NEt/CH$_2$Cl$_2$	84:16	73	97
BnO	Et$_3$N/Tol.-Hex.	15:85	78	97

Scheme 5.21

5.4 Triple asymmetric induction

We have seen examples in which the chiral ligands on boron have led to effective enantioselective synthesis, and we have seen how the chirality of the ligands on the boron has overcome any intrinsic facial preference of the aldehyde, but the most complex situation arises when we have the reaction of a chiral aldehyde with a chiral enolate bearing chiral ligands on the boron. There are few such examples in the literature and it is only those of methyl ketones which have given good yields.

One such example is seen in Masamune et al.'s[27] synthetic strategy towards the bryostatins. In this approach the chiral reagent **5.8** and ent-**5.8** are used to control the enantioselectivity of the reaction between chiral ketone **5.28** and the chiral aldehyde **5.29** (Scheme 5.22). There is considerable inherent selectivity for the anti-Felkin product for this reaction as seen from the 88:12 ratio of

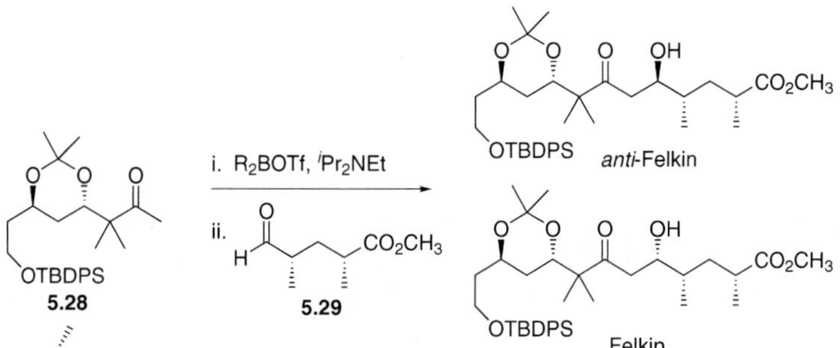

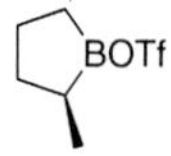

R$_2$BOTf	anti-Felkin:Felkin	Yield (%)
Et$_2$BOTf	8:1	–
5.8	1:1	92
ent-**5.8**	25:1	94

Scheme 5.22

products obtained when the achiral ethyl ligands are employed on the boron enolate. When **5.8** is used to form the enolate the amount of the Felkin product increases but the inherent preference of the substrates cannot be overcome and the product ratio is 1:1. On the other hand, when the *ent-***5.8** reagent is used the inherent selectivity of the substrates is enhanced significantly. Thus in this case the reagent control has been able to increase the inherent selectivity of the substrates, but has not been able to overturn that substrate control with any efficiency.

A more recent example of triple asymmetric induction is seen in the Paterson synthesis of the A,B-ring of the spongistatins.[6] Again, in this example the use of chiral ligands on the boron enhances the inherent selectivity of the substrates in an example of triple asymmetric inductions but fails to overcome the selectivity. In this synthesis of the A,B-ring of the spongastatins, two sequential aldol reactions starting with acetone are employed. First, acetone is enolised with (+)- or (−)-Ipc$_2$BCl and reacted with the chiral aldehyde **5.30** to give two isomeric products **5.31** and **5.32** in a double stereodifferentiating reaction (Scheme 5.23). In this reaction, the use of (−)-Ipc$_2$BCl enhances the inherent

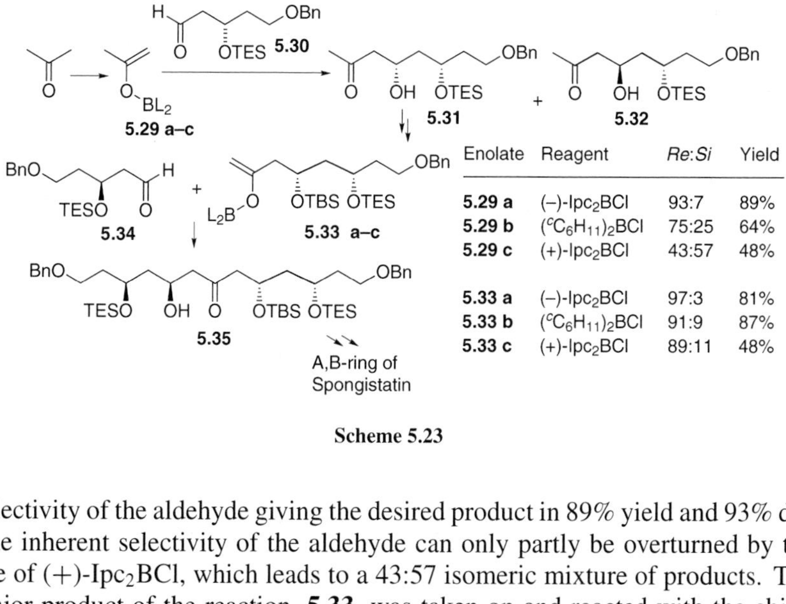

Enolate	Reagent	Re:Si	Yield
5.29 a	(−)-Ipc$_2$BCl	93:7	89%
5.29 b	(^{c}C$_6$H$_{11}$)$_2$BCl	75:25	64%
5.29 c	(+)-Ipc$_2$BCl	43:57	48%
5.33 a	(−)-Ipc$_2$BCl	97:3	81%
5.33 b	(^{c}C$_6$H$_{11}$)$_2$BCl	91:9	87%
5.33 c	(+)-Ipc$_2$BCl	89:11	48%

Scheme 5.23

selectivity of the aldehyde giving the desired product in 89% yield and 93% d.s. The inherent selectivity of the aldehyde can only partly be overturned by the use of (+)-Ipc$_2$BCl, which leads to a 43:57 isomeric mixture of products. The major product of the reaction, **5.33**, was taken on and reacted with the chiral aldehyde **5.34** to give **5.35**. The substrate selectivity for this reaction is very high as seen with reaction via the dicyclohexylboron enolate, which gives a 91:9 isomeric ratio of products. Use of a triple asymmetric reaction with (−)-Ipc$_2$BCl enhances the selectivity to 97:3, but the use of (+)-Ipc$_2$BCl cannot overturn the selectivity and gives a product mixture of 89:11.

One case where reagent control has been used successfully to overcome the inherent selectivity of the system is seen in the coupling of a chiral enolate **5.36** and an α,β-unsaturated chiral aldehyde **5.37**. This model system was studied as part of the Paterson total synthesis of discodermolide,[28] and use of achiral cyclohexyl ligands on the boron enolate **5.36 b** gives a mixture of products **5.38** and **5.39** in the ratio 11:89, with the unwanted isomer predominating. This ratio can be improved to 2:98 by using (−)-Ipc$_2$BCl, but more importantly it can be overturned to give predominantly the desired product **5.38** in the ratio 80:20 by the use of (+)-Ipc$_2$BCl (Scheme 5.24).

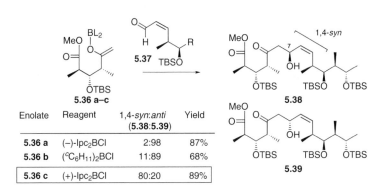

Enolate	Reagent	1,4-*syn*:*anti* (**5.38**:**5.39**)	Yield
5.36 a	(−)-Ipc$_2$BCl	2:98	87%
5.36 b	($^cC_6H_{11}$)$_2$BCl	11:89	68%
5.36 c	(+)-Ipc$_2$BCl	80:20	89%

Scheme 5.24

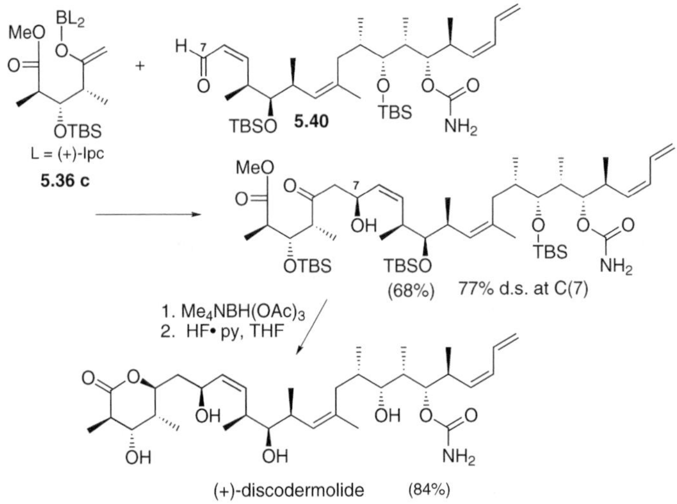

Scheme 5.25

Control of the selectivity of this reaction extends to the key step in the total synthesis of (+)-discodermolide where the coupling between enolate **5.36 c** and the very complex aldehyde **5.40** is achieved in 68% yield and 77% d.s. Hydroxyl-directed reduction and deprotection then gives (+)-discodermolide (Scheme 5.25).[28]

5.5 Conclusions

We have seen in this chapter how a wide range of aldol reactions can be controlled by use of chiral ligands attached to the metal of the enolate to give products with high yields and high enantiomeric excesses. Application of these reactions to the synthesis of complex natural products has also been demonstrated, establishing the utility of these reagents. However, it is apparent that the present day reagents are generally effective only where the substrates have little inherent facial selectivity or where they are reinforcing the inherent substrate facial selectivity. While we have seen that in some cases (e.g. Scheme 5.24 and 5.25) reagents can overcome the inherent substrate facial selectivity in the aldol reaction, increasing the scope and generality of these reactions remains a goal in this area.

References

1. Reetz, M.T., Rivadeneira, E. and Niemeyer, C., *Tetrahedron Lett.*, **1990**, *31*, 3863.
2. Paterson, I., Goodman, J.M., Lister, M.A., Schumann, R.C., McClure, C.K. and Norcross, R.D., *Tetrahedron*, **1990**, *46*, 4663.
3. Paterson, I. and Lister, M.A., *Tetrahedron Lett.*, **1986**, *27*, 4787.
4. Bonini, C., Racioppi, R., Righi, G. and Rossi, L., *Tetrahedron Asym.*, **1994**, *5*, 173.
5. Paterson, I. and Goodman, J.M., *Tetrahedron Lett.*, **1989**, *30*, 997.
6. Paterson, I., Oballa, R.M. and Norcross, R.D., *Tetrahedron Lett.*, **1996**, *37*, 8581.
7. Paterson, I. and Oballa, R.M., *Tetrahedron Lett.*, **1997**, *38*, 8241.
8. Cowden, C.J. and Paterson, I., in *Organic Reactions*, Ed. L.A. Paquete, John Wiley, New York, **1997**, vol. 51, p. 1.
9. Paterson, I., Goodman, J.M. and Isaka, M., *Tetrahedron Lett.*, **1989**, *30*, 7121.
10. Paterson, I. and Lister, M.A., *Tetrahedron Lett.*, **1988**, *29*, 585.
11. Paterson, I., McClure, C.K. and Schuman, R.C., *Tetrahedron Lett.*, **1989**, *30*, 1293.
12. Gennari, C., Hewkin, C.T., Molinari, F., Bernardi, A., Comotti, A., Goodman, J.M. and Paterson, I., *J. Org. Chem.*, **1992**, *57*, 5173.
13. Gennari, C., Moresca, D., Vieth, S. and Vulpetti, A., *Angew. Chem., Int. Ed. Engl.*, **1993**, *32*, 1618.
14. Gennari, C., Vulpetti, A., Moresca, D. and Pain, G., *Tetrahedron Lett.*, **1994**, *35*, 4623.
15. Gennari, C., Pain, G. and Moresca, D., *J. Org. Chem.*, **1995**, *60*, 6248.
16. Gennari, C., Bernadi, A., Cardani, S. and Scolastico, C., *Tetrahedron*, **1984**, *40*, 4059.
17. Gennari, C., Moresca, D. and Vulpetti, A., *Tetrahedron Lett.*, **1994**, *35*, 4857.
18. Riediker, M. and Duthaler, R.O., *Angew. Chem., Int. Ed. Engl.*, **1989**, *28*, 494.

19. Duthaler, R.O., Herold, P., Lottenbach, W., Oertle, K. and Riediker, M., *Angew. Chem., Int. Ed. Engl.*, **1989**, *28*, 495.
20. Masamune, S., Sato, T., Kim, B.M.W. and Wollmann, T.A., *J. Am. Chem. Soc.*, **1986**, *108*, 8279.
21. Masamune, S., *Pure Appl. Chem.*, **1988**, *60*, 1607.
22. Reetz, M.T., Kunisch, F. and Heitmann, P., *Tetrahedron Lett.*, **1986**, *27*, 4721.
23. Reetz, M.T., *Pure Appl. Chem.*, **1988**, *60*, 1607.
24. Duplantier, A.J. and Masamune, S., *J. Am. Chem. Soc.*, **1990**, *112*, 7079.
25. Corey, E.J. and Lee, D.-H., *Tetrahedron Lett.*, **1993**, *34*, 1737.
26. Corey, E.J., *Pure Appl. Chem.*, **1990**, *62*, 1209.
27. Duplantier, A.J., Nantz, M.H., Roberts, J.C., Short, R.P., Somfai, P. and Masamune, S., *Tetrahedron Lett.*, **1989**, *30*, 7357.
28. Paterson, I., *Angew. Chem., Int. Ed. Engl.*, **2000**, *39*, 377.

6   Allylation and crotylation reactions

6.1   Background

The stereoselective addition of allylmetal reagents to carbonyl compounds provides an alternative to the aldol reaction that can often be superior.[1-4] Both *syn* and *anti* type products can be produced by using γ-substituted allylmetal reagents, and the alkene functionality can be further functionalised to provide carbonyl compounds (Scheme 6.1). For example, the synthesis of *anti* type

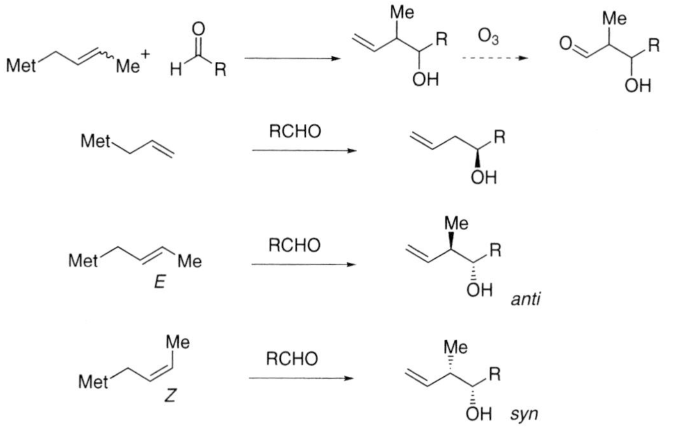

Scheme 6.1

products by an aldol condensation is difficult, and the addition of an (*E*)-γ-substituted allylmetal compound to an aldehyde can provide the analogous *anti* type products with high stereocontrol. *Syn* products can arise from addition of (*Z*)-γ-substituted allylmetal compounds which is analogous to the case for (*Z*)-*O*-enolates (see Chapters 4 and 5). For simplicity, this chapter will focus only on addition of allyl metals to aldehydes.

6.1.1 *Stereochemistry*

The configurational stability of a γ-substituted allylmetal compound is important if the (*E*)- or (*Z*)-isomer is to be utilised for a stereoselective addition reaction.[1] Allylmetal compounds can exist as either monohapto-(η^1) or trihapto-(η^3) bound forms (Scheme 6.2). These should be configurationally stable but E/Z isomerisation is possible for the monohapto-(η^1) compounds

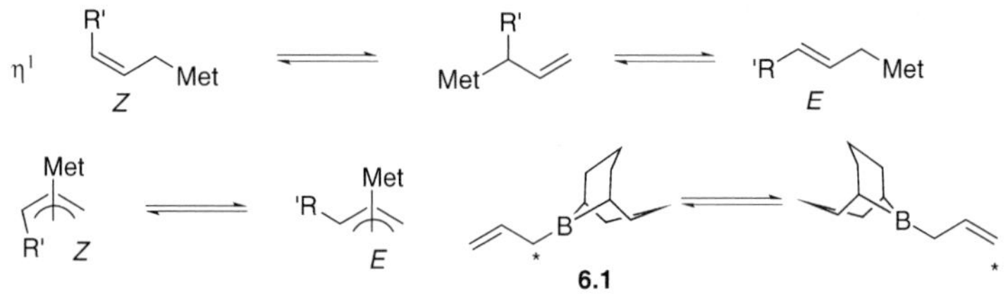

Scheme 6.2

via metallotropic rearrangement to the intermediate (η^1)-allylmetal, and the transition states for this rearrangement can be represented by the trihapto-(η^3)-allylmetal (Z)- and (E)-complexes, respectively. For example, β-allyl-9-borabicyclo[3.3.1]nonane **6.1** undergoes allyl rearrangement with a Gibbs energy ($\Delta G^{\ddagger}$) of 13.3 kcal/mol.[5] Allyllithiums also undergo rapid E/Z isomerisation and exist as slightly distorted π-complexes which can rearrange via the monohapto-intermediate, while allylmagnesium compounds exist as configurationally unstable monohapto-structures.[6,7]

Therefore, for the stereoselective formation of C—C bonds only allylmetal compounds in which this rearrangement is suppressed are useful and only monohapto-(η^1)-allylmetal compounds will be discussed in this chapter. Monohapto-allylmetal compounds which show the lowest tendency for metallotropic isomerisation are the allyl silanes, but these are somewhat less reactive than corresponding lithium and magnesium counterparts.[8] Allyl tin compounds are less stable and sometimes undergo E/Z isomerisation at temperatures less than 100°C or in the presence of Lewis bases.[9]

The most commonly used allylmetal compounds of the third-group elements are those of boron. The tendency of allyl boron compounds to undergo metallotropic rearrangement can be controlled to an extent by the proper choice of substituents on the boron atom. Although dialkyl(allyl)boranes rearrange rapidly at room temperature,[5] they are somewhat stable below −78°C.[10,11] The placement of π-donor atoms, such as oxygen or nitrogen, on boron raise the energy of the vacant boron p-orbital to such an extent that the trihapto-structure does not readily form and thus rearrangement is often entirely suppressed. (E)- and (Z)-allylboron derivatives with two oxygen atoms on boron (i.e. boronate esters) are sufficiently stable to be handled at room temperature without isomerisation.[12,13] Monohapto-allylchromium[14] and titanium compounds[15] are also stable and useful in stereoselective addition to aldehydes.

The addition of γ-substituted allylmetal or crotylmetal reagents to aldehydes can be classified into three main mechanistic types (1, 2 or 3) depending on the type of metal and the conditions of the reaction.[3,16] In type 1 crotylmetallation reactions the stereochemical outcome (*syn:anti* ratio) is dependant on the $Z:E$ ratio in the crotyl moiety and this type of addition is usually observed for boron-, aluminium and tin-based reagents. For type 1 additions, a cyclic six-membered transition state is evoked where the Lewis acidic metal atom can coordinate to the carbonyl oxygen to form an ate-complex.[17]

The four possible transition states for the reaction of a (Z)-crotylmetal reagent with an aldehyde (chair-like: C1 and C2, boat-like: B1 and B2) are depicted in Scheme 6.3 and require a synclinal orientation of the double bonds of the alkene

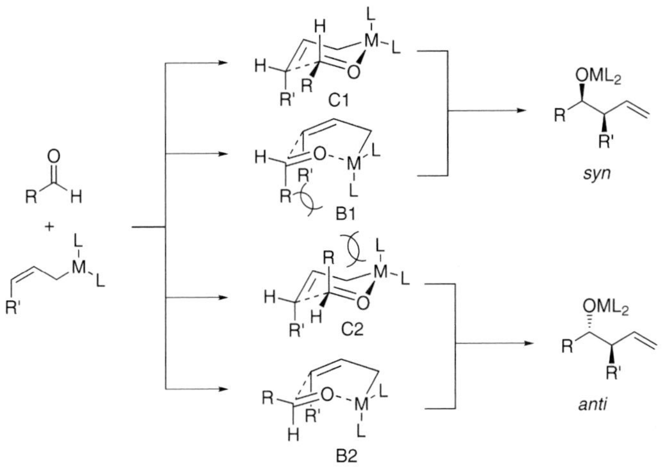

Scheme 6.3

and aldehyde (i.e. the dihedral angle between double bonds must be between 30° and 90°).[1] In the case of the C2 transition state a major steric interaction between the R group of the aldehyde and the L group on the metal atom is present. This 1,3-diaxial interaction is not present in C1 where the R group of the aldehyde is placed equatorially. Since the corresponding R–R' *gauche* interaction is not as large, C1 is the preferred chair-like transition state. In the boat-like transition states, B1 has two steric interactions between the R and R' groups as well as between the R and L groups. In B2, these interactions are absent with the *gauche* R–R' being present. However, boat-like transition states are usually less stable than chair-like transition states, as is the case for aldol reactions[18] and Claisen rearrangements, which suggests that C1 would be preferred. Therefore formation of the *syn* isomer is favoured with a (Z)-crotylmetal agent.

The selectivities for type 1 additions can be increased by using metal atoms with shorter M—O bond lengths [l(M—O)] which increases the R–L interaction in the C2 transition state. Therefore, diastereoselectivity should be highest for boron $[l(B{-}O) \approx 1.4\,\text{Å}]$ and lower for both titanium $[l(Ti{-}O) \approx 1.7\,\text{Å}]$ and tin $[l(Sn{-}O) \approx 2.1\,\text{Å}]$.[18] Ligands of increasing steric bulk can also increase stereoselectivity owing to an increase in the R–L interaction. These factors make boron the metal of choice because of the short B—O bond length and the ability to place large ligands on boron. Furthermore, as stated earlier, crotylboranes and boronates are configurationally stable at lower temperatures.

Some early examples of stereoselective crotylboron reagents are the (E)- and (Z)-crotylboronates **6.2** and **6.3** which are easily synthesised and which are configurationally stable.[12] Compounds **6.2** and **6.3** react with aldehydes via a type 1 process to give *anti* and *syn* diastereoisomers, respectively (Scheme 6.4).

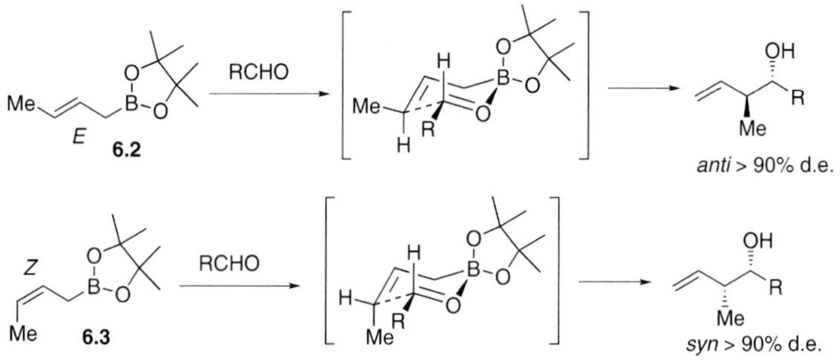

Scheme 6.4

The outcome of these additions can again be rationalised by a tight chair-like transition state which places the aldehyde R group in an equatorial orientation.

In type 2 additions, the stereochemical outcome is independent of crotyl geometry and the *syn* product is the major isomer in most cases. This type of addition is observed for the Lewis-acid-catalysed addition of the less nucleophilic silicon-based,[19,20] tin-based[21,22] and titanium-based[15] reagents and involves more complex acyclic transition states because of the fact that the carbonyl oxygen is no longer available for coordination to the allyl metal centre to form a cyclic ate-complex. The preference for the *syn* isomer can be attributed to addition occurring via an open-chain transition state (Scheme 6.5).[16,23]

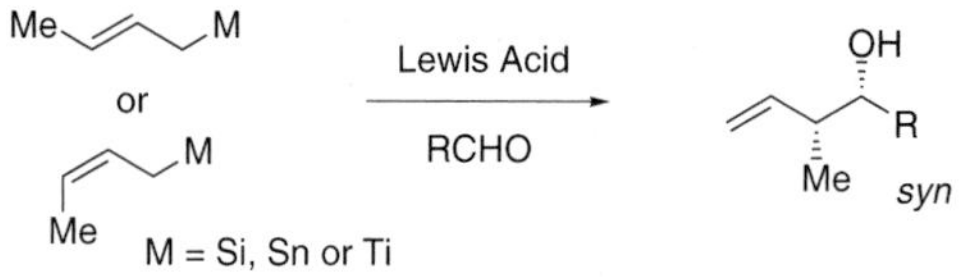

Scheme 6.5

The topicity of Lewis-acid-induced additions of allylsilanes and stannanes has been postulated to be either antiperiplanar[22,24] or synclinal.[23] In the anti-periplanar case, the double bond of the alkene is aligned with the carbonyl double bond in the transition state such that the dihedral angle is close to 180° (Scheme 6.6). It is assumed that the Lewis acid is bound to the oxygen of

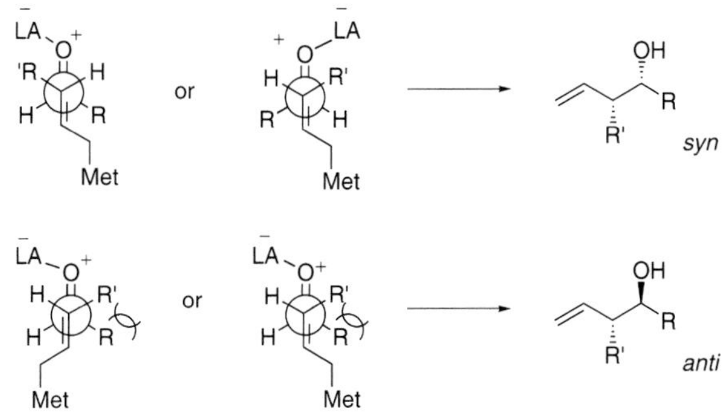

Scheme 6.6

the carbonyl to form an (E)-complex.[23] The key steric interactions are along the forming bond and could involve the R and R' substituents, as shown in Scheme 6.6. Therefore, both the (E)- and (Z)-crotylstannane would give the *syn* product, the *anti* product being disfavoured because of the R–R' interactions present in each transition state.[22]

In some cases the *syn* selectivity for allylsilanes is less than expected and this is not effectively explained by considering antiperiplanar transition states alone. For example, the $TiCl_4$-catalysed addition of (E)- or (Z)-crotyltrimethylsilane to aldehydes gives the *syn* isomer preferentially but the stereoselectivity for the (Z)-silane is much lower (Scheme 6.7). However, in the case of the $BF_3 \cdot OEt_2$-catalysed addition of (E)- or (Z)-crotyltributylstannane to benzaldehyde, high *syn* selectivity for both crotyl isomers is observed.

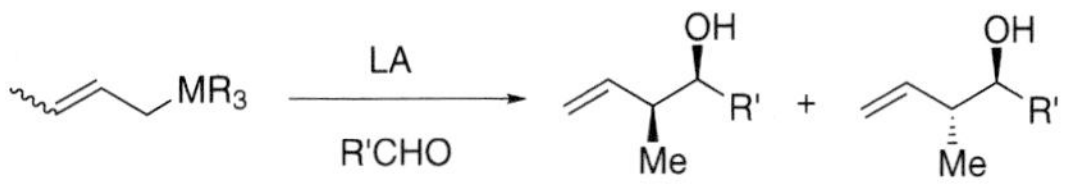

Isomer	MR_3	R'	LA	syn:anti
(E)	$SiMe_3$	Et	$TiCl_4$	95:5
(Z)	$SiMe_3$	Et	$TiCl_4$	69:31
(E)	$SnBu_3$	Ph	BF_3	98:2
(Z)	$SnBu_3$	Ph	BF_3	99:1

Scheme 6.7

An alternative to the antiperiplanar transition state is the open-chain synclinal mode of attack which is depicted in Figure 6.1. This involves synclinal

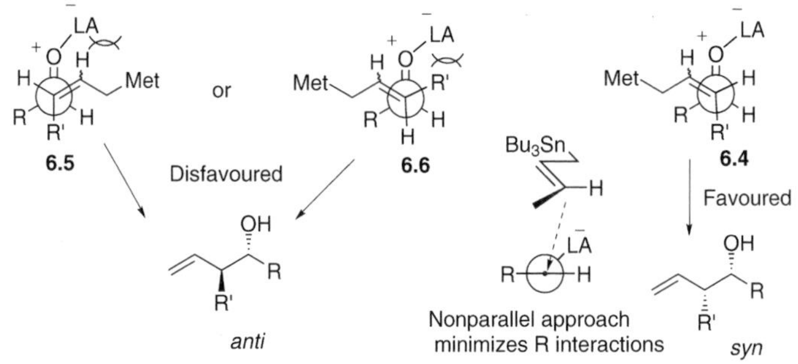

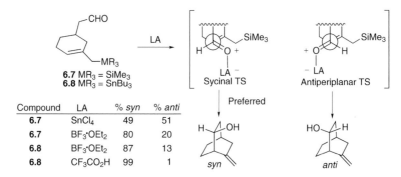

Figure 6.1

alignment of the double bonds where the torsional angle between them is 60°, which is similar to the alignment in the chair-like transition states. It has been suggested that the important steric interactions are between the Lewis acid bound to the carbonyl (E-complex) and the R' group or metal-methylene group[3] which favours the *syn* outcome since the transition state **6.4** would be preferred over **6.5** or **6.6**. This model can be further rationalised by assuming that the two reactants approach in a nonparallel manner which minimises interactions between the allyl metal reagent and R (Figure 6.1).

An early investigation into the preferred topicity for intramolecular Lewis-acid-catalysed additions of allylsilanes and stannanes was conducted by Denmark and Weber.[16,23] In their study, the silane **6.7** and stannane **6.8** were synthesised and allowed to undergo an intramolecular allylmetallation reaction catalysed by various Lewis acids (Scheme 6.8). In most cases there was a strong

Compound	LA	% syn	% anti
6.7	SnCl$_4$	49	51
6.7	BF$_3$·OEt$_2$	80	20
6.8	BF$_3$·OEt$_2$	87	13
6.8	CF$_3$CO$_2$H	99	1

6.7 MR$_3$ = SiMe$_3$
6.8 MR$_3$ = SnBu$_3$

Scheme 6.8

preference for the formation of the *syn* isomer which would result from a synclinal transition state geometry. Furthermore, stannane **6.8** shows high selectivity for the *syn* isomer which was not as dependent on the type of Lewis acid utilised, whereas for the silane case the selectivity was lower when the steric bulk of the Lewis acid was increased. In one example, a cyclisation induced by a proton gave the highest selectivity which led to the suggestion that the steric interaction with the Lewis acid was not important and that electronic effects play a major role in this type of addition.

Yamamoto *et al.* observed different outcomes in the intramolecular addition of a stannane to an aldehyde when using Lewis acids (MX_n) verses Brønsted acids (HA).[25] These results were attributed to a 'push–pull' mechanism where the HA acid serves to bridge the aldehyde and tin moieties (synclinal) whereas the Lewis acid gives the product via antiperiplanar attack. Later, performing deuterium labelling experiments, Denmark and Almstead noted that for silanes the silicon electrofuge is located away (*anti*) from the approaching electrophile regardless of the Lewis acid or double-bond orientation and that both synclinal and antiperiplanar geometries can occur.[26] For stannanes, the synclinal-anti S'_E geometry was preferred for both Lewis and Brønsted acids but the size of the Lewis acid could offset this preference towards an antiperiplanar geometry. It should be noted that the intramolecular case might differ from the intermolecular reactions as a result of steric congestion. Therefore, either the synclinal or antiperiplanar models can be applied in intramolecular Lewis-acid-catalysed additions of allylsilanes and stannanes to aldehydes.

Type 3 additions to aldehydes involve mainly titanium,[27] chromium[14] and zirconium[28] crotylmetal reagents and the stereochemical outcome is again independent of the crotyl geometry, with the *anti* product being preferred. Typically, type 3 crotylmetal reagents are generated *in situ* and then equilibrate to the more stable (*E*)-isomer which then reacts via a type 1 cyclic transition state to give the *anti* product (Scheme 6.9). The reaction of (*E*)- or (*Z*)-1-bromo-2-butene with benzaldehyde in the presence of $CrCl_2$ gave the *anti* product as the sole isomer in both cases.[14]

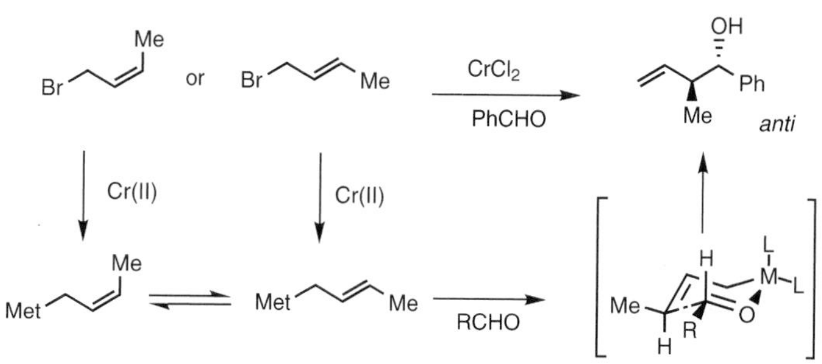

Scheme 6.9

An example of where a type 2 addition can be changed to a type 3 addition was reported by Keck *et al.* in 1984.[29] 'Normal' addition of (*E*)-crotyltributylstannane to a mixture of cyclohexanecarboxyaldehyde and $TiCl_4$ produced the *syn* isomer in a ratio of almost 93:7 whereas 'inverse' addition of the aldehyde to a solution prepared from the Lewis acid and stannane gave the *anti* isomer as the major product in a 95:5 ratio (Scheme 6.10). This result can be

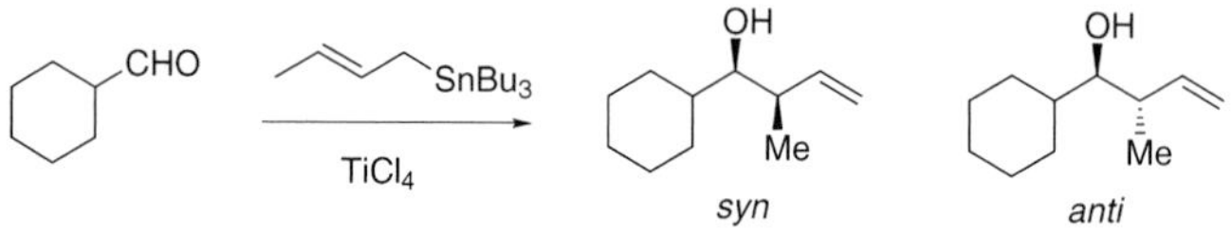

Scheme 6.10

explained by considering the 'normal' addition to be a type 2 addition where the Lewis acid is complexed to the aldehyde thus giving the *syn* isomer as the major product, whereas in the 'inverse' addition a crotyl titanium species may be formed which then reacts with the aldehyde via a type 3 mode of addition which would favour the *anti* product.

6.2   Addition of achiral allyl and crotylmetal reagents to chiral aldehydes

6.2.1   Boron-based reagents

Of the large number of allylmetal compounds available, boronic acid esters are amongst the most useful for asymmetric synthesis. Stereochemically defined (*Z*)- and (*E*)-crotylboronates are readily accessible and configurationally stable at or slightly above room temperature. Some early studies on additions of allyl and crotylboronates to chiral aldehydes were conducted independently by the groups of Hoffmann and Roush.[30–32] The pinacol-derived ester **6.9** can be synthesised from allylmagnesium bromide, and the configurationally stable (*E*)- and (*Z*)-crotyl derivatives **6.2** and **6.3** are available from (*E*)- and (*Z*)-but-2-ene by metallation under Schlosser conditions followed by borylation, hydrolysis and ester formation (Scheme 6.11).[32]

A number of additions of the allylboronate **6.9** to chiral aldehydes have been reported by Hoffmann and the majority proceed preferentially by the Felkin–Anh mode of attack (Scheme 6.12). Addition of **6.9** to the chiral aldehyde **6.10** gave a low selectivity for what was assumed to be the Felkin–Anh product whereas addition to aldehyde **6.11** provided an almost 1:1 selectivity.[30]

Roush also examined the addition of boronate **6.9** to α-oxygenated chiral aldehydes **6.12** and **6.13** and found that the reactions again proceeded under Felkin–Anh control; however, in these cases, the α-oxygen is the large group

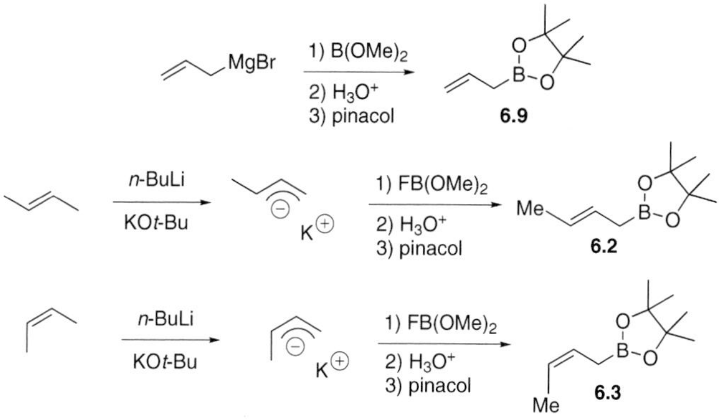

Scheme 6.11

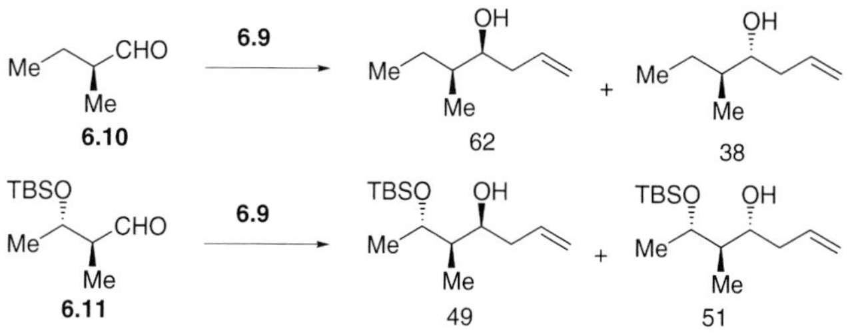

Scheme 6.12

in the transition state. Better selectivities were observed as exemplified by the reactions shown in Scheme 6.13.[32]

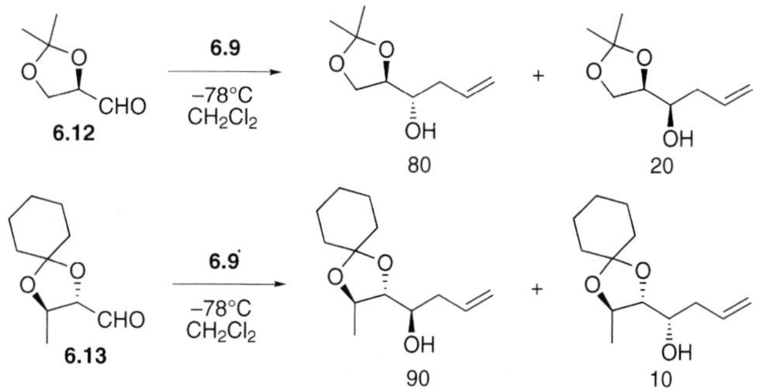

Scheme 6.13

More striking results were obtained when the additions of the crotyl boronates where investigated. Treatment of aldehyde **6.11** with the (*E*)-boronate **6.2** gave the 3,4-*anti*-4,5-*syn* isomer as the major compound (Scheme 6.14).[30] This product was a result of Felkin–Anh attack (*Si* face) on the aldehyde, whereas

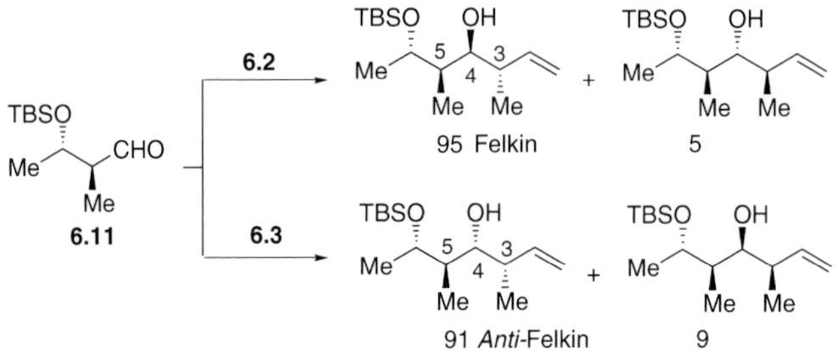

Scheme 6.14

the 3,4-*anti* stereochemistry is that predicted by a type 1 cyclic transition state. The reaction between **6.11** and (*Z*)-boronate **6.3** also proceeded with good stereo-selectivity but gave the 3,4-*syn*-4,5-*anti* isomer which is the *anti*-Felkin–Anh product (i.e. *Re* face attack).

The addition of boronate **6.2** to the α-oxygenated aldehyde **6.12** was much less selective, giving the 3,4-*anti*-4,5-*anti* product as the major isomer which is a result of Felkin–Anh attack (Scheme 6.15).[32] Similarly, addition of

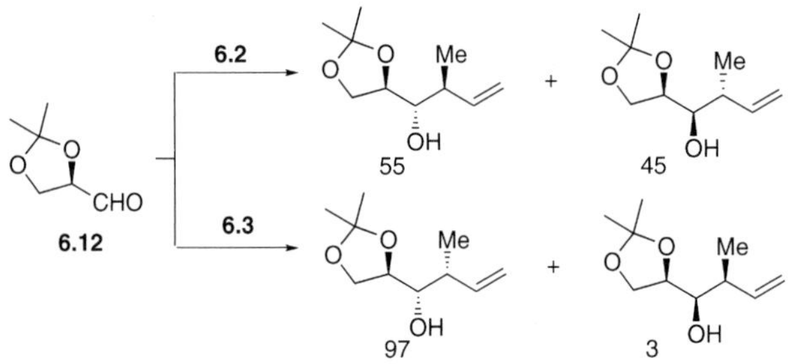

Scheme 6.15

(*Z*)-boronate **6.3** to **6.12** also gave the product resulting from Felkin–Anh attack but with excellent selectivity for the 3,4-*syn*-4,5-*anti* product. It is interesting that this last observation was opposite to that for the addition of **6.3** to **6.11** (Scheme 6.14) which proceeded via an *anti*-Felkin mode of attack.

Roush has proposed separate models to rationalise the results for addition of crotyl boronates to α-methyl chiral aldehydes[33] and α-oxygenated chiral aldehydes.[32] In the case of α-methyl chiral aldehydes, the addition of (*Z*)-crotyl boronate **6.3** proceeds by an *anti*-Felkin mode of attack because in the Felkin mode there are 1,5-nonbonding Me–Me or Me–R interactions in the type 1 chair-like transition states **6.13** and **6.14** (Scheme 6.16). In the transition state

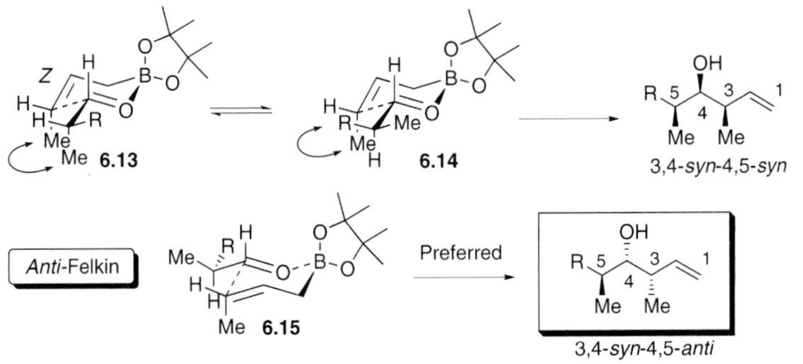

Scheme 6.16

6.15 there are fewer interactions between the crotyl unit and the α-substituents on the aldehyde and this is favoured even though it is an *anti*-Felkin mode of attack on the aldehyde.

In the other case, the addition of (*E*)-crotyl boronate **6.2** gives product **6.17** via a Felkin mode of attack where the transition state **6.16** is preferred over the *anti*-Felkin transition states **6.18** and **6.19** where there are again interactions between the R and Me groups on both the crotyl unit and C2 of the aldehyde (Scheme 6.17).

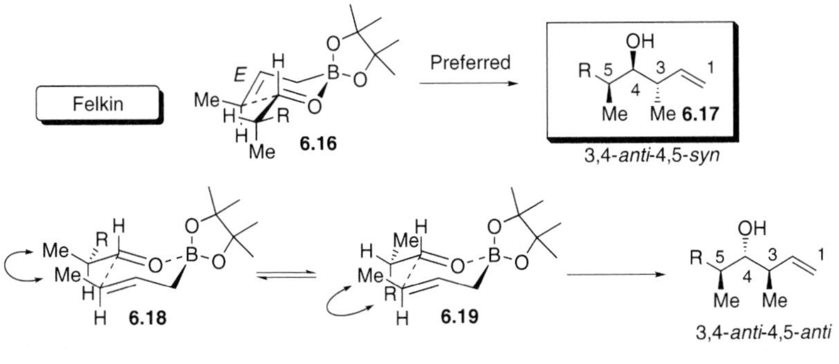

Scheme 6.17

In the case of α-oxygenated chiral aldehydes, such as acetonide **6.12**, dia-stereoselectivity is still dependent on the geometry of the crotyl unit but the addition of (Z)-crotyl boronate occurs from the same face of the aldehyde as Felkin–Anh attack. This has been attributed to electronic stabilisation of the transition state **6.20** which corresponds to the Cornforth model.[32] The correct Felkin–Anh transition state **6.21** and the *anti*-Felkin transition state **6.22** both have stronger steric interactions which destabilise them relative to **6.20**. For the (E)-crotylborane the selectivity for addition to **6.12** is low and this was attributed to the fact that the transition states for Felkin, Cornforth and *anti*-Felkin **6.25** modes of attack are similar in energy. Although **6.25** has lesser steric interactions, the destabilisation of **6.23** and **6.24** owing to steric interactions is offset by electronic stabilisation.

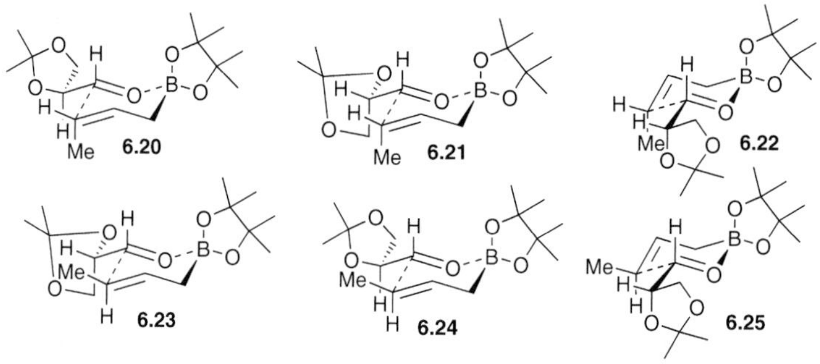

However, more complex chiral aldehydes give different results, with an example being the addition of the (E)-crotylboronate **6.2** to chiral aldehyde **6.26** (Scheme 6.18).[34] This reaction proceeds with very high stereocontrol possibly

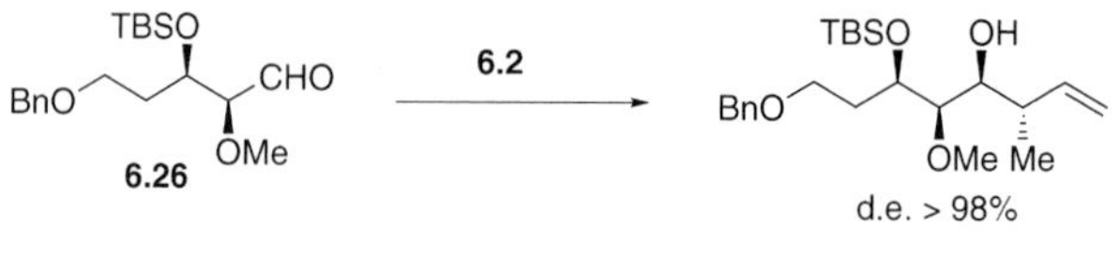

Scheme 6.18

by an *anti*-Felkin–Anh transition state similar to **6.25**. Remote steric interactions may also be involved.

Additions of other γ-substituted boronates to chiral aldehydes have been reported. In the example shown in Scheme 6.19, addition of the (Z)-allylboronate **6.27** to the chiral aldehyde **6.28** proceeds with modest selectivity to give the 3,4-*syn*-4,5-*anti* product.[35]

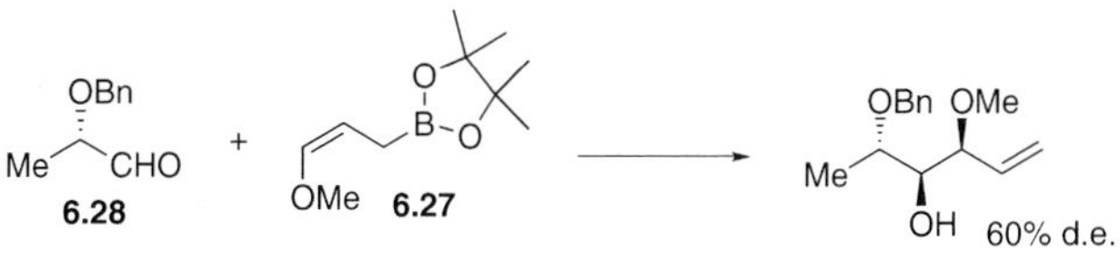

Scheme 6.19

6.2.2 Chromium-based reagents

The addition of crotylchromium compounds to aldehydes is often highly selective and, as mentioned earlier, gives the *anti* isomer regardless of their configuration (type 3 reagent). Allylchromium reagents are usually generated *in situ*
from the corresponding allylic halides and $CrCl_2$ in THF or $CrCl_3/LiAlH_4$
in THF. For example addition of the allylchromium species generated from
iodides **6.30** or **6.31** to aldehyde **6.29** affords the product with the 3,4-*anti*-4,5-
syn stereochemistry at the newly formed asymmetric centres (Scheme 6.20).[36]

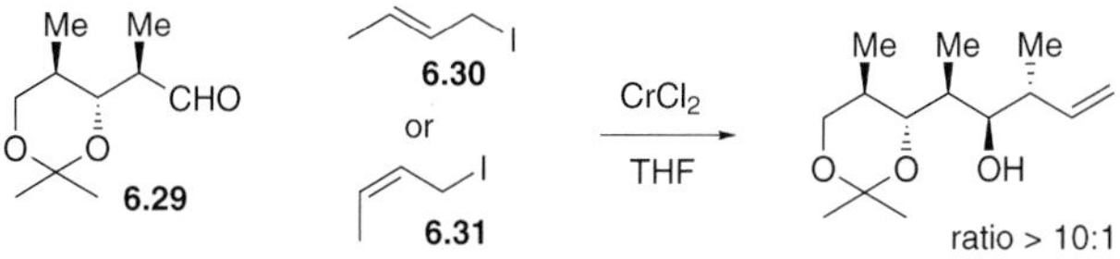

Scheme 6.20

The 3,4-*anti* stereochemistry arises from a cyclic transition state whereas the
4,5-*syn* stereochemistry is a result of Felkin–Anh attack on the aldehyde as
observed for (*E*)-crotylboronates in type 1 additions (see Scheme 6.17).

An example using the method where $CrCl_2$ is generated by $LiAlH_4$ reduction of $CrCl_3$ is presented in Scheme 6.21. The (*E*)-crotylchromium reagent

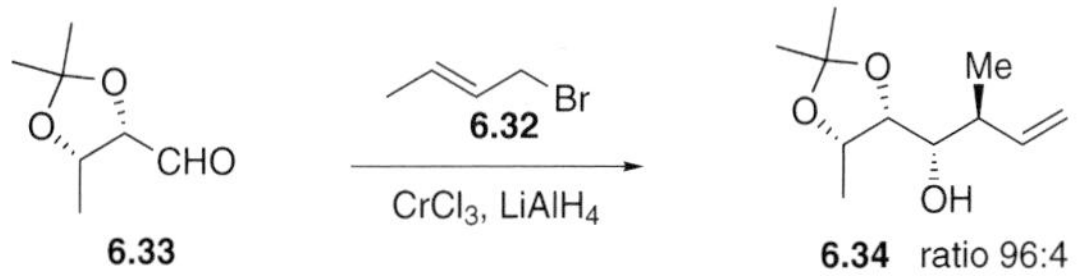

Scheme 6.21

generated from bromide **6.32** was allowed to react with chiral aldehyde **6.33** to
provide the isomer **6.34** in high diastereoisomeric excess.[37]

6.2.3   Silicon- and tin-based reagents

The Lewis-acid-promoted type 2 reactions of allylsilanes and stannanes also often proceed with good stereocontrol. Heathcock *et al.* investigated the reaction of allylsilanes with simple chiral aldehydes in the presence of Lewis acids and found that additions to α- and β-alkoxy aldehydes show exceptional diastereofacial preference.[38] The SnCl$_4$-mediated addition of allyltrimethylsilane to aldehyde **6.35** gives the alcohol **6.36** in high d.e. (Scheme 6.22). Presumably,

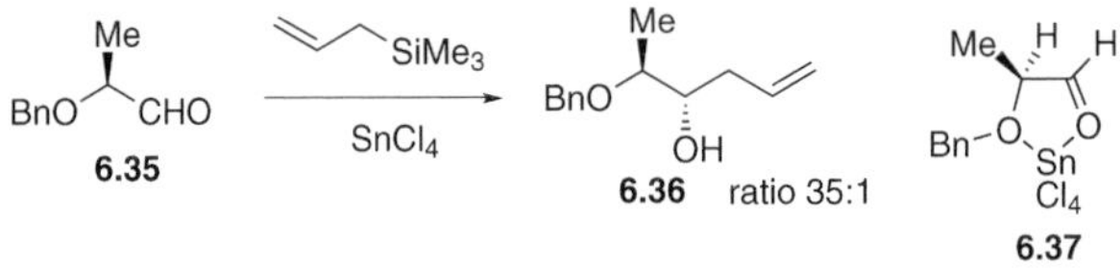

Scheme 6.22

the intermediate five-membered rigid chelate **6.37** undergoes selective attack from the face opposite to the methyl group (Cram chelate model).

The addition of allyl and crotylstannanes to chiral aldehydes can often give higher selectivities than those observed for silanes. Keck and Boden reported that the addition of allyl tributylstannane to aldehyde **6.38** catalysed by MgBr$_2$ proceeded with high selectivity for the *syn* isomer, which is a result of chelation control (Scheme 6.23).[39] On the other hand, addition to the TBS-protected

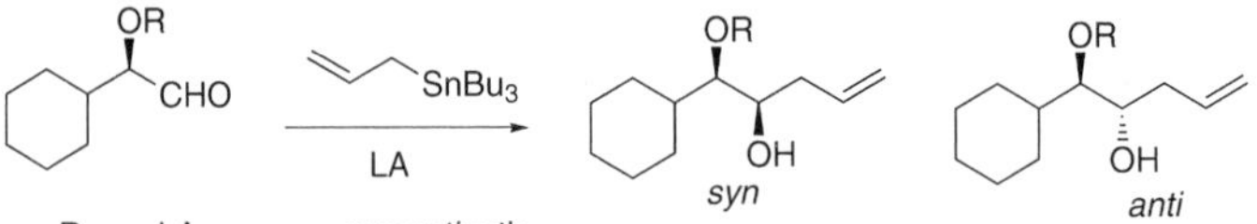

	R	LA	syn:anti ratio
6.38	Bn	MgBr$_2$	> 250:1
6.39	TBS	BF$_3$·OEt$_2$	5:95

Scheme 6.23

analogue **6.39** catalysed by BF$_3$ · OEt$_2$ switched the selectivity to the *anti* Felkin–Anh product. The TBS group is known to reduce the chelation ability of the ether oxygen thus causing the addition to occur by a Felkin–Anh mode of attack (see Chapter 2).

Additions of crotyl stannanes also proceed with excellent stereocontrol, and the facial selectivity on β-oxygenated aldehydes can again be controlled by choice of protecting group and Lewis acid. When aldehyde **6.40** was allowed to react with (*E*)-crotyltributylstannane in the presence of MgBr$_2$ the 3,4-*syn*-4,5-*anti* product was the preferred outcome (Scheme 6.24).[40] Switching

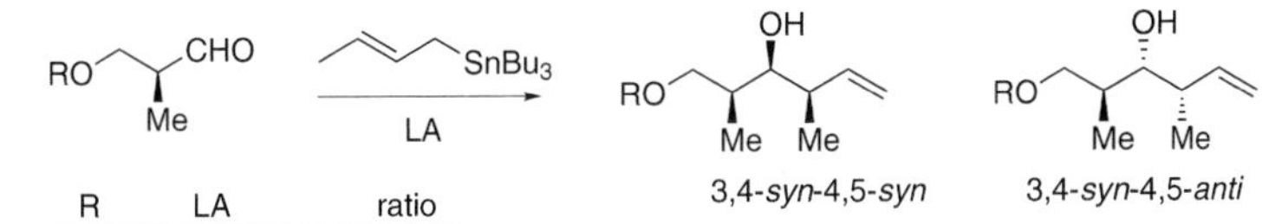

	R	LA	ratio
6.40	Bn	MgBr$_2$	7:81
6.41	TBS	BF$_3$·OEt$_2$	95:5

Scheme 6.24

to the TBS-protected aldehyde **6.41** as substrate, one obtains the 3,4-*syn*-4,5-*syn* isomer as the major product.

6.3 Addition of chiral allyl and crotylmetal reagents to achiral aldehydes

6.3.1 Chiral allylboron compounds

A number of chiral allylmetal reagents have been developed, but the most effective are based on boron, and only this type of reagent will be discussed here. Allylboron reagents can deliver products in high stereoselectivity, and the ability to easily introduce chiral groups on the boron atom makes this the metal of choice for chiral allylmetal reagents. Early studies focused on chiral boronate esters because of the stability and ease of preparation of these esters.

An early example of a chiral boronate ester derived from camphor was reported by Hoffmann and co-workers in the early 1980s.[41–43] The allylboronate **6.42** was synthesised from the camphor derivative **6.43** by the sequence outlined in Scheme 6.25. Reduction of ketone **6.43** followed by ketal hydrolysis

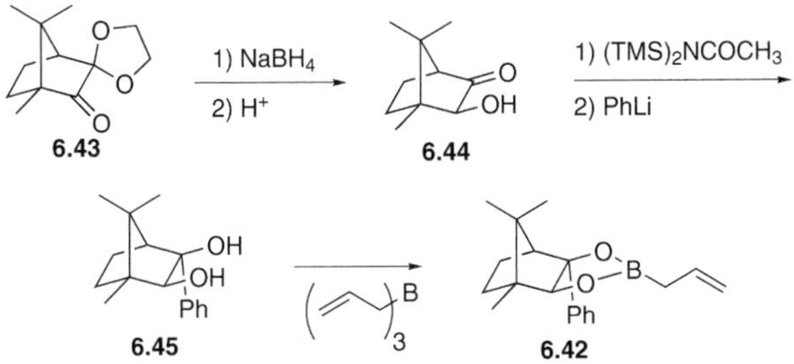

Scheme 6.25

provided hydroxyketone **6.44** which on temporary protection and treatment with phenyllithium provided crystalline diol **6.45**. Exposure of diol **6.45** to triallyl borane gave boronate **6.42** in high yield. Condensation of boronate **6.42** with acetaldehyde at low temperature followed by triethanolamine work-up provided 4-penten-2-ol in good e.e. (Scheme 6.26).[41] Other aldehydes such as butanal

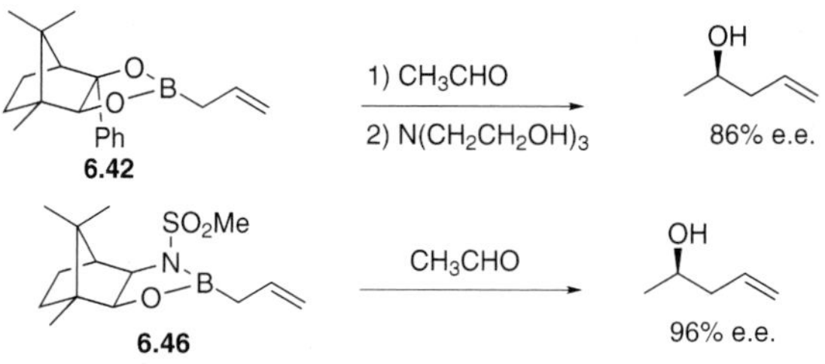

Scheme 6.26

and benzaldehyde gave lower e.e. values. Later, Reetz and Zierke reported that the related allylborane **6.46** reacts with a range of aldehydes to give products with e.e. values of 88%–96%.[44]

Hoffmann and Landmann have also reported a series of α-chiral allyl-boronates which react with aldehydes to give products in high e.e.[45] The chiral chloroboronate **6.47** was synthesised by using the enantioselective α-haloalkylboronate alkylation method described by Matteson. Reagent **6.47** reacts with number of achiral aldehydes to provide optically active (Z)-chloroallylic alcohols in good yield and excellent e.e. (Scheme 6.27). This result

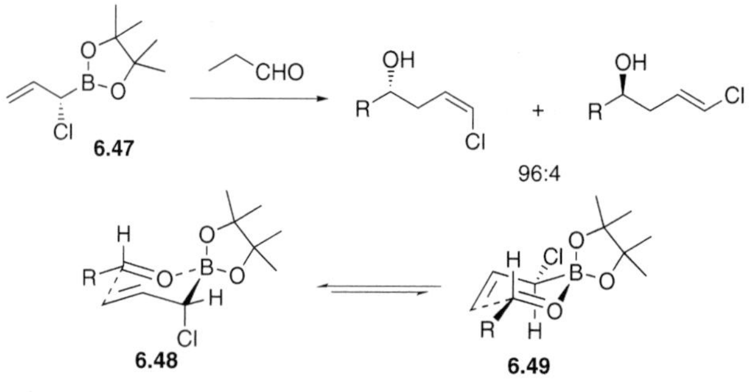

Scheme 6.27

is in accord with a type 1 reaction and it was postulated that transition state **6.48** would be preferred over **6.49** because of stereoelectronic effects.[46] In conformer

6.49 the C—Cl bond is parallel to the π-orbital of the alkene and such overlap might diminish the electron density in the double bond. Conversely, the C—Cl bond in **6.48** is essentially orthogonal to the π-orbital of the alkene, which would preclude any delocalisation of electron density from the double bond, making conformation **6.48** more reactive than **6.49**.

Optically active α-chiral (*E*)- and (*Z*)-crotylboronates have also been synthesised but not exclusively by the method described by Matteson. For example, the (*E*)-crotylboronate **6.50** was synthesised by an allylic rearrangement conducted on silyl ether **6.51**.[47] The addition of **6.50** to propanal proceeded in high e.e. and gave the expected *anti*-(*Z*)-product **6.52** (Scheme 6.28).

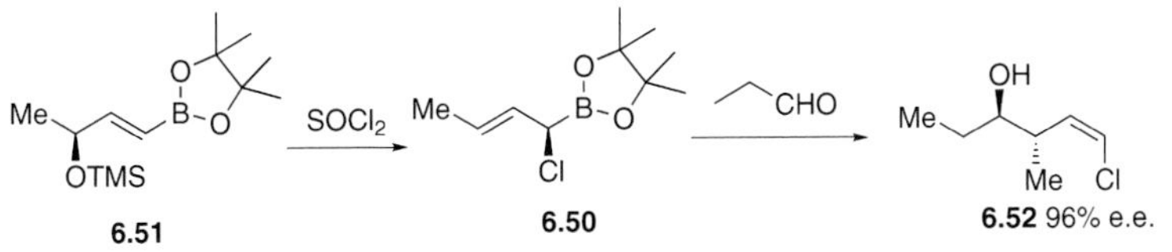

Scheme 6.28

The (*Z*)-α-methylboronate **6.53** was synthesised by using the Matteson asymmetric alkylation procedure and condensed with benzaldehyde to give the (*E*)-*syn*-alcohol **6.54** in excellent optical yield (Scheme 6.29).[48] In this case it was

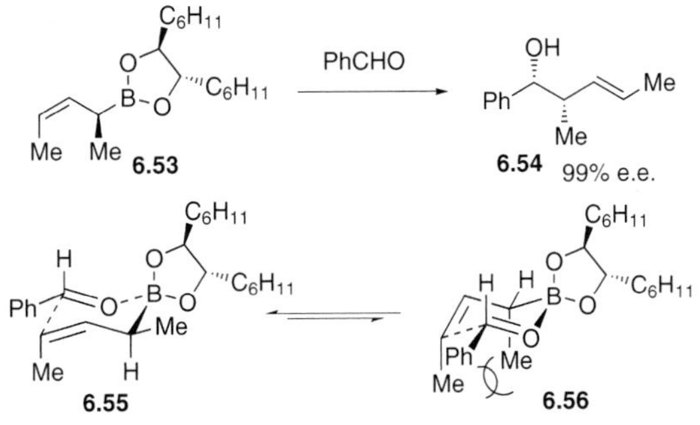

Scheme 6.29

proposed that the aldehyde facial selectivity arises from a preference for the transition state **6.55** over **6.56**. In **6.56** there is a 1,3-steric interaction between the α-methyl and the (*Z*)-methyl groups, making **6.55** the preferred conformer.

Some of the most synthetically useful chiral allylmetal reagents are the dialkylboranes reported by Brown. Although alkylboranes are less stable than their boronate counterparts they are configurationally stable at low temperatures and are more reactive. As opposed to most of the chiral boronate esters, the asymmetric centres can be directly attached to the boron atom, which can provide higher enantioselectivities. Furthermore, their preparation is simple, short and economical. The most common chiral dialkylboranes are those derived from the chiral terpene α-pinene which is readily available in both the (+)-form and the (−)-form. The so-called diisopinocampheylboranes [derived from (+)-α-pinene] **6.57**,[49,50] **6.58/6.59**[51,52] and **6.60**[53] all react readily with a range of aldehydes at −78°C to provide the corresponding condensation products in very high e.e.

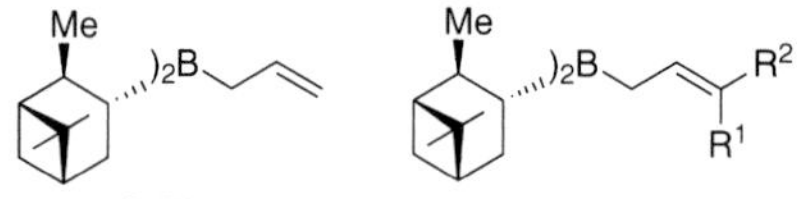

6.58 R¹ = Me; R² = H
6.59 R¹ = H; R² = Me
6.60 R¹ = OMe; R² = H

The preparation of allyldiisopinocampheylborane **6.57** is described in Scheme 6.30.[50] Hydroboration of (+)-α-pinene **6.61** with borane dimethylsulfide complex followed by equilibration at 0°C over 3 days in the presence of

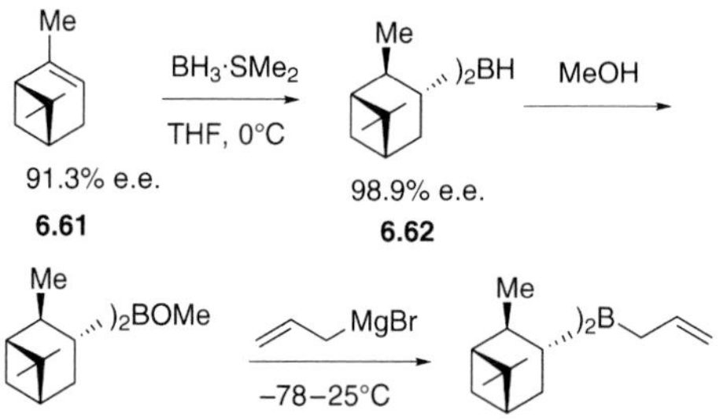

Scheme 6.30

15% excess (+)-α-pinene gives crystalline diisopinocampheylborane **6.62** in high yield and optical purity.[54] It should be noted that in this reaction (+)-α-pinene in 91.3% e.e. can be utilised to produce diisopinocampheylborane in almost 99% e.e. after equilibration. Addition of methanol to **6.62** gives methoxydiisopinocampheylborane **6.63** which on treatment with allylmagnesium bromide yields allyldiisopinocampheylborane (ᵈIpc₂B-allyl) **6.57**.

Acetaldehyde, propionaldehyde and benzaldehyde all react readily with **6.57** at −78°C to give, after hydrogen peroxide oxidation and alkaline hydrolysis, the corresponding homoallylic alcohols with high enantiomeric purities (Scheme 6.31).[49]

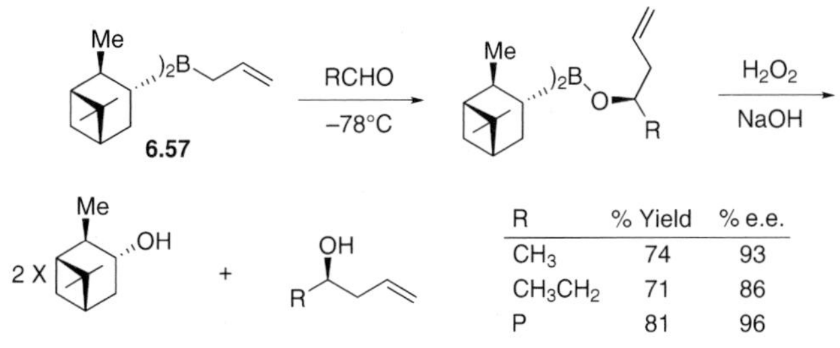

Scheme 6.31

R	% Yield	% e.e.
CH_3	74	93
CH_3CH_2	71	86
P	81	96

The (Z)- and (E)-crotyldiisopinocampheylboranes **6.58** [$^dIpc_2B$-(Z)-crotyl] and **6.59** [$^dIpc_2B$-(E)-crotyl] can also be synthesised from methoxy-diisopinocampheylborane **6.63** (Scheme 6.32).[51,52] Metallation of (E)- and

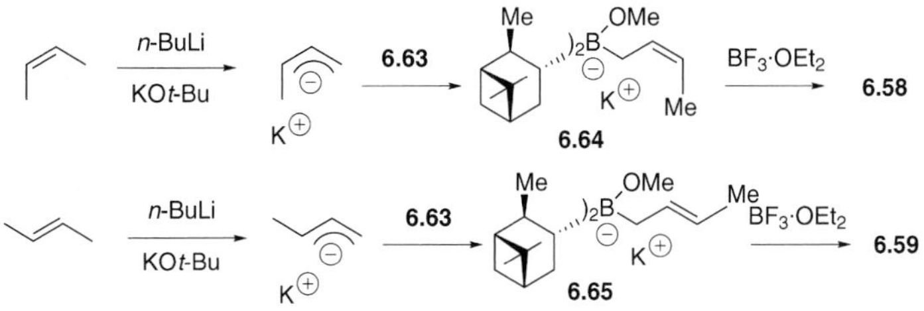

Scheme 6.32

(Z)-but-2-ene under Schlosser conditions provides the corresponding anions which upon treatment with chiral methoxyborane **6.63** gives the ate complexes **6.64** and **6.65**. Boranes **6.58** and **6.59** are formed by treatment of the ate complexes with boron trifluoride etherate and then immediately allowed to react with the desired aldehyde at $-78°C$ (Scheme 6.33).

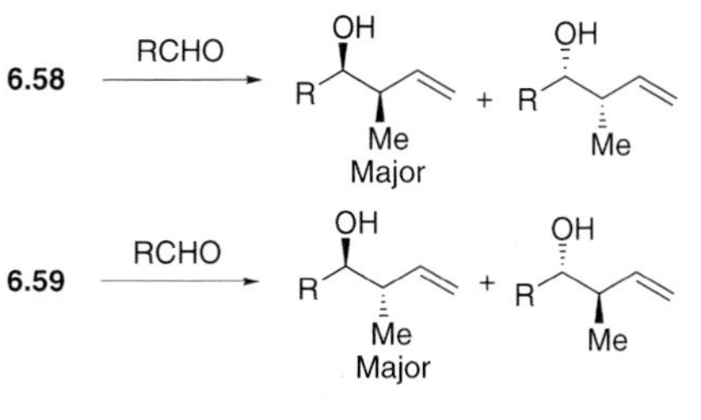

R	% Yield	ratio
CH_3	75	95:5
CH_3CH_2	70	95:5
Bn	72	94:6

R	% Yield	ratio
CH_3	78	95:5
CH_3CH_2	70	95:5
Bn	79	94:6

Scheme 6.33

(*Z*)-Borane **6.58** reacts with various aldehydes to give *syn* products in greater than 99% diastereoselectivity and 90% enantioselectivity. The (*E*)-borane gives the corresponding *anti* isomers in similar diastereo- and enantioselectivities. It is noteworthy that in all cases the chiral boranes derived from (+)-α-pinene give the products which are a result of attack by the reagent on the *Si* face of the aldehyde.

The exact nature of the stereoselectivity of diisopinocampheylboranes derived from (+)-α-pinene is not fully understood but it has been proposed that they react by a cyclic type 1 transition state such as those represented by conformers **6.66** and **6.67** (Scheme 6.34).[3,51] One possible reason why

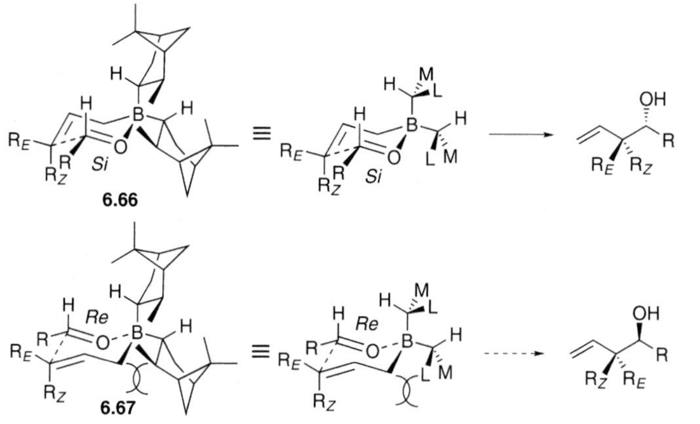

Scheme 6.34

transition state **6.66** is preferred (*Si* face attack on the aldehyde) is because there is a developing steric interaction between the allyl moiety and the L group on the asymmetric carbon atom attached to boron in the alternative conformer **6.67**.

The simple *R,R*-tartrate-modified boronate esters **6.68**, **6.69** and **6.70** described by Roush's group are alternatives to the alkylborane reagents.[55,56]

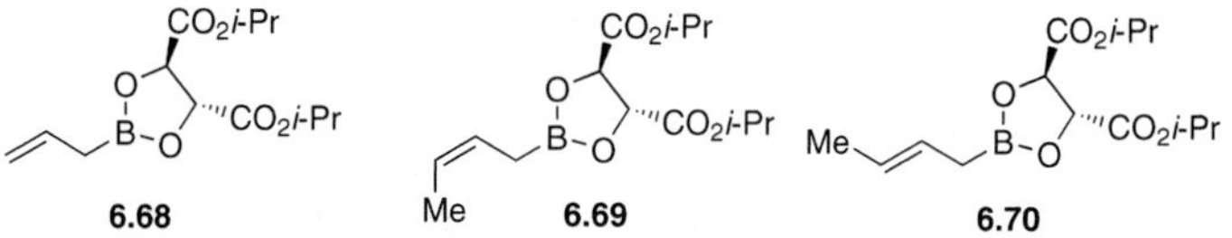

They are easily prepared in either enantiomeric form and have the advantage that they can be stored at −10°C without apparent decomposition. Treatment of allylboronic acid with L-(+)-diisopropyltartrate provides boronate **6.68** which can be purified by distillation (Scheme 6.35).

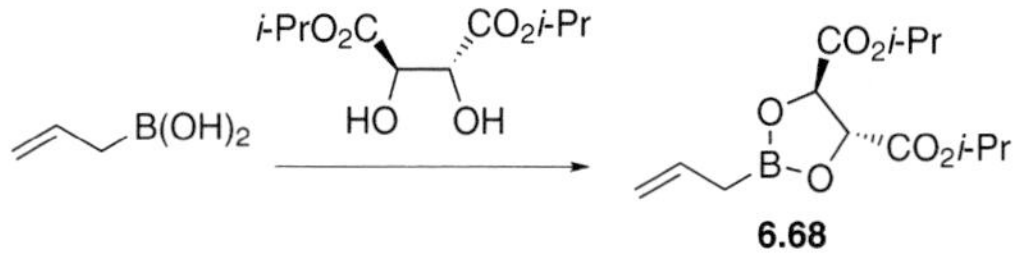

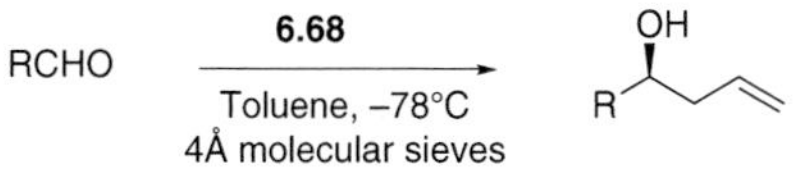

R	% Yield	% e.e.
C_6H_{11}	72	87
Ph	78	71
$CH_3(CH_2)_8$	86	79

Scheme 6.35

Boronate **6.68** reacts with a range of aldehydes to give homoallylic alcohols in good e.e.[55,57] Reactions are usually conducted in toluene as solvent, and the addition of molecular sieves improves enantioselectivity. As opposed to the Ipc reagents, reaction of tartrate-modified boronates with hindered aliphatic, aromatic and α- and β-alkoxy aldehydes give lower enantioselectivities.

Tartrate-modified crotylboronates can also be prepared, and they react with aliphatic aldehydes with useful levels of diastereo- and enantiocontrol.[56,58] Treatment of either metallated (*E*)- or (*Z*)-but-2-ene with B(*i*-PrO)$_3$ followed by acid and tartrate ester provides the chiral crotylboronates **6.69** and **6.70** in excellent yield with high isomeric purity (Scheme 6.36). Simple

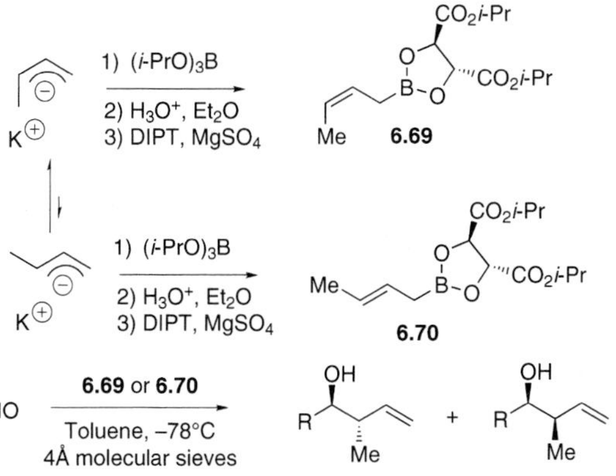

R	Reagent	*anti*:*syn*	% e.e.
n-C_9H_{19}	**6.69**	3:97	82
C_6H_{11}	**6.70**	96:4	87
Ph	**6.70**	99:1	66

Scheme 6.36

diastereoselectivities obtained on addition to achiral aldehydes are often > 97%, and enantioselectivities range from 70% e.e. to 86% e.e.[58]

The origins of the asymmetric induction realised with tartrate-ester-modified boronates such as **6.68** cannot be explained by simple steric interactions alone. To explain this selectivity a novel and previously undocumented stereo-electronic effect has been proposed.[55,59] The transition states **6.69** and **6.71** interconvert via the intermediate **6.70** where the dioxaborolane ring would exist in the conformation shown in Figure 6.2. B—O bond rotation to form transition

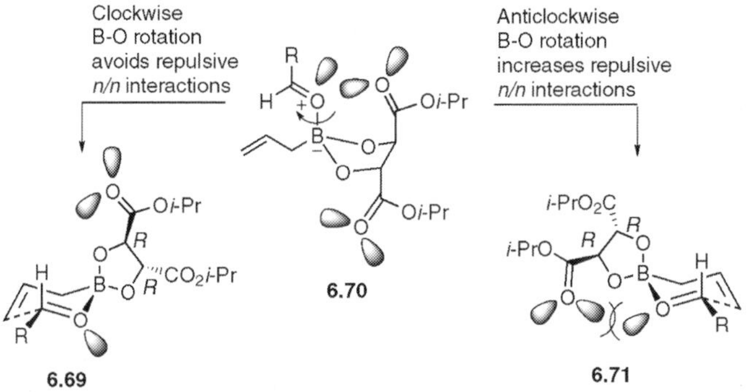

Figure 6.2

state **6.69** is preferred over the alternative **6.71** because there is a destabilising electronic interaction that develops in **6.71** between the lone pairs of the alde-hydic oxygen and one of the ester group oxygens. These interactions can arise as a favoured conformation of α-heteroatom-substituted carbonyl systems in one where the carbonyl group and heteroatom are *syn*-coplanar.

This proposal is supported by the fact that the allylboronate **6.72** contain-ing the rigid tartramide auxiliary reacts with aldehydes to give substantially improved enantioselectivities relative to **6.68**.[60] For example, **6.72** condenses with cyclohexanecarboxaldehyde to give the (*S*)-configured product in high e.e. (Scheme 6.37). However, reagent **6.72** is less reactive and less soluble

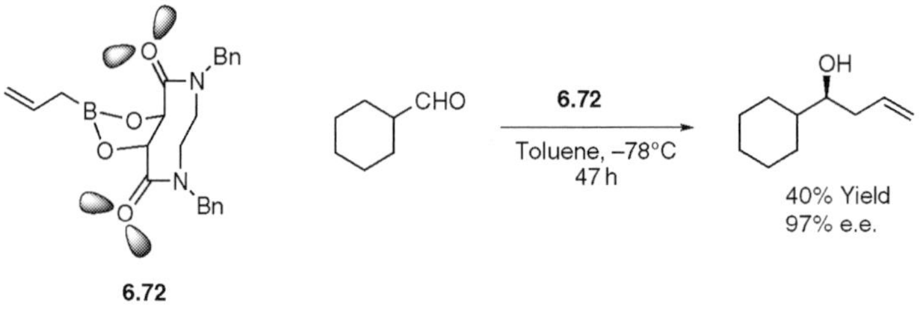

Scheme 6.37

than the tartrate-ester-modified boronates and is consequently not as useful for large-scale work.

6.4 Addition of chiral allyl and crotylmetal reagents to chiral aldehydes

6.4.1 Chiral allylboranes

In many cases significant improvements in diastereoselection can be obtained by allowing a chiral allylmetal reagent to react with a chiral aldehyde in what is commonly known as double asymmetric synthesis.[61] There are two types of double asymmetric reaction: one where the intrinsic diastereofacial preferences of the aldehyde and chiral allyl metal reagent both favour the formation of the same diastereoisomeric product, that is, a 'matched' double asymmetric synthesis; the other where the diastereofacial preferences of each reactant oppose each other, that is, a 'mismatched' case. It is therefore more difficult to obtain high diastereoselectivities in a mismatched case compared with a matched case. If a matched pair is utilised, the stereoselectivity can be very high; however, even in a mismatched case, one reactant can still dominate the outcome.

Chiral boranes react with α-chiral aldehydes to provide outcomes which are for the most part dependent of the chirality of the borane and not the aldehyde.[62–64] In this case, the diisopinocampheylborane (Ipc) reagent can dominate over the intrinsic facial (i.e. Felkin–Anh) preference of the aldehyde, and the isomer produced is that predicted from the chirality on the allylborane. Addition of borane **6.57** or the enantiomeric borane **6.73** [derived from (−)-α-pinene] to an α-methyl chiral aldehyde can therefore provide both enantiomeric *syn* and *anti* isomer pairs with high selectivity (Scheme 6.38).

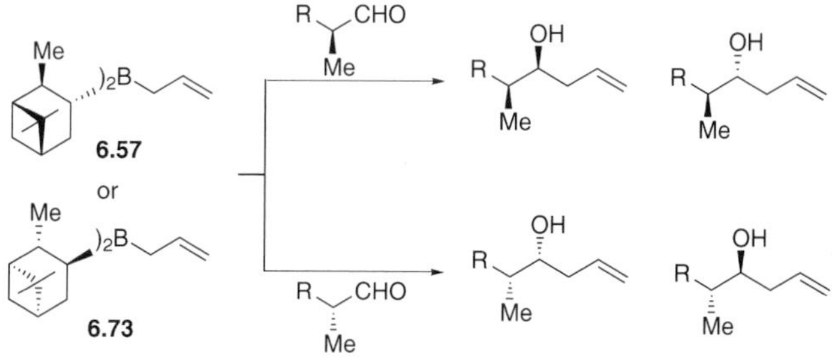

Scheme 6.38

The additions of boranes **6.57** or **6.73** to chiral aldehyde **6.74** gave products with high *syn* or *anti* selectivities (Scheme 6.39).[62,64] However, the addition of **6.73** to aldehyde **6.75** gives a low selectivity. In this case, the facial preference

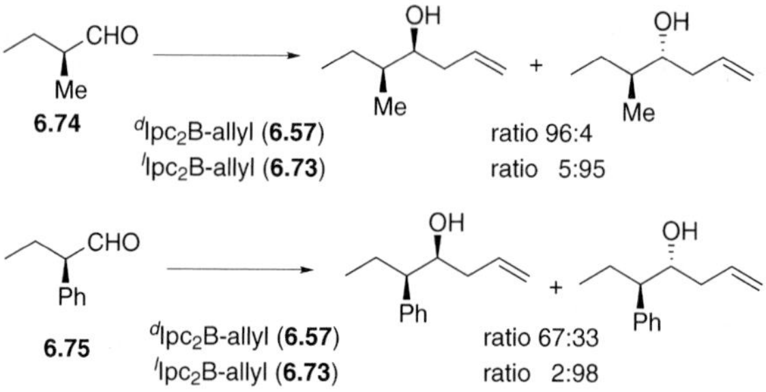

Scheme 6.39

for aldehyde **6.75** (*Re* face) is stronger than for **6.74** and fights effectively against the facial preference of the reagent (*Si* face) giving a mismatched case in which neither reactant dominates. On the other hand, in the matched case of the addition of **6.73** to aldehyde **6.75** the selectivity is very high.

The (*E*)- and (*Z*)-crotyldiisopinocampheylboranes **6.59** and **6.58** as well as their enantiomeric counterparts derived from (−)-α-pinene all react with chiral aldehyde **6.76** to provide all possible diastereoisomeric outcomes in excellent diastereoisomeric excess (Scheme 6.40).[64] In this manner any propionate

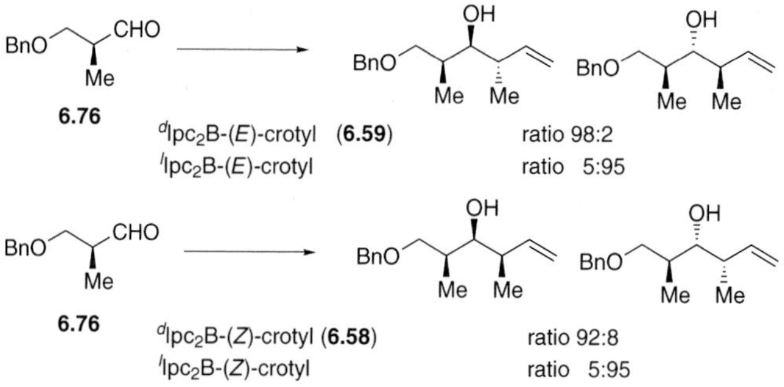

Scheme 6.40

stereo-triad can be constructed with high control by using the Brown Ipc ligand. α-Alkoxy, aromatic and α,β-unsaturated chiral aldehydes also react with Ipc$_2$B-crotyl reagents to give excellent results.

6.4.2   Chiral allylboronates

The chiral boronate esters developed by Hoffmann's group were amongst the first chiral allyl metal reagents to be utilised in double asymmetric synthesis. Varied results were obtained with use of the crotyl metal agents **6.77** and **6.78** derived from camphor (Scheme 6.41).[43] Addition of the (E)-crotylboronate **6.77** to optically pure aldehyde **6.74** gave good selectivity for the 3,4-*anti*-4,5-*syn*

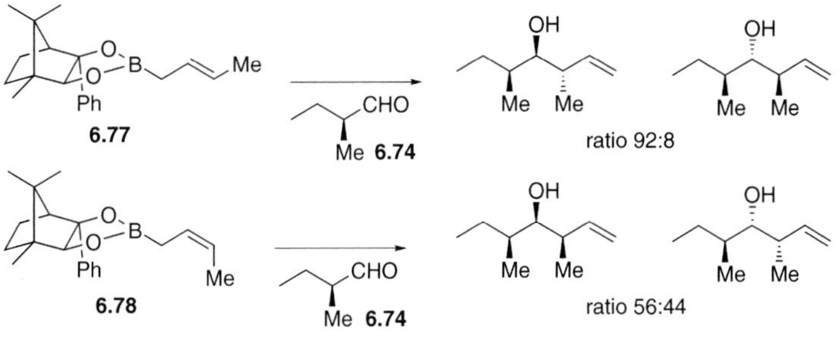

Scheme 6.41

product as would be expected given the rules for addition of achiral (E)-boronates to chiral aldehydes (see Scheme 6.17). The other example presented is the addition of (Z)-crotylboronate **6.78** to the same aldehyde, **6.74**. In this reaction, the selectivity is poor and the Felkin product (3,4-*syn*-4,5-*syn*) is slightly preferred. This is not in accord with the behaviour of achiral boronates which add to chiral aldehydes by an *anti*-Felkin mode of attack as outlined in Scheme 6.16. The chiral auxiliary on the boron atom in **6.78** is therefore effecting the outcome and favours a Felkin mode of attack on **6.74** which is a mismatched case.

The tartrate-ester-modified boronates can often provide higher selectivities than the camphor-derived reagents described above. The addition of the (R,R)-boronate **6.68** to chiral aldehyde **6.12** is a matched case, providing an excellent ratio in favour of the *anti* product (Scheme 6.42).[55,59] When the enantiomeric

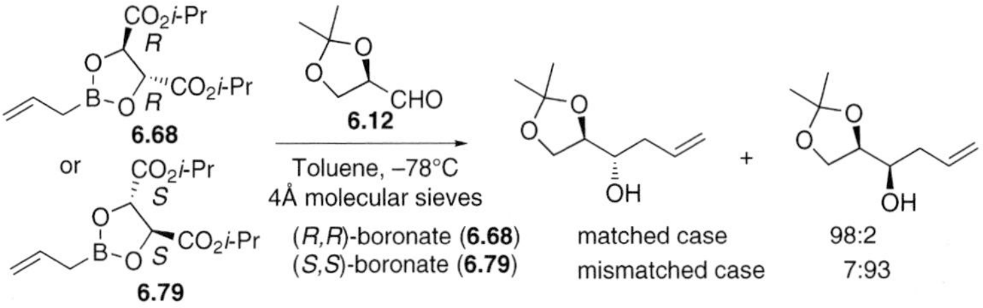

Scheme 6.42

(S,S)-boronate **6.79** was utilised the *syn* product was favoured in a slightly lower ratio as this is a mismatched case.

Tartrate-modified (E)- and (Z)-crotyl boronates **6.70** and **6.69** react with the chiral α-methyl branched aldehyde **6.80** to provide three of the four possible stereoisomers with selectivities greater than 90%.[65] As shown in Scheme 6.43

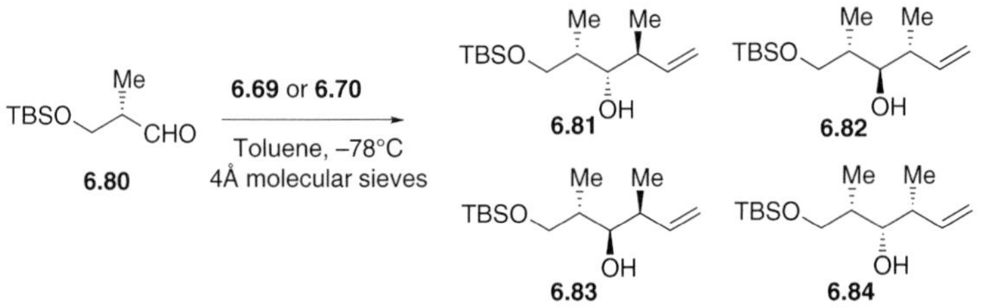

Scheme 6.43

the (R,R)-boronate **6.70** reacts with **6.80** to give the 3,4-*anti*-4,5-*syn* isomer **6.81** and the 3,4-*anti*-4,5-*anti* isomer **6.82** in a ratio if 97:3. Switching to the (S,S)-boronate, which is enantiomeric with **6.70**, gives 90% of diastereoisomer **6.82** and 10% isomer **6.81**. The (Z)-boronate (R,R)-**6.69** reacts with **6.80** to give the 3,4-*syn*-4,5-*anti* isomer **6.83** and the 3,4-*syn*-4,5-*syn* isomer **6.84** as well as a little of isomer **6.82** in a ratio of 95:1:4. The worst case is the mismatched reaction of the (S,S)-isomer (enantiomer of **6.69**) with aldehyde **6.80** which gives all four products **6.81**, **6.82**, **6.83** and **6.84** in a ratio of 12:2:45:41. The 3,4-*syn*-4,5-*syn* isomer **6.84** is therefore the most difficult to access by this method.

6.4.3   Chiral allylsilanes

Chiral allylsilanes can be easily prepared and add effectively to aldehydes in the presence of Lewis acids.[66,67] Panek and co-workers have demonstrated that chiral crotylsilanes can add to chiral α-methyl-β-alkoxy aldehydes to provide high selectivities where the stereochemical outcome is dependent on the configuration of the silane, the Lewis acid used and the protecting group on the aldehyde.[68,69]

Treatment of the aldehyde **6.85** with the (S)-crotylsilane **6.86** gives the 5,6-*anti*-6,7-*syn* isomer in good yield and d.e. (Scheme 6.44).[69] On the other hand, condensation of aldehyde **6.85** with the (R)-crotylsilane **6.87** gives the 5,6-*syn*-6,7-*syn* isomer as the major product.

The stereochemical outcomes depicted in Scheme 6.44 can be explained by considering the open transition state (TS) models shown in Scheme 6.45. The addition of **6.86** to aldehyde **6.85** might proceed by the synclinal transition state **6.88** to give the 5,6-*anti*-6,7-*syn* isomer which is a Felkin–Anh mode of attack by the crotylsilane on the *Si* face of the aldehyde. The antiperiplanar transition state **6.89** explains the 5,6-*syn*-6,7-*syn* outcome arising from Felkin–Anh

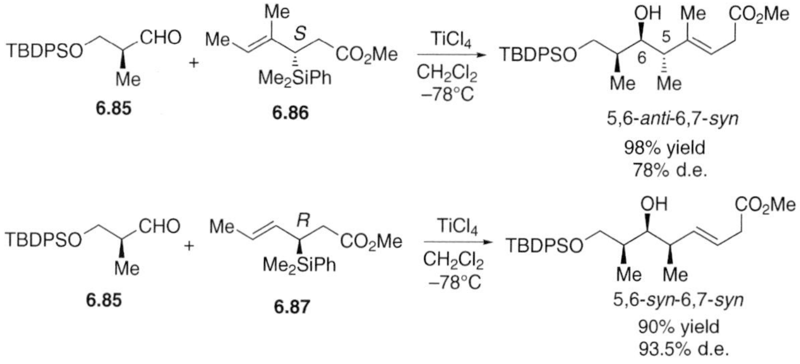

Scheme 6.44

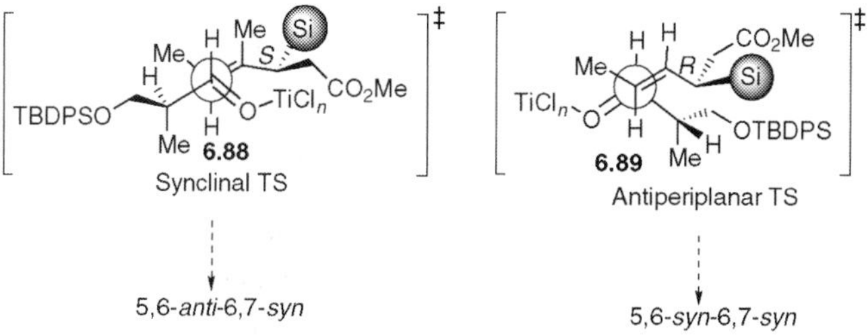

Scheme 6.45

attack of **6.85** by the (R)-crotylsilane **6.87**. The configuration of the crotylsilane reagent controls the stereochemistry of the bond construction, and the nonchelating bulky silyl protecting group controls the facial (Felkin–Anh) attack of the aldehyde.

Changing the protecting group of the aldehyde to a benzyloxy group reverses the facial selectivity of the aldehyde. Treatment of aldehyde **6.90** with (S)-silane **6.86** gives the 5,6-*syn*-6,7-*anti* isomer in high d.e., whereas addition of the (R)-silane **6.87** to aldehyde **6.90** provides the 5,6-*anti*-6,7-*anti* isomer in excellent yield and d.e. (Scheme 6.46).[69]

In the above cases, chelation-controlled or Cram chelate attack on the *Re* face of the aldehyde is preferred.[68] The antiperiplanar transition state **6.91** can be postulated for addition of (S)-crotylsilane **6.86** to aldehyde **6.90** where the titanium Lewis acid chelates to both the aldehyde carbonyl and the benzyl oxygen, forcing the crotylsilane to attack on the *Re* face of the aldehyde, providing the 5,6-*syn*-6,7-*anti* isomer (Scheme 6.47). In the case of the (R)-silane **6.87**, the synclinal transition state **6.92** is proposed where aldehyde **6.90** again reacts on the *Re* face via a Cram chelate to give the 5,6-*anti*-6,7-*anti* isomer as the major product.

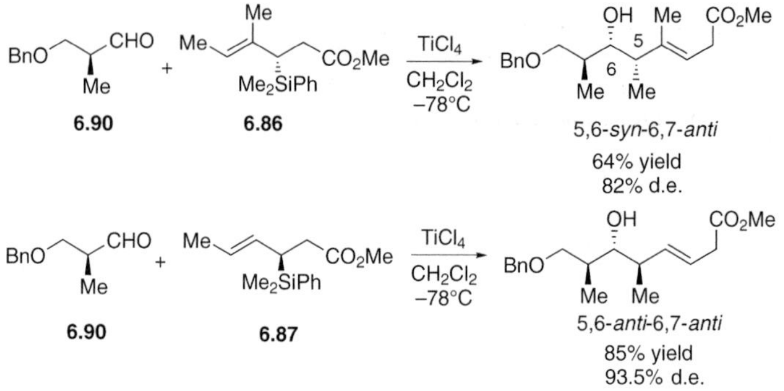

Scheme 6.46

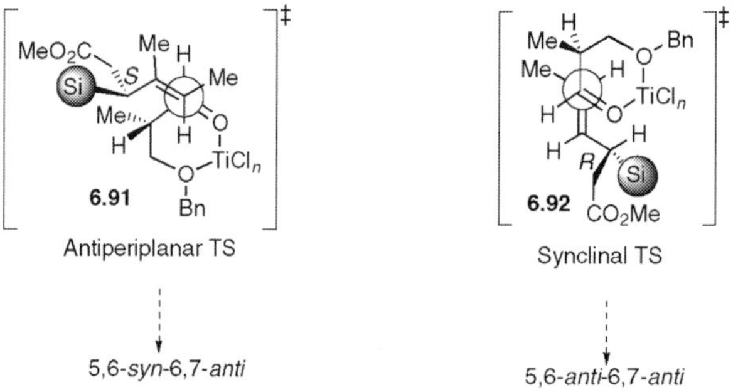

Scheme 6.47

References

1. Hoffmann, R.W., *Angew. Chem., Int. Ed. Engl.*, **1982**, *21*, 555.
2. Yamamoto, Y. and Maruyama, K., *Heterocycles*, **1982**, *18*, 357.
3. Roush, W.R., in *Comprehensive Organic Synthesis, vol 2. Additions to C—X Bonds, Part 2*, Ed. C.H. Heathcock, Pergamon, Oxford, **1991**, p. 1.
4. Yamamoto, Y. and Asao, N., *Chem. Rev.*, **1993**, *93*, 2207.
5. Kramer, G.W. and Brown, H.C., *J. Organomet. Chem.*, **1977**, *132*, 9.
6. Stähle, M. and Schlosser, M., *J. Organomet. Chem.*, **1981**, *220*, 277.
7. Neugebauer, W. and Schleyer, P.V.R., *J. Organomet. Chem.*, **1980**, *198*, C1.
8. Chan, T.H. and Fleming, I., *Synthesis*, **1979**, 761.
9. Verdone, J.A., Mangravite, J.A., Scarpa, N.M. and Kuivila, H.G., *J. Am. Chem. Soc.*, **1975**, *97*, 843.
10. Yamaguchi, M. and Mukaiyama, T., *Chem. Lett.*, **1980**, 993.
11. Schlosser, M. and Rauchschwalbe, G., *J. Am. Chem. Soc.*, **1978**, *100*, 3258.
12. Hoffmann, R.W. and Zeiß, H., *J. Org. Chem.*, **1981**, *46*, 1309.
13. Brown, H.C., DeLue, N.R., Yamamoto, Y., Maruyama, K., Kasahara, T., Marahashi, S. and Sonoda, A., *J. Org. Chem.*, **1977**, *42*, 4088.

14. Hiyama, T., Kimura, K. and Nozaki, H., *Tetrahedron Lett.*, **1981**, *22*, 1037.
15. Reetz, M.T. and Sauerwald, M., *J. Org. Chem.*, **1984**, *49*, 2292.
16. Denmark, S.E. and Weber, E.J., *Helv. Chim. Acta*, **1983**, *66*, 1655.
17. Tochtermann, W., *Angew. Chem., Int. Ed. Engl.*, **1966**, *5*, 351.
18. Evans, D.A., Nelson, J.V. and Taber, T.R., in *Topics in Stereochemistry*, Eds. N.L. Allinger, E.L. Eliel and S.H. Wilen, John Wiley, New York, **1982**, vol. 13, p. 1.
19. Hosomi, A. and Sakurai, H., *Tetrahedron Lett.*, **1976**, *16*, 1295.
20. Hiyashi, T., Kabeta, K., Hamachi, I. and Kumada, M., *Tetrahedron Lett.*, **1983**, *24*, 2865.
21. Yamamoto, Y., Yatagai, H., Naruta, Y. and Maruyama, K., *J. Am. Chem. Soc.*, **1980**, *102*, 7109.
22. Yamamoto, Y., Yatagi, H., Ishihara, Y., Maeda, N. and Maruyama, K., *Tetrahedron*, **1984**, *40*, 2239.
23. Denmark, S.E. and Weber, E.J., *J. Am. Chem. Soc.*, **1984**, *106*, 7970.
24. Yamamoto, Y., *Acc. Chem. Res.*, **1987**, *20*, 243.
25. Gevorgyan, V., Kadota, I. and Yamamoto, Y., *Tetrahedron Lett.*, **1993**, *34*, 1313.
26. Denmark, S.E. and Almstead, N.G., *J. Org. Chem.*, **1994**, *59*, 5130.
27. Sato, F., Iida, K., Ijima, S., Moriya, H. and Sato, M., *J. Chem. Soc., Chem. Commun.*, **1981**, 1140.
28. Yamamoto, Y. and Maruyama, K., *Tetrahedron Lett.*, **1981**, *22*, 2895.
29. Keck, G.E., Abbott, D.E., Boden, E.P. and Enholm, E.J., *Tetrahedron Lett.*, **1984**, *25*, 3927.
30. Hoffmann, R.W. and Weidmann, U., *Chem. Ber.*, **1985**, *118*, 3966.
31. Hoffmann, R.W., Zeiß, H., Ladner, W. and Tabche, S., *Chem. Ber.*, **1982**, *115*, 2357.
32. Roush, W.R., Adam, M.A., Walts, A.E. and Harris, D.J., *J. Am. Chem. Soc.*, **1986**, *108*, 3422.
33. Roush, W.R., *J. Org. Chem.*, **1991**, *56*, 4151.
34. Wuts, P.G.M. and Bigelow, S.S., *J. Org. Chem.*, **1988**, *53*, 5023.
35. Wuts, P.G.M. and Bigelow, S.S., *J. Org. Chem.*, **1983**, *48*, 3849.
36. Nagaoka, H. and Kishi, Y., *Tetrahedron*, **1981**, *37*, 3873.
37. Fronza, G., Fuganti, C., Grasselli, P., Pedrocchi-Fantoni, G. and Zirotti, C., *Chem. Lett.*, **1984**, 335.
38. Heathcock, C.H., Kiyooka, S. and Blumenkopf, T.A., *J. Org. Chem.*, **1984**, *49*, 4214.
39. Keck, G.E. and Boden, E.P., *Tetrahedron Lett.*, **1984**, *25*, 265.
40. Keck, G.E. and Abbott, D.E., *Tetrahedron Lett.*, **1984**, *25*, 1883.
41. Herold, T., Schrott, U., Hoffmann, W., Schnelle, G., Ladner, W. and Steinbach, K., *Chem. Ber.*, **1981**, *114*, 359.
42. Hoffmann, R.W. and Herold, T., *Chemie Briechte*, **1981**, *114*, 375.
43. Hoffmann, R. and Zeiß, H.-J., *Angew. Chem., Int. Ed. Engl.*, **1980**, *19*, 218.
44. Reetz, M.T. and Zierke, T., *Chem. Ind.*, **1988**, 663.
45. Hoffmann, R.W. and Landmann, B., *Angew. Chem., Int. Ed. Engl.*, **1984**, *23*, 437.
46. Hoffmann, R.W. and Landmann, B., *Chem. Ber.*, **1986**, *119*, 1039.
47. Hoffmann, R.W. and Dresley, S., *Angew. Chem., Int. Ed. Engl.*, **1986**, *25*, 189.
48. Ditrich, K., Bube, T., Stürmer, R. and Hoffmann, R.W., *Angew. Chem., Int. Ed. Engl.*, **1986**, *25*, 1028.
49. Brown, H.C. and Jadhav, P.K., *J. Am. Chem. Soc.*, **1983**, *105*, 2092.
50. Jadhav, P.K., Bhat, K.S., Perumal, P.T. and Brown, H.C., *J. Org. Chem.*, **1986**, *51*, 432.
51. Brown, H.C. and Bhat, K.S., *J. Am. Chem. Soc.*, **1986**, *108*, 5919.
52. Brown, H.C. and Bhat, K.S., *J. Am. Chem. Soc.*, **1986**, *108*, 293.
53. Brown, H.C., Jadhav, P.K. and Bhat, K.S., *J. Am. Chem. Soc.*, **1988**, *110*, 1535.
54. Brown, H.C., Desai, M.C. and Jadhav, P.K., *J. Org. Chem.*, **1982**, *47*, 5065.
55. Roush, W.R., Walts, A.E. and Hoong, L.K., *J. Am. Chem. Soc.*, **1985**, *107*, 8186.
56. Roush, W.R. and Halterman, R.L., *J. Am. Chem. Soc.*, **1986**, *108*, 294.
57. Roush, W.R., Hoong, L.K., Palmer, M.A.J. and Park, J.C., *J. Org. Chem.*, **1990**, *55*, 4109.
58. Roush, W.R., Ando, K., Powers, D.B., Palkowitz, A.D. and Halterman, R.L., *J. Am. Chem. Soc.*, **1990**, *112*, 6339.

59. Roush, W.R., Hoong, L.K., Palmer, M.A.J., Straub, J.A. and Palkowitz, A.D., *J. Org. Chem.*, **1990**, *55*, 4117.
60. Roush, W.R. and Banfi, L., *J. Am. Chem. Soc.*, **1988**, *110*, 3979.
61. Masamune, S., Choy, W., Peterson, J.S. and Sita, L.R., *Angew. Chem., Int. Ed. Engl.*, **1985**, *24*, 1.
62. Brown, H.C., Bhat, K.S. and Randad, R.S., *J. Org. Chem.*, **1987**, *52*, 319.
63. Brown, H.C., Bhat, K.S. and Randad, R.S., *J. Org. Chem.*, **1987**, *52*, 3701.
64. Brown, H.C., Bhat, K.S. and Randad, R.S., *J. Org. Chem.*, **1989**, *54*, 1570.
65. Roush, W.R., Palkowitz, A.D. and Ando, K., *J. Am. Chem. Soc.*, **1990**, *112*, 6348.
66. Panek, J.S., Yang, M. and Solomon, J.S., *J. Org. Chem.*, **1993**, *58*, 1003.
67. Masse, C.E. and Panek, J.S., *Chem. Rev.*, **1995**, *95*, 1293.
68. Jain, N.F., Cirillo, P.F., Pelletier, R. and Panek, J.S., *Tetrahedron Lett.*, **1995**, *36*, 8727.
69. Jain, N.F., Takenaka, N. and Panek, J.S., *J. Am. Chem. Soc.*, **1996**, *118*, 12475.

7 Pericyclic processes

7.1 Asymmetric Diels–Alder reactions

7.1.1 Background

The Diels–Alder reaction has been widely utilised in asymmetric synthesis because of its ability to control stereochemistry effectively while forming two new C—C bonds at the same time.[1] The concerted thermal [4+2] cycloaddition between an electron-rich diene and an electron-poor dienophile is known as a normal electron demand Diels–Alder reaction in which the lowest unoccupied molecular orbital (LUMO) of the dienophile interacts with the highest occupied molecular orbital (HOMO) of the diene (Figure 7.1).[2] Any factor that lowers the

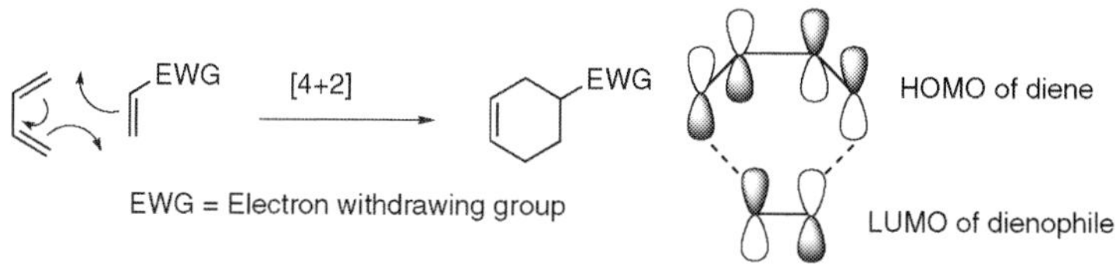

EWG = Electron withdrawing group

HOMO of diene

LUMO of dienophile

Figure 7.1

LUMO of the dienophile can enhance the rate of the reaction. Electron withdrawing groups or activation of a carbonyl substituent on the dienophile by Lewis acids can reduce the LUMO energy of the dienophile, which reduces the energy-separation between the HOMO of the diene and the LUMO of the dienophile and increases the rate of reaction.[2]

Stereocontrol in [4+2] cycloaddition reactions arises from the secondary interaction of the frontier molecular orbitals, and *endo* type products are favoured in most cases. For example, *cis*-1,2-disubstituted cyclohexanes can be prepared by a Diels–Alder reaction between a *cis* olefin and diene, and *endo* substituted [2.2.1]-bicycloheptanes can be formed by condensation of a cyclopentadiene with a (*Z*)-dienophile (Scheme 7.1). An all-*cis*-substituted cyclohexane is produced when an (*E,E*)-diene reacts with a (*Z*)-dienophile via an *endo* rather than an *exo* type transition state, as depicted in Scheme 7.1. An *exo* type transition state would produce the *trans-cis-trans* cyclohexane product but this is not preferred for the case where R or R′ is a carbonyl-containing group because of secondary π-interactions which stabilise the *endo* transition state.

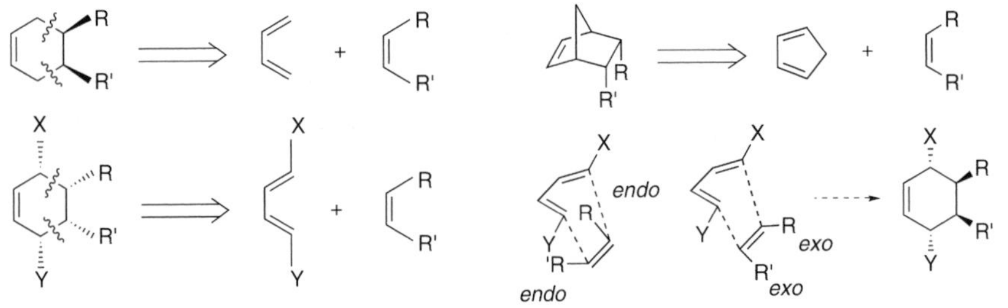

Scheme 7.1

7.1.2 *Chiral dienophiles: type II chiral acrylates*

One way by which the facial selectivity of a Diels–Alder reaction and thus the absolute stereochemistry of a chiral product can be controlled is by utilising a chiral dienophile. One of the most common types of acyclic chiral dienophile is where the electron-withdrawing group is a carbonyl functionality which is then covalently attached to a chiral group or auxiliary.[3,4] In so-called type I dienophiles (Figure 7.2), the chiral moiety is directly attached to the carbonyl

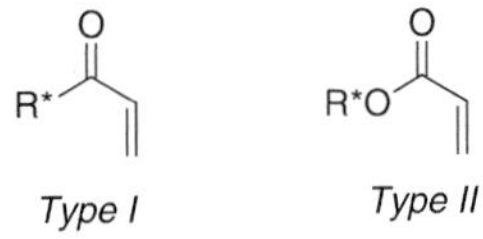

R* = Chiral group

Type I   *Type II*

Figure 7.2 Dienophiles.

carbon, whereas in type II dienophiles the chirality is attached via an oxygen atom (i.e. a chiral acrylate ester) which is easily removed after the cycloaddition reaction.[5] Type I dienophiles are more difficult to synthesise and the chiral auxiliary is not as easily removed as in the type II dienophiles. There are numerous examples of type II chiral dienophiles and only a few will be dealt with in this section.

One problem with type II chiral dienophiles is that they can adopt a number of conformations. There are X-ray data and spectroscopic evidence that there are two preferred conformations, **7.1** and **7.2**, where the C—H and C=O are synperiplanar in both. There is an equilibrium between **7.1** and **7.2** and conformer **7.2**, where the C=O and C=C bonds are synplanar, is preferred but the difference is only of the order of 0.23 kcal/mol (Scheme 7.2).[3] Each conformer displays a different topicity, with attack of a diene occurring opposite to the L group, and this explains why the thermal Diels–Alder reactions have never proceeded to greater than 65% d.e.

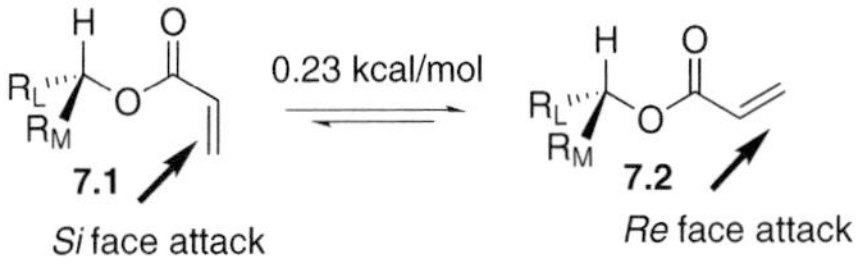

Scheme 7.2

The situation changes dramatically under the influence of Lewis acids. The carbonyl is now coordinated to the metal of the Lewis acid *anti* to the ester oxygen and there is again X-ray and spectroscopic evidence for the preference for this conformation. Conformation **7.3** is now preferred over **7.4** on steric grounds (Scheme 7.3) and the Lewis acid coordination also increases the *endo*-selectivity as well as the rate of the Diels–Alder reaction as noted earlier.

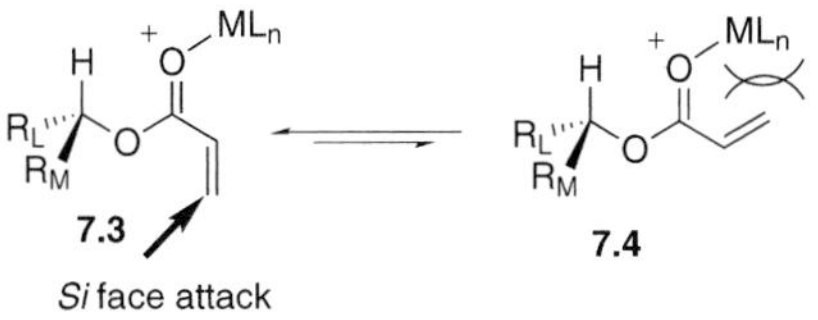

Scheme 7.3

The earliest example of a chiral dienophile was dimenthyl fumarate **7.5**, which reacted with butadiene in the presence of $AlCl_3$ to provide the (S,S)-adduct **7.6** in 57% d.e. (Scheme 7.4).[6] In the absence of a Lewis acid, the

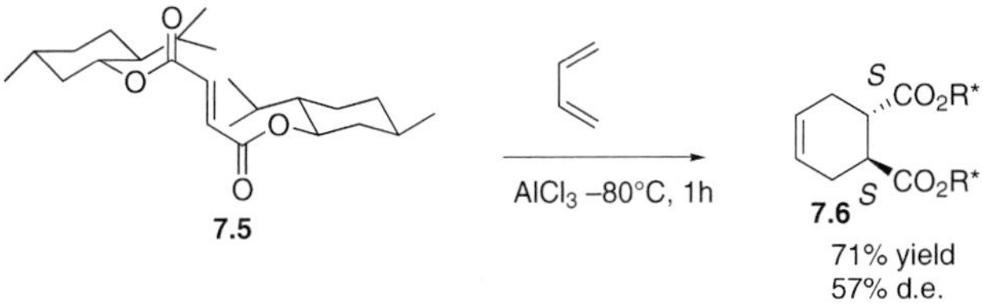

Scheme 7.4

cycloaddition proceeded slowly at 67°C and gave a product in only 2.7% d.e., favouring the (S,S)-isomer. Yamamoto *et al.* later showed that by employing a homogeneous aluminium catalyst *i*-Bu$_2$AlCl the d.e. was improved to 95%.[7]

Yamamoto *et al.* suggested that the Lewis-acid-coordinated fumarate exists in the *s-trans* form predominantly, and the difference in activation energy leading to the (R,R) and (S,S) product is estimated to be 2.3 kcal/mol.[7] Thus the diene approaches the Lewis-acid-complexed dienophile almost exclusively from the face depicted in Scheme 7.5.

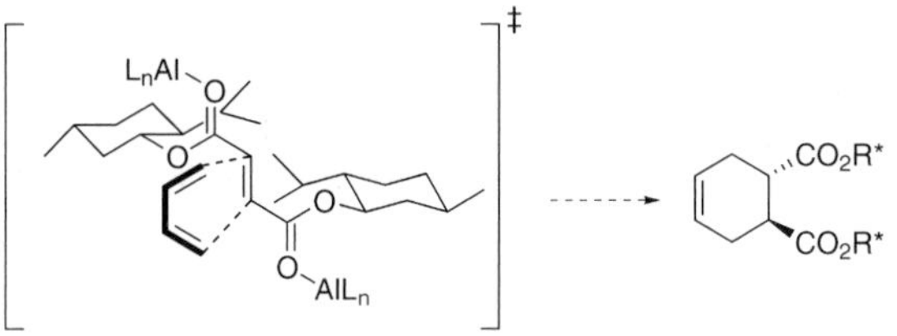

Scheme 7.5

In a later example, Corey *et al.* utilised chiral acrylate **7.7** derived from (+)-8-phenylmenthol which in turn was synthesised from (*S*)-(−)-pulegone.[8] The phenyl group of the chiral auxiliary effectively shields the *Si* face of the alkene and addition of the cyclopentadiene occurred from the *Re* face to provide key prostagladin intermediate **7.8** in 99% d.e. (Scheme 7.6). This example stands as the first synthetically useful application of asymmetric induction in a Diels–Alder reaction.

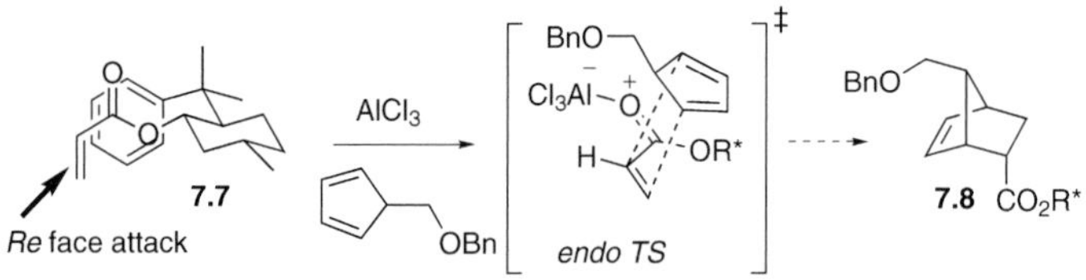

Scheme 7.6

Oppolzer *et al.* have investigated the Lewis-acid-catalysed Diels–Alder reactions of the enantiomeric (−)-8-phenylmenthol-derived acrylate **7.9**.[9] Dienophile **7.9** reacts with cyclopentadiene to give the *endo* product **7.10** along with a small amount of the *exo* product (Scheme 7.7). The chiral auxiliary is then

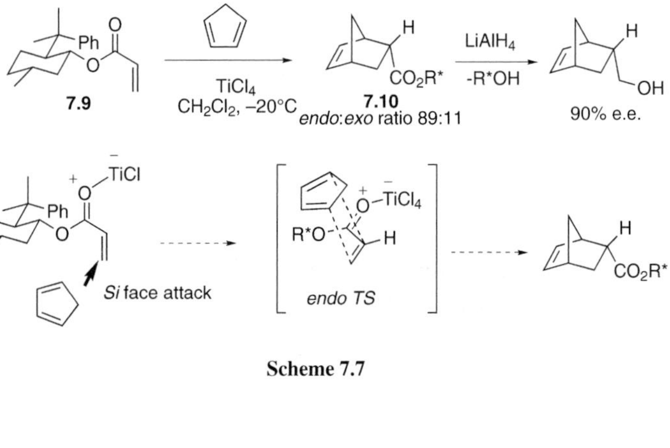

Scheme 7.7

removed by hydride reduction and can easily be recycled. The stereochemistry observed for the product is again a result of attack on the Lewis-acid-coordinated acrylate from the face opposite to the α-1-methyl-1-phenylethyl substituent. The high stereoselection observed for this chiral auxiliary is probably the result of a combination of complementary steric and aryl-acrylate π-stacking effects.

Camphor-derived chiral auxiliaries have also been utilised effectively in asymmetric Diels–Alder reactions. The chiral acrylate **7.11** reacts with cyclopentadiene under the influence of a mild Lewis acid to give the bicycle **7.12** in excellent d.e. and high *endo* selectivity (Scheme 7.8).[10] In this case, it was

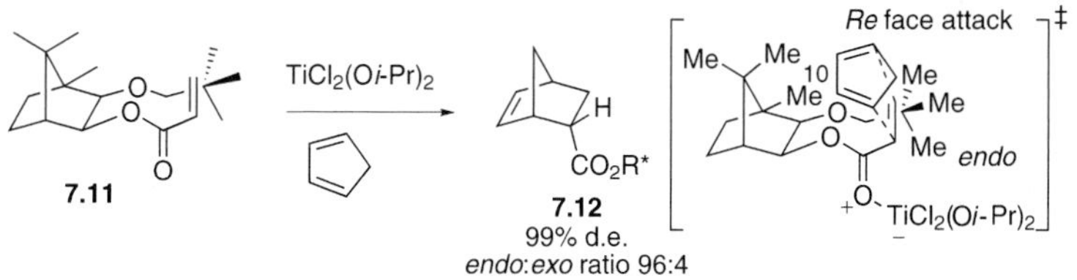

Scheme 7.8

proposed that the C(10) methyl group probably plays a role in providing the observed high selectivity as it pushes the neopentyl ether side-chain closer to the acrylate which effectively blocks the *Si* face from attack.

Simple chiral auxiliaries which contain an ester functionality for additional chelation to a Lewis acid are also effective in asymmetric Diels–Alder reactions. Helmchen *et al.* utilised commercially available D-pantolactone to synthesise chiral acrylate **7.13** which provided high selectivities upon condensation with dienes in the presence of titanium-based Lewis acids (Scheme 7.9).[11] It has

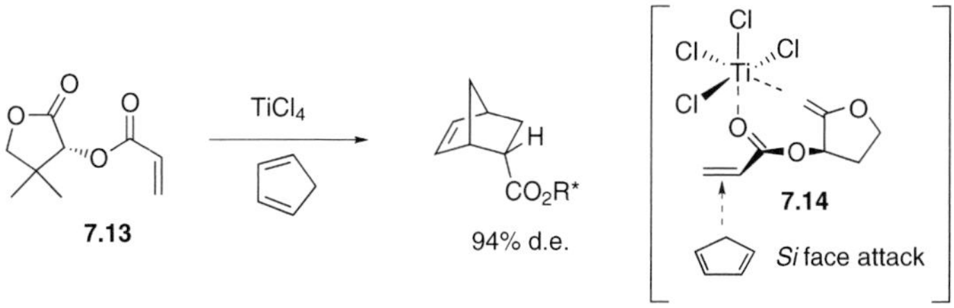

Scheme 7.9

been suggested that the seven-membered chelate[12] **7.14** is formed, and it is clear that the *Re* face is effectively blocked. In this case it should be noted that the titanium is chelated to the acrylate carbonyl *syn* to the ester oxygen atom owing to additional chelation to the lactone carbonyl and thus the titanium atom is now *anti* to the alkene. The C=C bond then lies in the preferred synperiplanar orientation to the C=O bond and approach of the diene via the *Si* face is preferred.

7.1.3   Chiral dienophiles: acyclic chiral N-acyl derivatives

A number of chiral *N*-acyl dienophiles have been utilised with great success in asymmetric Diels–Alder reactions. The *N*-acyl sultam **7.15**, derived from (+)-camphorsulfonic acid, undergoes highly selective addition to various dienophiles in the presence of mild Lewis acids.[13] Both butadiene and cyclopentadiene react with it to provide the corresponding adducts **7.16** and **7.17** with (*S*)-stereochemistry at the newly created asymmetric centre (Scheme 7.10).

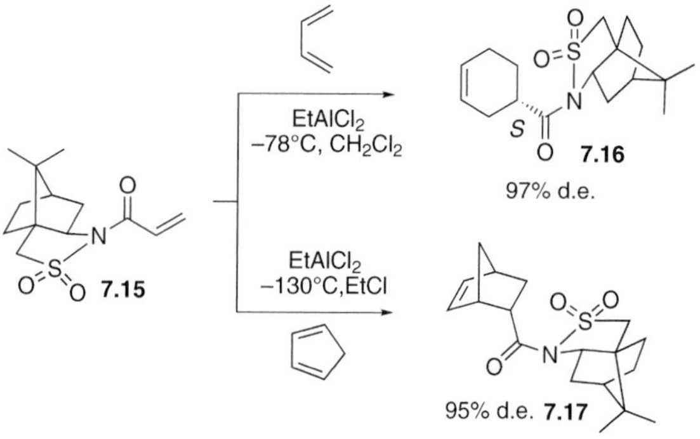

Scheme 7.10

The facial selectivity arises from *Re* face addition of the diene to the alkene. Evidence for the chelation of the Lewis acid to both a sulfonyl and a carbonyl oxygen arose from examination of the infrared (IR) spectrum of a 1:1 mixture of the acylsultam and $TiCl_4$ where changes in both the C=O band and the SO_2 band were observed. Apparently, the preferred conformer is **7.18** where the C=C bond is synperiplanar to the C=O bond (Scheme 7.11).[13] In the alternative

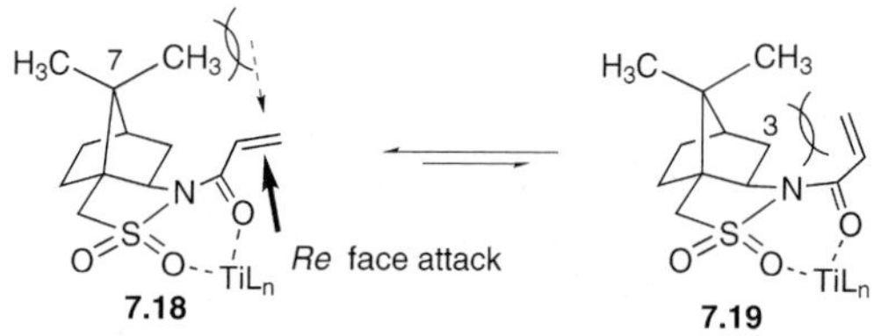

Scheme 7.11

conformer **7.19**, where the C=C and C=O bonds are antiperiplanar, there is a steric interaction developing between the β-carbon of the alkene and C(3) of the sultam. The diene then undergoes *endo* attack from the less-hindered bottom *Re* face as attack from the top face is hindered by one of the C(7) methyl groups.

The α,β-unsaturated *N*-acycloxazolidinones reported by Evans and co-workers also give impressive results in asymmetric Diels–Alder reactions. It was envisaged that this type of auxiliary would provide excellent control since high selectivities were observed for the alkylation of enolates derived from the corresponding *N*-acycloxazolidinones. Once activated by a Lewis acid, the dienophiles **7.20** and **7.21** react rapidly with a number of dienes to give adducts with excellent *endo* and *exo* and diastereoisomeric selectivity (Scheme 7.12).[14]

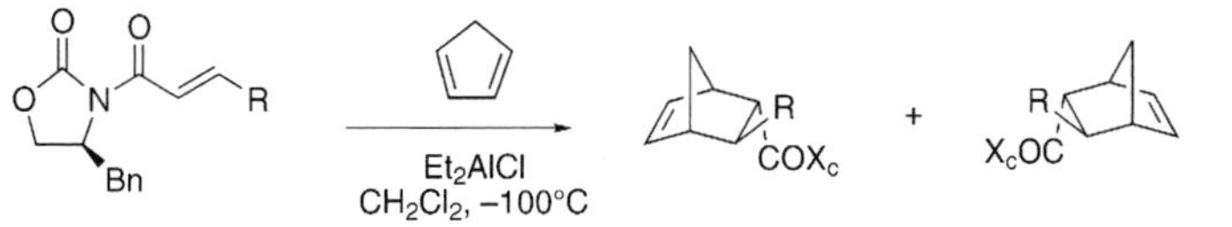

7.20 R = H
7.21 R = Me

Et₂AlCl
CH₂Cl₂, –100°C

Dienophile	*endo/exo*	Ratio	Yield (%)
7.20	>100:1	93:7	81
7.21	100:1	95:5	82

Scheme 7.12

The Lewis-acid-complexed ionic dienophile is very reactive and highly organised, with the major conformer being **7.22** (Scheme 7.13). The alternative

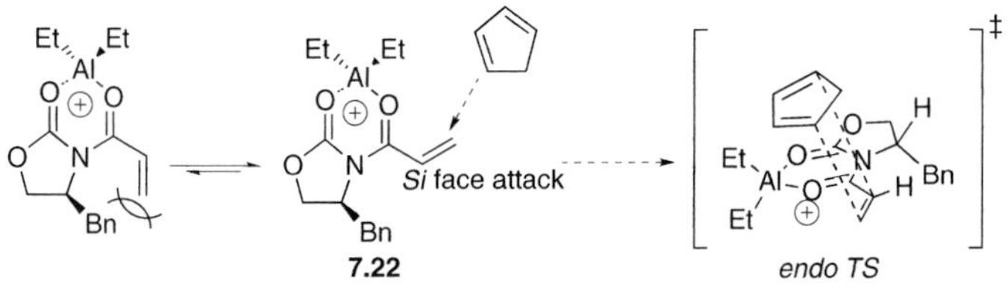

Si face attack

7.22

endo TS

Scheme 7.13

conformer suffers form a steric interaction, as indicated, and therefore reaction occurs preferentially via the C_α-*Si* face in conformer **7.22**. The high *endo* preference observed for this dienophile makes the oxazolidinone auxiliary amongst the most useful for asymmetric Diels–Alder reactions.

The importance of the C(4) substituent on the oxazoline ring was noted, and it was found that a benzyl substituent (oxazolidinone derived from phenylalanine) gave the highest diastereoselectivities even though this is not the most sterically demanding group (Scheme 7.14).[14] It was proposed that a charge-transfer interaction exists between the aromatic ring and the dienophilic moiety

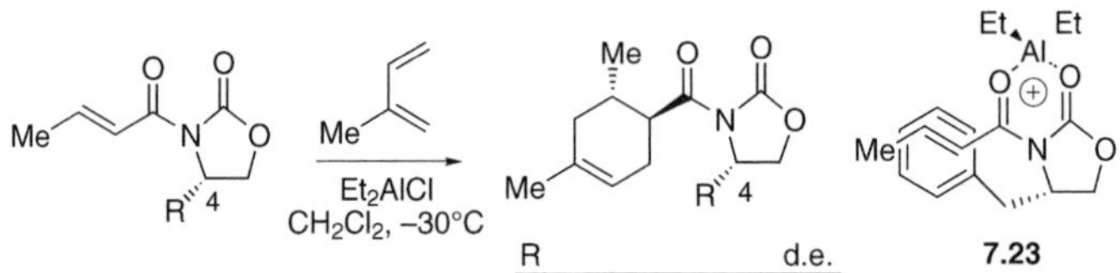

R	d.e.
CHMe$_2$	68%
CH$_2$cyclohexyl	82%
CH$_2$Ph	90%

Scheme 7.14

(π-stacking interaction) to give conformer **7.23** and this manifests itself in an increase in both diastereoselectivity and reactivity.

Asymmetric intramolecular Diels–Alder reactions are also effected by using the oxazolidinone auxiliary as shown in Scheme 7.15.[14] Triene **7.24** undergoes

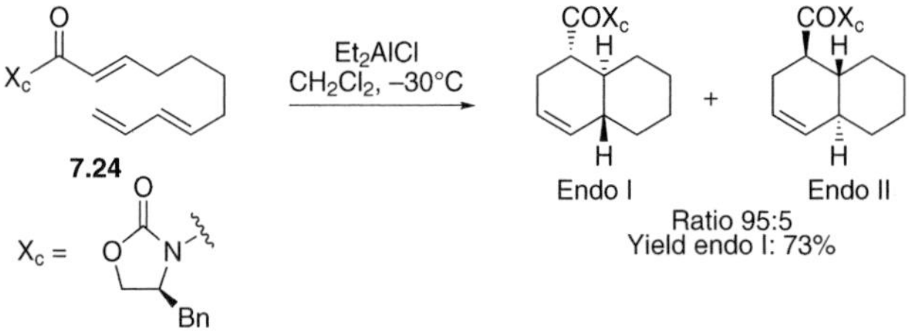

Scheme 7.15

cycloaddition upon treatment with dimethylaluminium chloride to give the *endo* I isomer as the major product in high d.e. Similar results have been reported by Roush *et al.* who used 8-phenylmenthol as the chiral auxiliary (see Schemes 7.6 and 7.7).[15]

7.1.4 Chiral dienophiles: type I chiral acrylates

As mentioned, type I dienophiles have the chiral moiety directly attached to the carbonyl group. Although not as common as type II dienophiles, there are some examples where high selectivities can be obtained with the simplest of auxiliaries. In one elegant example, Masamune *et al.* reported that the chiral acrylate **7.25** reacts with cyclopentadiene to give the adduct **7.26** in high d.e.

and with good *endo/exo* selectivity (Scheme 7.16).[5] It is noteworthy that the reaction proceeds in the absence of Lewis acid catalyst and it was proposed that

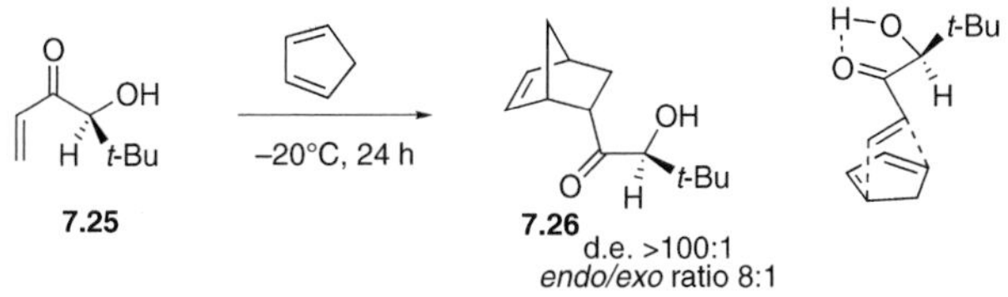

Scheme 7.16

the strong intramolecular hydrogen-bond in **7.25** freezes the free rotation along the carbonyl-C–asymmetric-C axis making the two diastereotopic faces easily distinguishable. Furthermore, the hydrogen-bonding also increases the reaction rate owing to stabilisation of the transition state.

Chiral sulfoxides have also been utilised as chiral inducing groups in Diels–Alder reactions. The (*S*)-*p*-tolylsufinylmalate **7.27** reacts with butadiene in the presence of titanium tetrachloride to provide adduct **7.28** that undergoes spontaneous sulfinyl elimination at room temperature to give diene **7.29** in 96% e.e. (Scheme 7.17).[16]

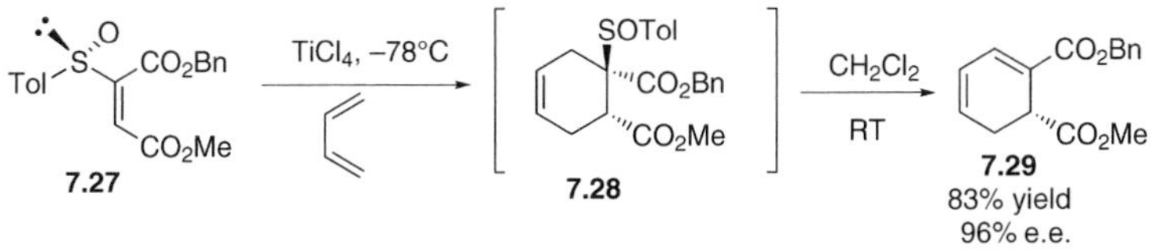

Scheme 7.17

The high selectivity observed in the above cycloaddition can be attributed to the transition state depicted in Figure 7.3 where the *endo* approach of the diene

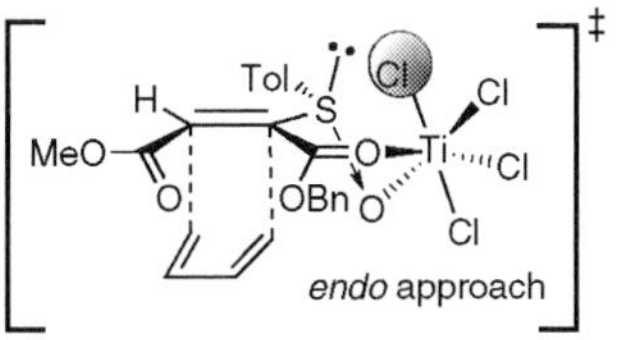

Figure 7.3

occurs from the less-hindered face of the titanium-complexed vinyl sulfoxide opposite the apical chlorine atom.

7.1.5   Chiral dienes

Chiral dienes are not as common as chiral dienophiles in asymmetric Diels–Alder reactions; however, high selectivities are often obtained with use of these inducers. An early example of a chiral diene with a removable auxiliary was reported by Trost *et al.* in 1980.[17] The (*S*)-mandelate ester **7.30**, available via a retro-Diels–Alder reaction of a norbornene derivative, reacts with acrolein under Lewis acid catalysis to give the adducts **7.31** and **7.32** in a ratio of 94:6 favouring **7.31** (Scheme 7.18).[18]

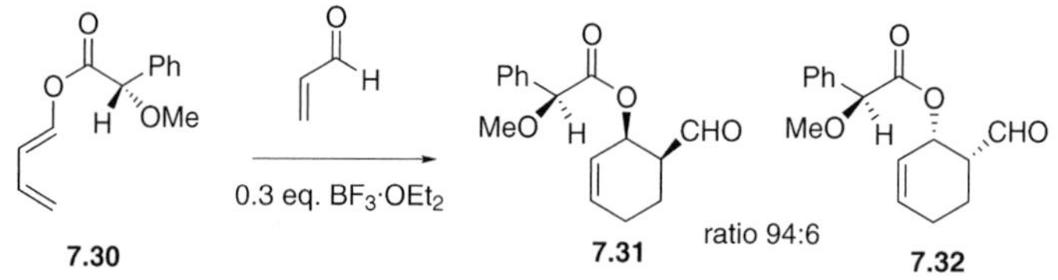

Scheme 7.18

The chiral diene **7.30** also reacts with juglone **7.33** in the presence of boron triacetate to yield the adduct **7.34** in 98% yield as the only detectable isomer (Scheme 7.19).[17] In the second example shown in Scheme 7.19 the chiral acrylate

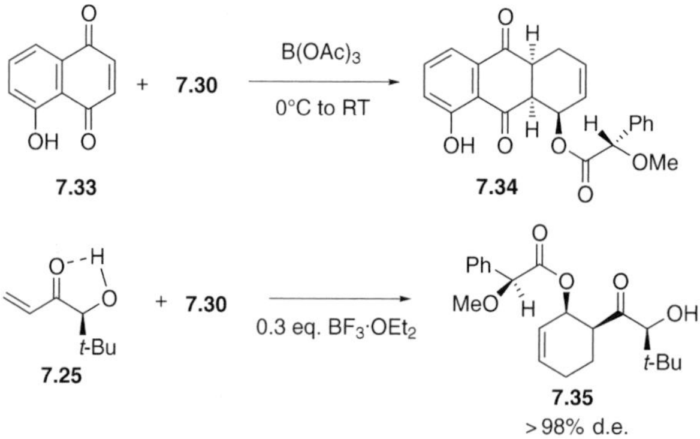

Scheme 7.19

7.25 discussed earlier reacts with **7.30** in a matched asymmetric reaction to provide the cyclohexene **7.35** in very high d.e.[19]

Two possible explanations for the selectivity shown by chiral diene **7.30** have been proposed. The first is based on a π-stacking model where the phenyl ring and diene system are parallel and the conformer preferred **7.36** is based on the nonbonding interaction between the methoxy group and diene in alternative conformer **7.37** (Scheme 7.20).[17] The other model proposed is represented by

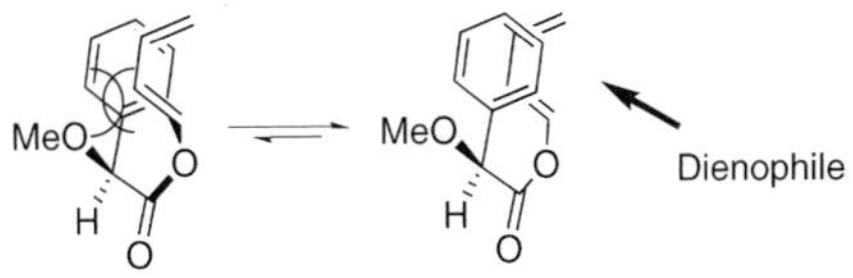

Scheme 7.20

conformer **7.38** and this is based on X-ray studies of various adducts.[18] In the alternative model **7.38** the phenyl ring is perpendicular to the diene blocking the

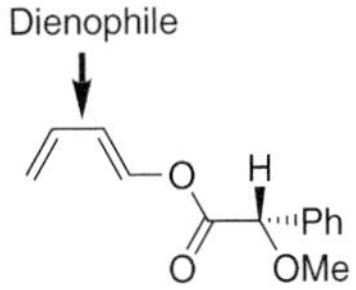

underside and the dienophile approaches from the top face as shown. This model also explains why a cyclohexyl rather than a phenyl ring gives the same stereo-selectivities and it was therefore suggested that π-stacking is not the controlling interaction.

More elaborate chiral dienes have also been utilised in asymmetric Diels–Alder reactions. The D-glucose-derived chiral diene **7.39** reacts with dienophile **7.40** to provide the demethoxydaunomycinone precursor **7.41** in reasonable d.e. (Scheme 7.21).[20]

The selectivity in the above cycloaddition was attributed to an *exo* anomeric effect. The conformer **7.42** is preferred over **7.43** and attack of the dienophile occurs from the top face as shown in Scheme 7.21.

A recent example of a very effective chiral diene is the simple ether **7.44**. The diene **7.44** reacts with electron-deficient dienophiles under thermal conditions to give high chemical yields and diastereoisomeric excesses greater than 95%.[21] In the example shown in Scheme 7.22, a thermal Diels–Alder reaction between diene **7.44** and diethyl fumarate gives the adduct **7.45** in excellent yield and diastereoselectivity. The selectivity displayed by diene **7.44** was again attributed to a π-stacking effect where the conformer **7.46** is preferred and the dienophile attacks in an *endo* fashion from the top face. It is noteworthy that the diene **7.44** is

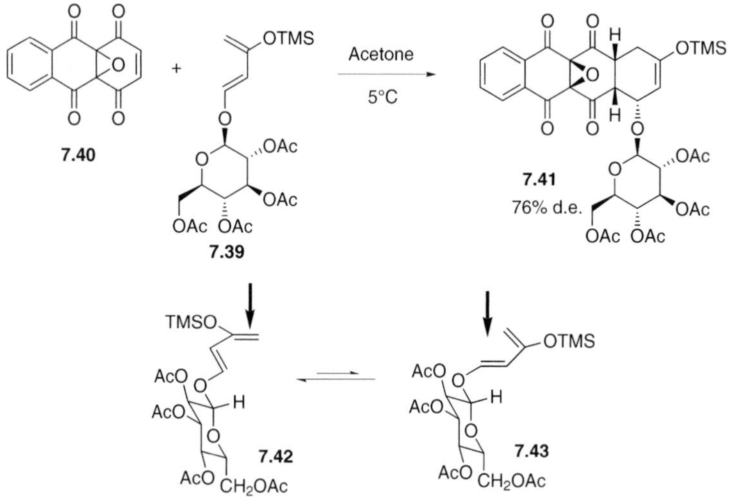

Scheme 7.21

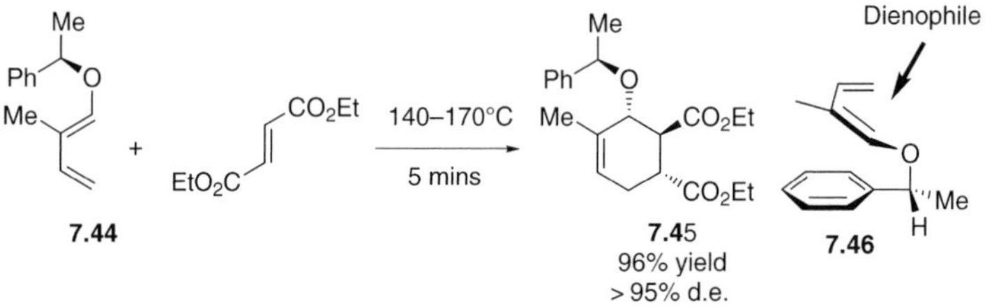

Scheme 7.22

extremely stereoselective and does not require Lewis acids or low temperatures to be effective.

7.2 Asymmetric 1,3-dipolar cycloadditions

7.2.1 Background

1,3-Dipolar cycloaddition (1,3-DC) reactions provide a simple method for the synthesis of five-membered heterocycles. A 1,3-dipole is defined as an a-b-c structure and can be divided into two main categories: allyl anion type and propargyl anion type (Scheme 7.23).[22] In the allyl anion type, the 1,3-dipole has four electrons in three parallel p_z orbitals and is bent. The central atom b can be nitrogen, oxygen or sulfur. For the propargyl anion type of 1,3-dipole, the central atom b is limited to nitrogen. The terminal atom a is usually carbon,

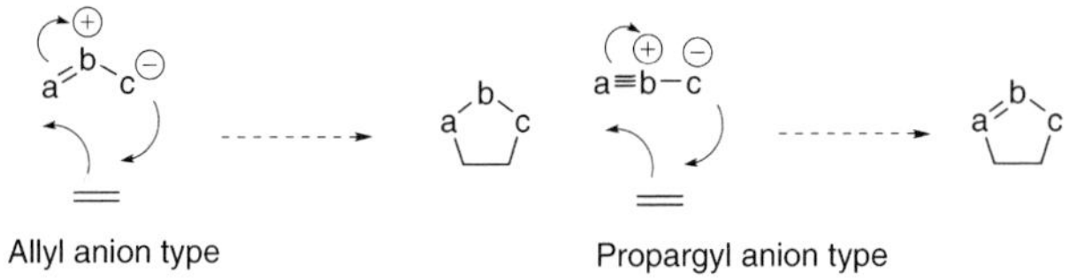

Scheme 7.23

and atom c is either nitrogen, oxygen or carbon. A cycloaddition between an allyl anion type 1,3-dipole and an alkene produces a saturated five-membered ring, and a 1,3-DC between a propargyl anion type dipole and alkene gives a monounsaturated five-membered ring.

Some of the more common 1,3-dipoles, where the b atom is nitrogen, are depicted in Scheme 7.24 and only the asymmetric versions of these will be

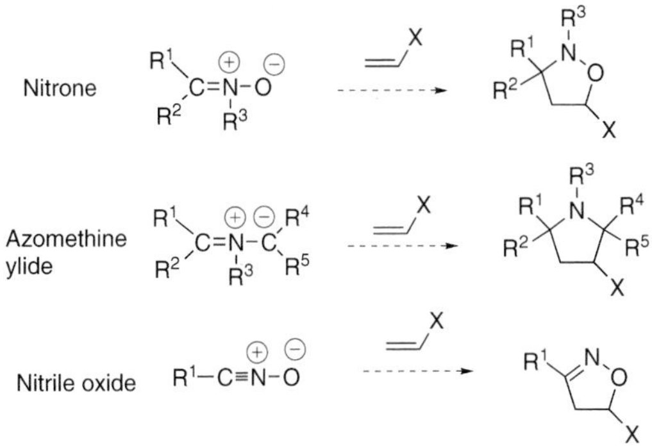

Scheme 7.24

dealt with in this section. A nitrone dipole can be generated from a carbonyl compound and a hydroxylamine, and these condense readily with substituted alkenes. Azomethine ylides are unstable species that can be generated *in situ* from methods such as proton abstraction of imines and thermolysis of aziridines. Nitrile oxides are also somewhat unstable and are also generated *in situ* from an aldoxime by treatment with a chlorinating agent and base.[22]

1,3-Dipoles react with functionalised alkenes in a manner analogous to dienes in Diels–Alder reactions. The selectivity of the addition is determined from reaction via an *endo* or *exo* transition state as depicted in Scheme 7.25. In the Diels–Alder reaction, the *endo* transition state is preferred because of the secondary π-orbital interactions. However, in the case of 1,3-DC reactions the stabilisation of the *endo* transition state is much smaller and the *endo/exo* selectivity is primarily controlled by the structure of the substrates. In an asymmetric 1,3-DC reaction, facial selectivity can arise by the use of a chiral 1,3-dipole or

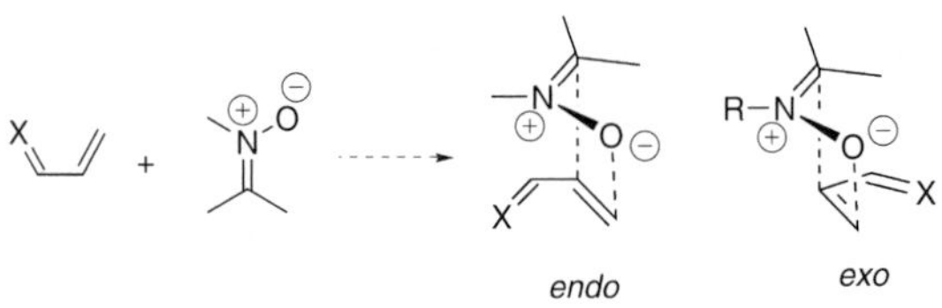

endo   *exo*

Scheme 7.25

alkene in a manner similar to that for asymmetric Diels–Alder reactions where both chiral dienes and dienophile can be utilised.

7.2.2   Asymmetric nitrone 1,3-dipolar cycloadditions: chiral nitrones

The majority of chiral nitrones utilised for asymmetric 1,3-DC reactions have the chiral substituent located on the nitrogen or carbon atom.[23] The most common nitrogen chiral substituent is the 1-phenylethyl group which arises from 1-phenylethylamine (Scheme 7.26). Oxidation to a hydroxylamine followed by

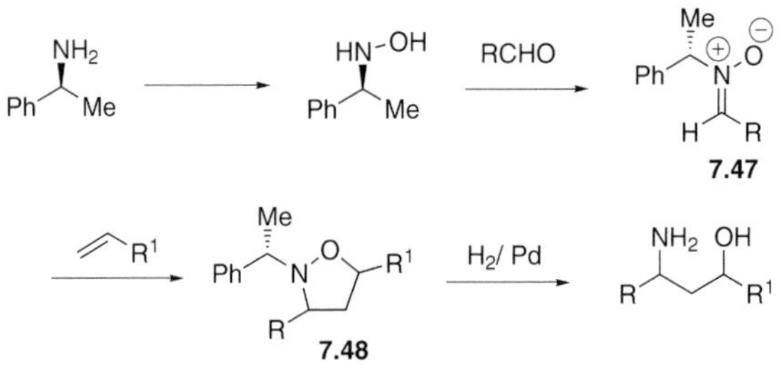

Scheme 7.26

condensation with an aldehyde provides chiral nitrone **7.47**. A 1,3-DC with a substituted alkene then provides isoxazolidine **7.48** as the major regioisomer which upon reduction gives an amino alcohol.

An early example of a nitrone bearing a 1-phenylethyl substituent is shown in Scheme 7.27. A 1,3-DC between styrene and the chiral nitrone **7.49** provided

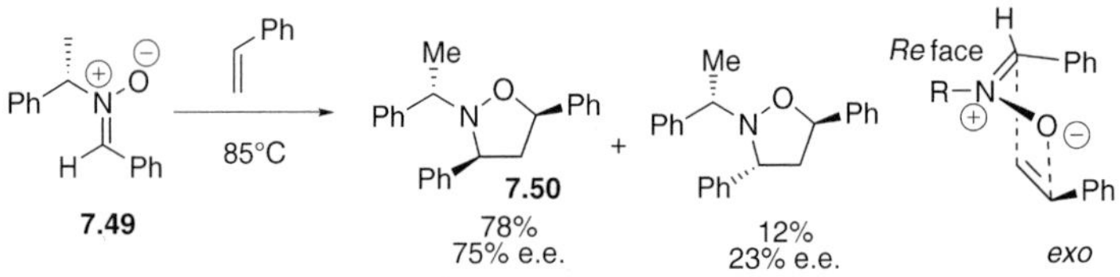

Scheme 7.27

the isoxazolidine **7.50** as the major isomer, in 75% e.e.[24] This outcome can be explained by an *exo* transition state in which attack occurs on the *Re* face of the nitrone. An *exo* mode of addition on the *Si* face of the nitrone results in the enantiomer of **7.50** so the e.e. of this addition is 75%. On some occasions the minor *endo* isomers were obtained in 100% e.e. In these cases, *endo* attack brings the chiral R group and the alkene substituent in proximity and thereby increases the facial selectivity of the chiral nitrone.

The chiral nitrone **7.51** has the chirality attached to the carbon atom and reacts with ethylvinyl ether in a highly stereoselective manner to provide the isoxazolidine **7.52** exclusively (Scheme 7.28).[25] The selectivity was rationalised

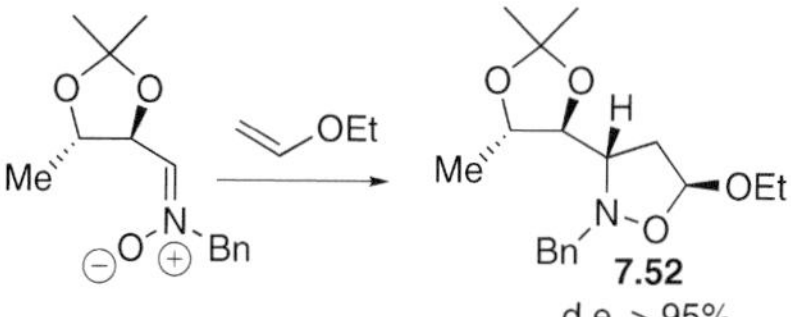

7.51

7.52
d.e. > 95%
93% yield

Scheme 7.28

as a Felkin–Anh type attack of the alkene to the *Si* face of the nitrone where the major conformer has the OR^1 substituent orientated perpendicular to the C=N bond (Figure 7.4). The alkene then reacts with nitrone **7.51** via the *endo* mode (Figure 7.5) to give the adduct **7.52** as the only product.

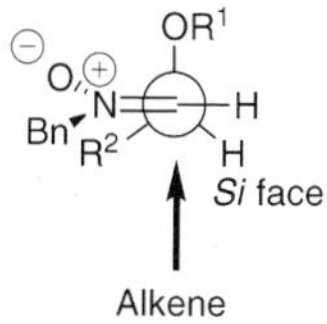

Figure 7.4

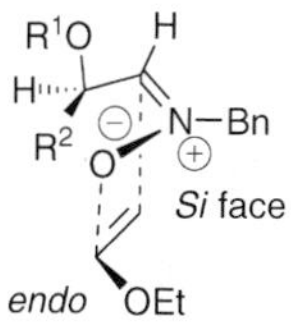

Figure 7.5

7.2.3   Asymmetric nitrone 1,3-dipolar cycloadditions: chiral alkenes

Chiral alkenes have also been utilised in asymmetric 1,3-DC reactions with nitrones. The simple chiral alkene **7.53** reacts with cyclic nitrone **7.54** to give the C2, C3a-*trans*-cycloadducts **7.55** and **7.56**.[26] The size of the alkyl group on the allylic stereocentre plays a major role in the selectivity, and in the case shown in Scheme 7.29 there is a strong preference for the formation of isomer **7.55**.

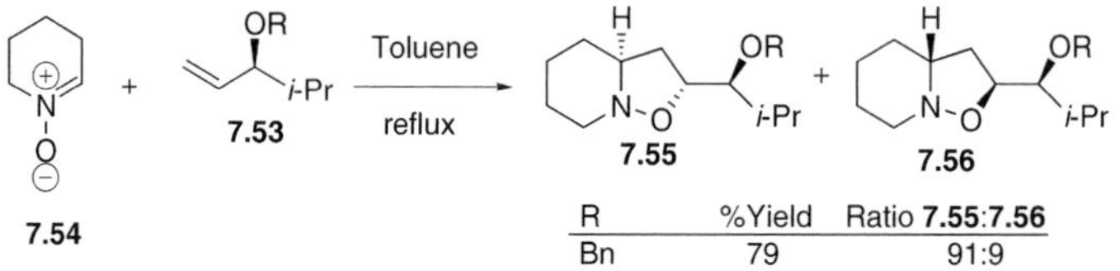

R	%Yield	Ratio 7.55:7.56
Bn	79	91:9
TBDPS	85	93:7

Scheme 7.29

The selectivity displayed by chiral alkene **7.53** can be explained by considering transition state model **7.57** as the preferred mode of attack (Figure 7.6). This places large R^1 group in the position *anti* to the approaching

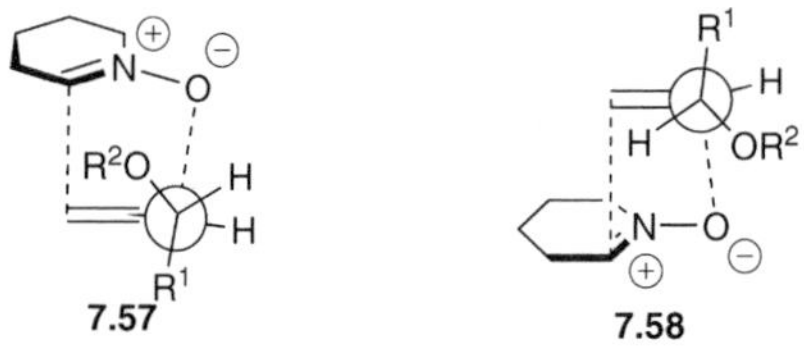

Figure 7.6

nitrone while the electronegative OR^2 group prefers the 'inside' rather than the 'outside' position in order to minimise electron withdrawal from the π system via σ^*_{C-O} overlap and to avoid repulsive interaction between the allylic oxygen and the nitrone oxygen. Transition state **7.58** suffers destabilisation as a result of an increase in the unfavourable polar interaction because of the 'outside' OR^2 group.

The chiral vinyl ether **7.59** displays excellent selectivity in the 1,3-DC reaction with cyclic nitrone **7.54** (Scheme 7.30).[27] The product **7.60** is obtained in high d.e. and this result is interesting since the chirality is one atom removed from the alkene. Removal of the chiral auxiliary results in the destruction of one of the new asymmetric centres.

The chiral sultams developed by Oppolzer have given mixed results in 1,3-DC reactions.[28] Condensation of alkene **7.61** with nitrone **7.54** provided two

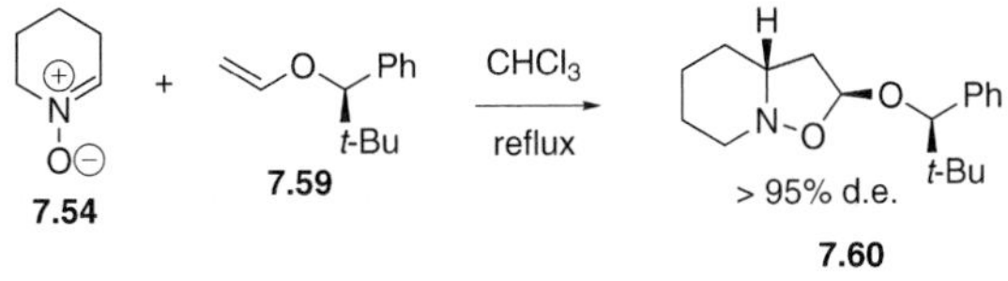

Scheme 7.30

major adducts in a low ratio, each a result of *endo* attack (Scheme 7.31). Other minor products isolated were a result of *exo* attack.

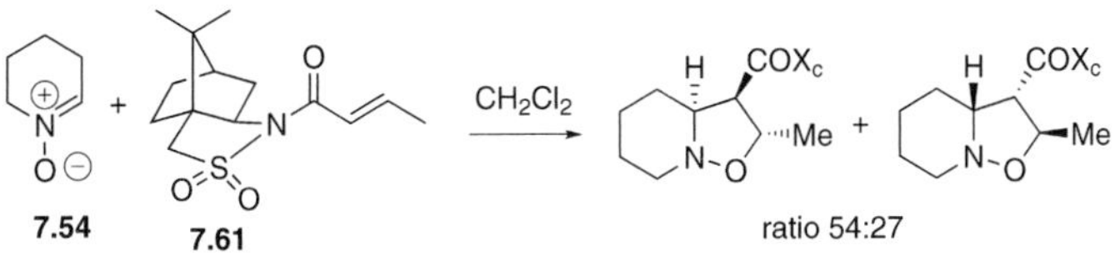

Scheme 7.31

7.2.4 Asymmetric azomethine ylide 1,3-dipolar cycloadditions

As mentioned earlier, azomethine ylides are generated *in situ* in the presence of an alkene which then undergoes the 1,3-DC. The majority of asymmetric versions involve a chiral azomethine ylide with the chirality located on the nitrogen atom. The chiral ylide **7.62** generated from the oxazolidine **7.63** reacts with *N*-phenylmaleimide to give the *exo* product **7.64** in high d.e. and yield (Scheme 7.32).[29] The result implies an *exo* transition state in which one face

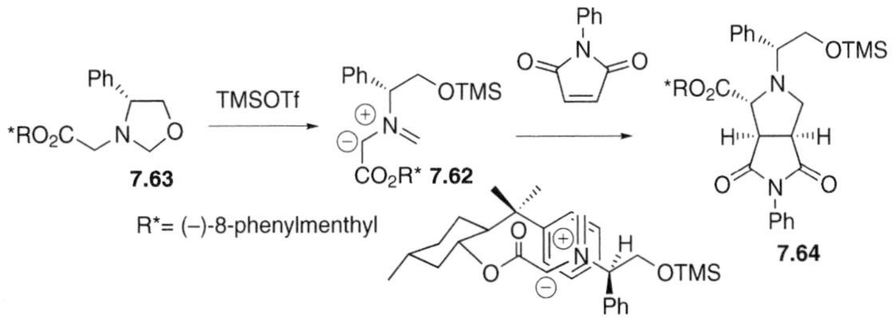

Scheme 7.32

of the ylide is blocked by the phenyl group of the 8-phenylmenthol auxiliary as shown.

In another example, the photochemically generated chiral azomethine **7.65** ylide reacted with chiral alkene **7.15** from the face indicated by an *exo* mode of attack to give the adduct **7.66** in 55% yield based on recovered starting material

(Scheme 7.33).[30] This methodology was later utilised in the total synthesis of quinocarcin **7.67**.[31]

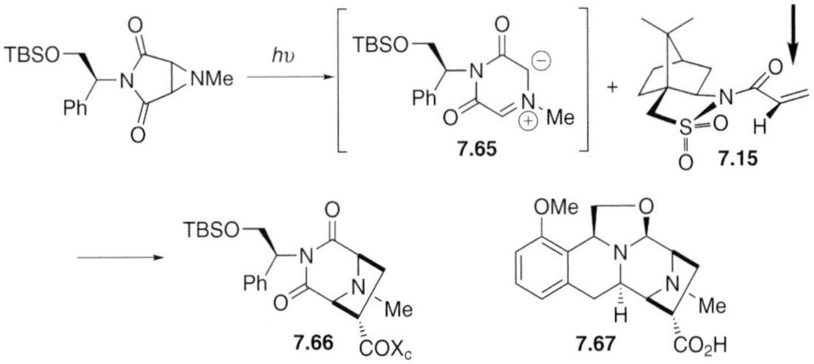

Scheme 7.33

7.2.5   Asymmetric nitrile oxide 1,3-dipolar cycloadditions: chiral alkenes

The vast majority of asymmetric 1,3-DC reactions involving nitrile oxides utilise a chiral alkene as opposed to a chiral nitrile oxide. These reactions supply adducts which can be converted into β-hydroxy carbonyl compounds or β-amino alcohols as shown in Scheme 7.34.

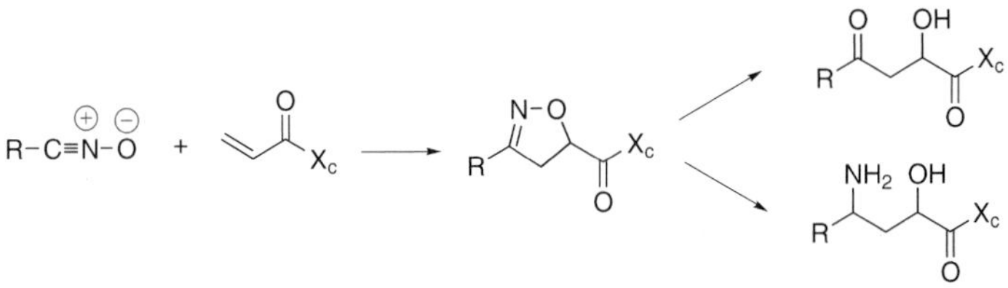

Scheme 7.34

Chiral acrylate ester derivatives have proven successful in asymmetric Diels–Alder reactions but only in the presence of Lewis acids as they normally show low rotameric preference for either the s-*cis* or s-*trans* conformer (Scheme 7.35). Lewis acids cannot be employed successfully for 1,3-DC reactions involving nitrile oxides presumably because of their Lewis basicity. Therefore, it not surprising that only modest selectivities have been observed in cycloadditions between chiral acrylate esters and nitrile oxides. Chiral tertiary acrylamides can provide a solution to the rotamer problem as they exist mainly in the planar s-*cis* form. However, acrylamides do have two low-energy rotamers about the C—N bond, also shown in Scheme 7.35.

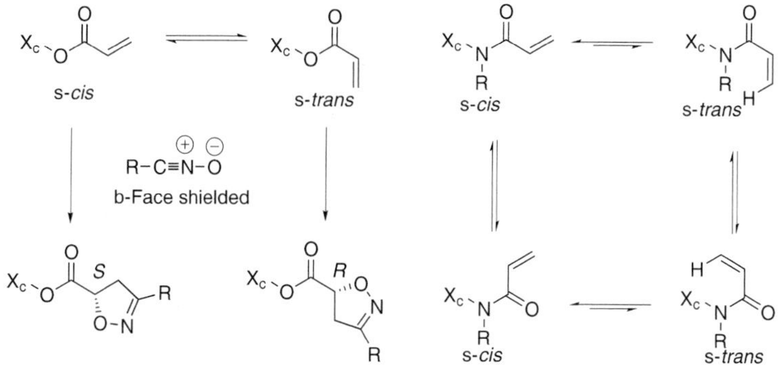

Scheme 7.35

Curran *et al.* reported that the Oppolzer sultam **7.15** reacts with nitrile oxide **7.68** on the top face of the conformer shown to give the adduct **7.69** as the major product with excellent selectivity (Scheme 7.36).[32] The chiral auxiliary is then easily removed by L-selectride reduction.

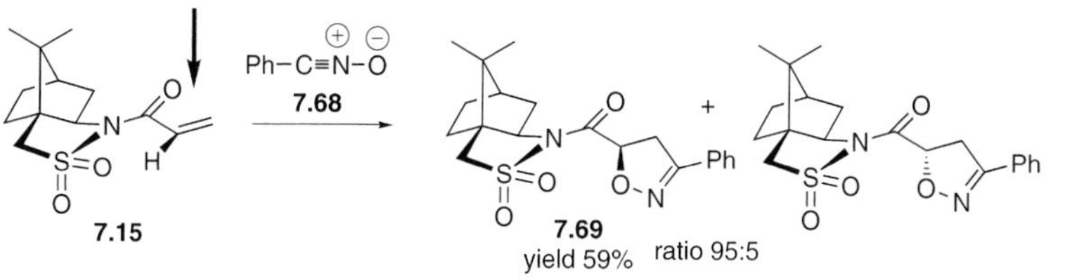

Scheme 7.36

The chiral acrylamide **7.70** developed by Kim *et al.* also displays high selectivity in 1,3-DC reactions with nitrile oxides.[33] Condensation of alkene **7.70** with nitrile oxide **7.68** at low temperature provided adduct **7.71** in 90% d.e. and 74% yield (Scheme 7.37). This is a result of *Si* face attack by the nitrile oxide on the conformer (Figure 7.7).

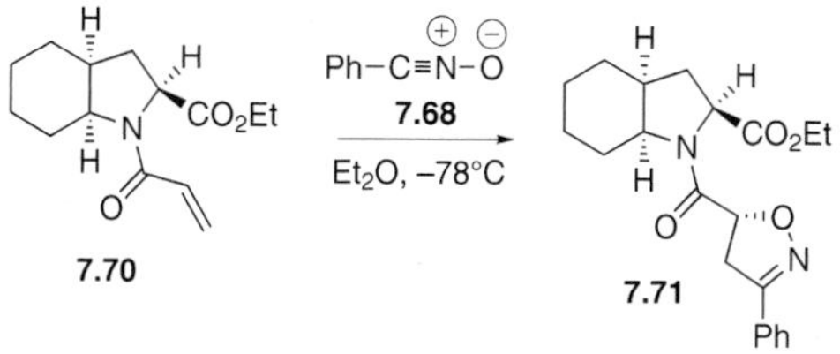

Scheme 7.37

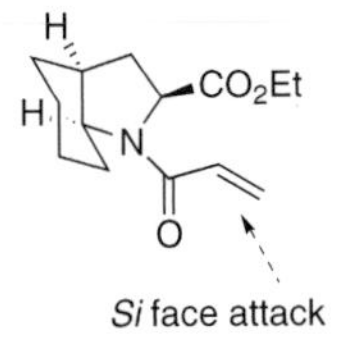

Si face attack

Figure 7.7

7.3 Asymmetric [3,3] rearrangements

7.3.1 Background

Sigmatropic [3,3]-rearrangements are suprafacial concerted reactions that often provide products in high stereoselectivity. The two major types that will be discussed in this section are the asymmetric versions of Cope and Claisen rearrangements (Scheme 7.38). Both the Cope rearrangement and the Claisen

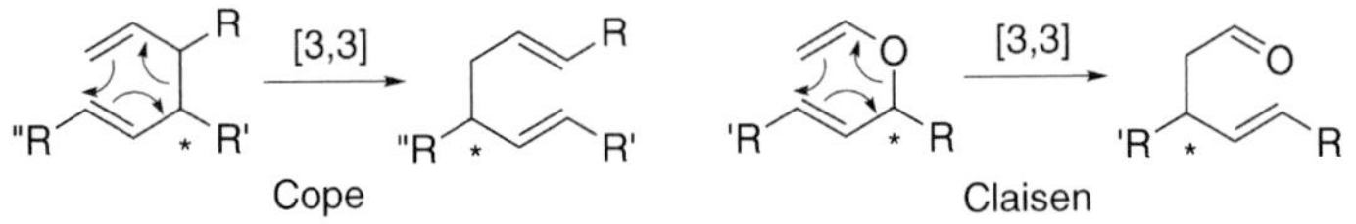

Scheme 7.38

rearrangement are characterised by large negative entropies of activation, and 1,3-chirality transfer can occur from the centres indicated via a rigid chair-like six-membered transition state to form the new σ-bond and asymmetric centre.

7.3.2 Asymmetric Cope rearrangement

The Cope rearrangement can proceed by a tightly ordered transition state to form new asymmetric centres with a high degree of control. Evidence for a cyclohexane chair-like transition state arose from an early investigation into the rearrangement of the chiral 1,5-diene **7.72** (Scheme 7.39).[34] Cope rearrangement of optically active **7.72** gave the optically active dienes **7.73** and **7.74** in a ratio of 87:13. This result suggested that rearrangement occurred by the cyclohexane transition states **7.75** and **7.76** (Scheme 7.39) to give the products in greater than 90% optical purity. This nearly quantitative transfer of asymmetry provided strong evidence for a concerted cyclic rearrangement. In this case there is a preference for the transition state **7.75** (equatorial phenyl) over

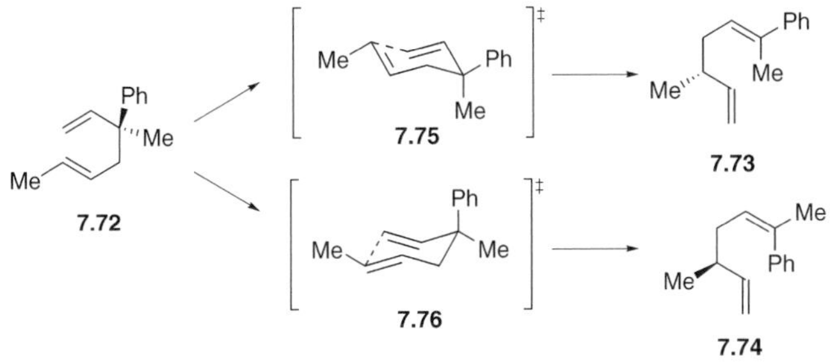

Scheme 7.39

7.76 (axial phenyl) which corresponds to a free-energy difference of around 2 kcal/mol.

7.3.3 Asymmetric anionic oxy-Cope rearrangement

The Cope rearrangement can also proceed in the presence of a C(3) hydroxyl substituent. Futhermore, the so-called oxy-Cope rearrangement occurs at a much faster rate upon conversion of the hydroxyl group to the alkoxide (Scheme 7.40).[35,36] The anionic oxy-Cope rearrangement can therefore proceed

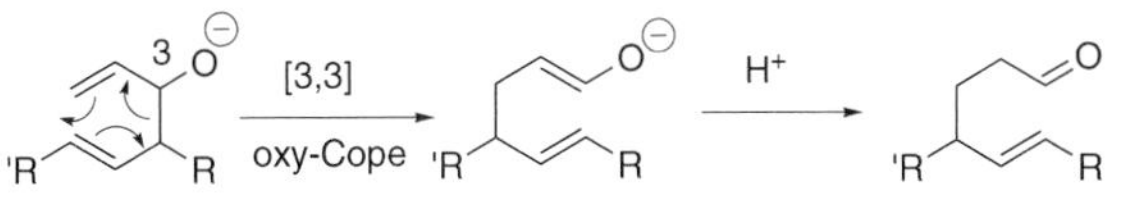

Scheme 7.40

at lower temperatures allowing for a higher control of stereochemistry. Also, the product is a resonance-stabilised enolate anion which causes the rearrangement to become essentially irreversible. The carbonyl group obtained upon protonation is then available for further synthetic manipulation.

Paquette reported that the acyclic oxy-Cope substrate **7.77** rearranges on treatment with base to provide the enantiomers **7.78** and **7.79**, favouring the former (Scheme 7.41). This suggests that the anion derived from **7.77** rearranges preferentially via the chair-like transition state **7.80** where the alkoxy group is orientated in the pseudoequatorial position.[37]

Cyclohexenes are known to show a strong preference for axial bond formation during [3,3] sigmatropy. Oxy-Cope rearrangement of the cyclohexene **7.81** proceeds almost exclusively via the transition state **7.82** in which axial bond formation and equatorial oxyanion orientation is exhibited (Scheme 7.42).[38] Reduction then provided the alcohols **7.83** and **7.84** in a ratio of 61:1.

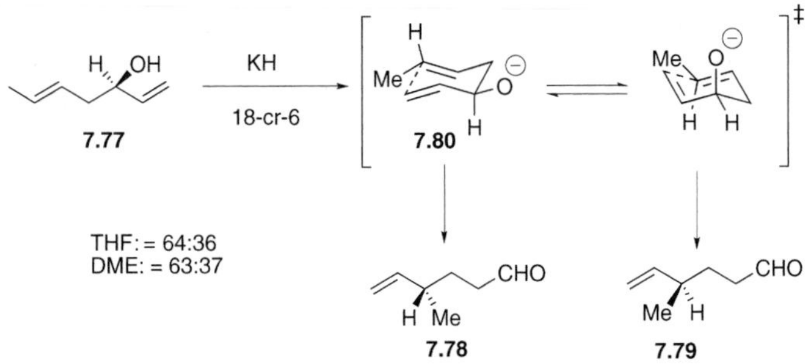

THF: = 64:36
DME: = 63:37

Scheme 7.41

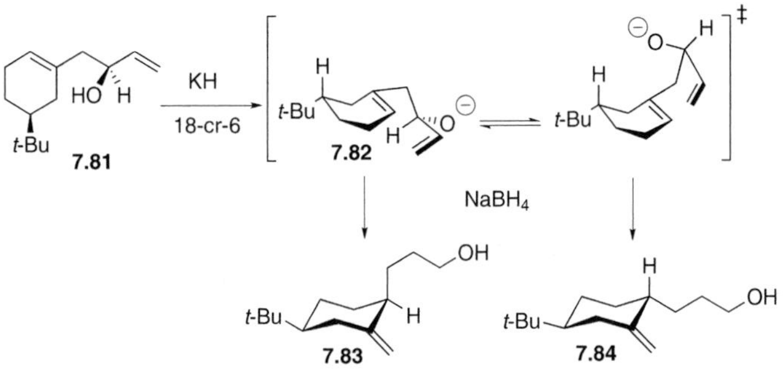

NaBH₄

Scheme 7.42

7.3.4 Asymmetric Claisen rearrangement

Claisen rearrangements also proceed with excellent stereocontrol. The enol
ether **7.85** undergoes a [3,3]-sigmatropic rearrangement to give the aldehyde
7.87 with high transfer of chirality (Scheme 7.43).[39] It was postulated that the

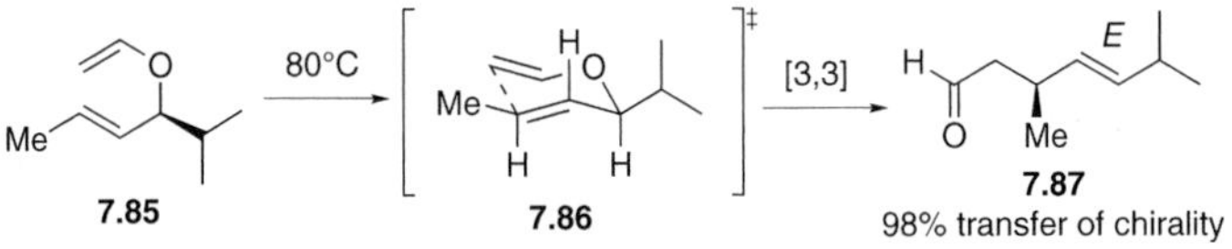

Scheme 7.43

reaction proceeded through the chair-like transition state **7.86** where all the
substituents are placed pseudoequatorial.

Denmark and Marlin have reported that a phosphorus-stabilised carbanion can accelerate the Claisen rearrangement, and the asymmetric version of

this reaction proceeds with high stereocontrol. Treatment of the 1,3,2-oxazaphosphorinane **7.88** with lithium chloride and the potassium anion derived from DMSO gives the rearranged product **7.89** in good d.e. (Scheme 7.44).[40]

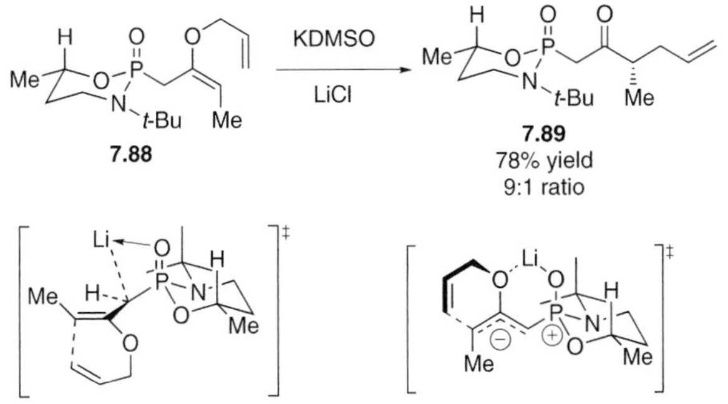

Scheme 7.44

The two transition state models depicted have been proposed for the lithium carbanion. The key features common to these structures are the planar carbanion and the strong polarisation of the phosphorus oxygen bond.

7.3.5 Asymmetric Ireland–Claisen rearrangement

Although a number of variants of the aliphatic Claisen rearrangement have been reported, most require harsh conditions and give low yields. The Ireland variant of the Claisen rearrangement involves an allylic ester enolate, generated from an allylic ester, which undergoes [3,3]-rearrangement at or below room temperature to provide carboxylic acid products with excellent levels of stereocontrol (Scheme 7.45).[41,42] However, enolates are often unstable and can

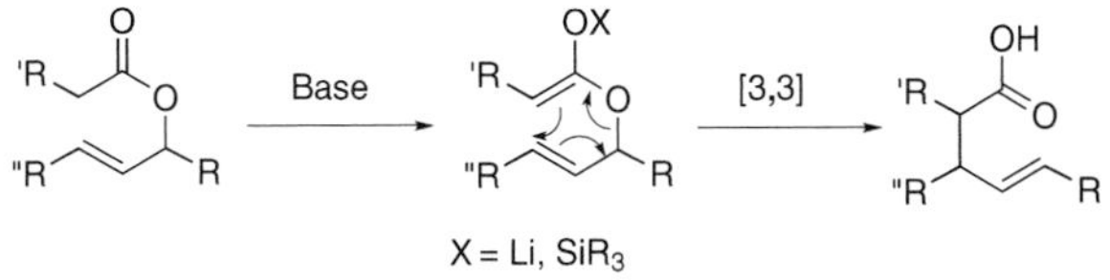

X = Li, SiR$_3$

Scheme 7.45

undergo unwanted side-reactions or extensive decomposition. A solution to this problem was to effect silylation of the intermediate ester enolate which provides a stable silyl-ketene-acetal that rapidly rearranges at similar rates.[42] This variant, know as the Ireland–Claisen rearrangement, has been extensively utilised in asymmetric synthesis because of the mild conditions and the ease of synthesis of the ester precursors.

In simple acyclic systems, the Ireland–Claisen rearrangement proceeds with high stereocontrol; however, the outcome is dependent on the stereochemistry of the enolate (and subsequent silyl-ketene acetal) generated from the ester. Ireland and co-workers demonstrated that ester enolates could be generated in a stereoselective manner by choice of solvent and base (see Chapter 3).[43,44]

Treatment of the propionate ester **7.90** with LDA in THF followed by silylation gives the (Z)-silylketene acetal **7.91** which rearranges by the chair-like transition state **7.92** to give the *threo* (*syn*) stereoisomer (Scheme 7.46).[43]

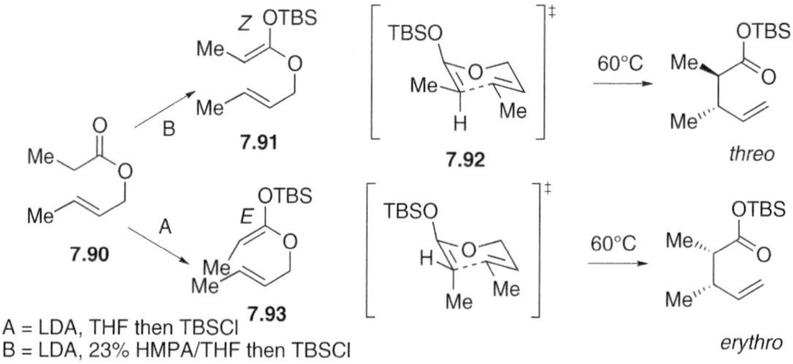

A = LDA, THF then TBSCl
B = LDA, 23% HMPA/THF then TBSCl

Scheme 7.46

Conversely, treatment of **7.90** with LDA in 23% HMPA/THF followed by TBSCl resulted in preferential formation of the (E)-silylketene acetal **7.93** which then rearranges again by a chair-like transition state to provide the *erythro* (*anti*) stereoisomer.

It was later found that enolate geometry could also be effectively controlled by chelation.[45,46] Treatment of the allylic glycolate ester **7.94** with LDA resulted in the formation of a stable chelated (Z)-enolate which on silylation provide the (Z)-silylketene acetal selectively (Scheme 7.47). Rearrangement via the chair-like transition state **7.95** followed by hydrolysis and methylation of the resultant acid provides the ester **7.96** in 82% d.e.[45]

Ireland *et al.* noted that for certain cyclic systems conformational constraints can override the inherent preference for chair-like transition states in the ester enolate Claisen rearrangement and lead to a dominance of boat-like transition state structures.[47] For example, rearrangement of the (Z)-silylketene acetal **7.97** gave the acid **7.98** on workup (Scheme 7.48). It was proposed that the chair-like transition state **7.99** suffers from an unfavourable interaction between the silyloxy group and the CHMe of the pyranoid glycal ring. Thus a preference for the boat-like transition state **7.100** was exhibited in this case. The high preference

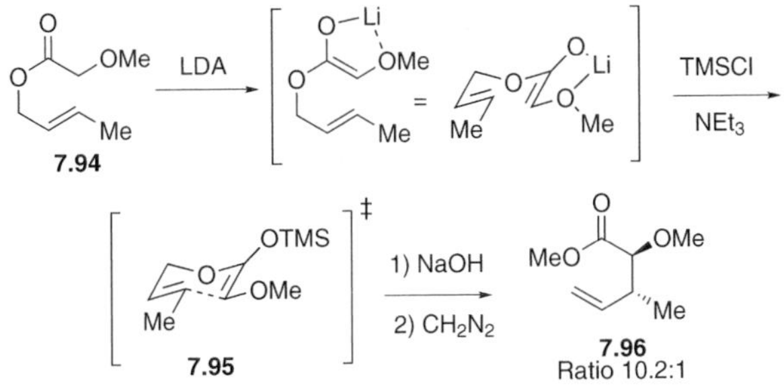

Scheme 7.47

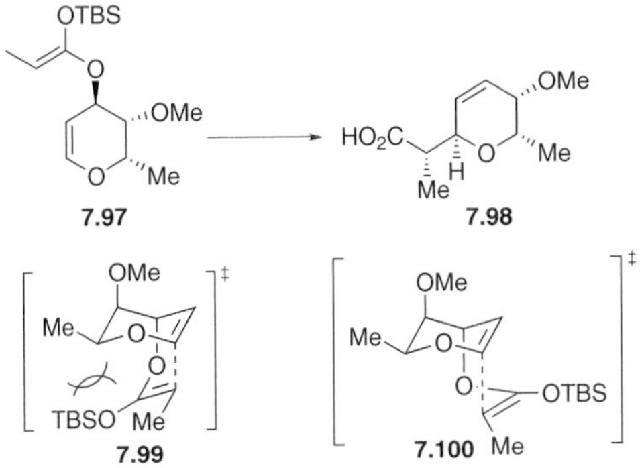

Scheme 7.48

(> 90%) shown for the boat-like transition state in the glycal example shown as opposed to related cyclohexenes (c. 70%) was explained by the fact that the boat-like transition state has a more product-like geometry which results in a higher degree of O—C bond cleavage. The resonance stabilisation by the C(6) oxygen in the glycal is significant for the more dissociated boat-like transition state.

The shift for a preference for boat-like transition states is reproduced in five-membered glycal systems. Rearrangement of glycal **7.101** gave acid **7.102** almost exclusively (Scheme 7.49).[47] This result again suggests a strong preference (> 90%) for the boat-like transition state **7.103** over the chair-like transition state **7.104**.

The ester enolate Claisen rearrangement has been successfully utilised in a large number of asymmetric syntheses. In the total synthesis of ebelactone A **7.105** reported by Paterson and co-workers, the (E)-silylketene acetal **7.106** rearranged via the chair-like transition state to provide the ester **7.107** in excellent d.e. (Scheme 7.50).[48,49] It should be noted that the (E)-O-ester enolate leading to acetal **7.106** was generated with high stereocontrol in the presence of an unprotected ketone functionality.

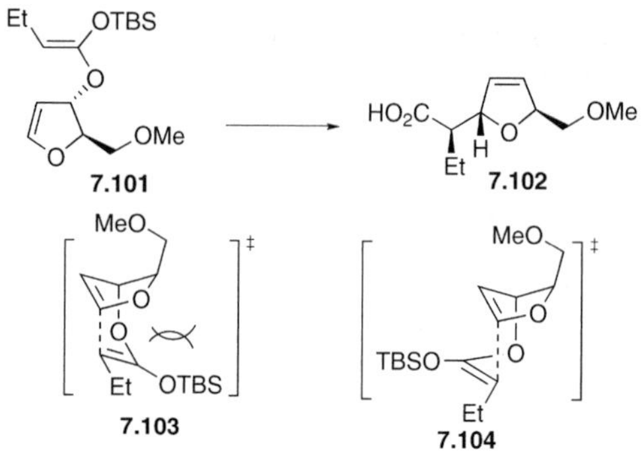

Scheme 7.49

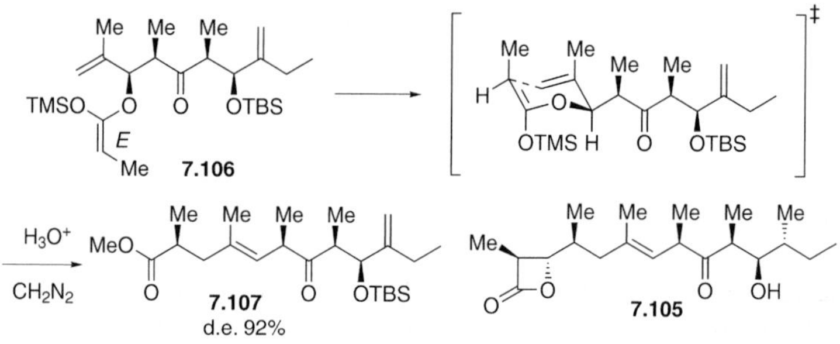

Scheme 7.50

The ester enolate Claisen rearrangement can also be carried out in the presence of a β-leaving group. Treatment of the tetrahydrofuronate **7.108** with base in the presence of HMPA and silylating reagent at low temperature forms the ketene acetal which then undergoes [3,3]-rearrangement stereoselectively (Scheme 7.51).[50] Under the conditions utilised, the formation of the

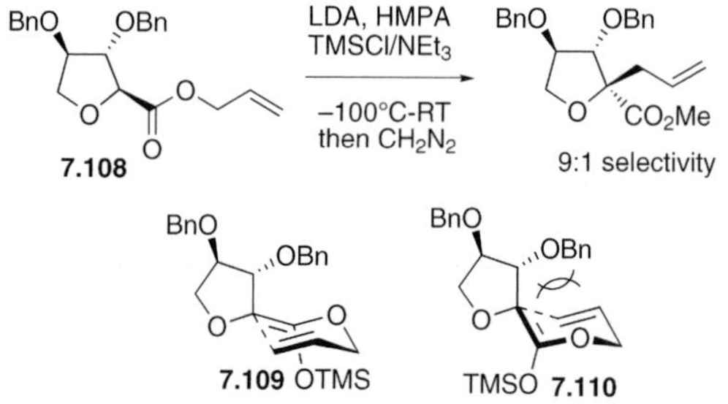

Scheme 7.51

(Z)-silylketene acetal should be favoured[44] leading to two major chair-like transition states. Transition state **7.109** is favoured owing to a developing steric interaction in the alternative transition state **7.110**.

7.3.6   Auxiliary-directed Ireland–Claisen rearrangement

Auxiliary-directed ester enolate Claisen rearrangements have been reported to give useful levels of diastereoselection. Chiral glycolate esters can be synthesised and the stereochemistry of the enolate is then easily controlled by chelation (Scheme 7.52). Treatment of the simple chiral glycolate **7.111** with KHMDS

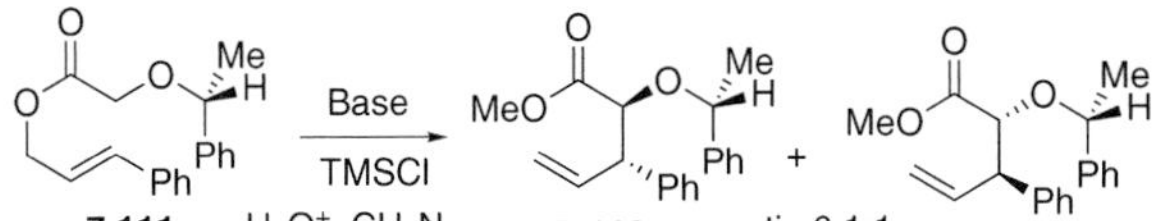

Scheme 7.52

and TMSCl at −78°C followed by warming to 0°C gave the rearranged product **7.112** as the major isomer.[51]

The enhanced auxiliary induction displayed by cinnamyl glycolates such as **7.111** was explained by examining the chair-like transition states shown in Scheme 7.53. A favourable π-stacking interaction of the vinyllic aryl substituent

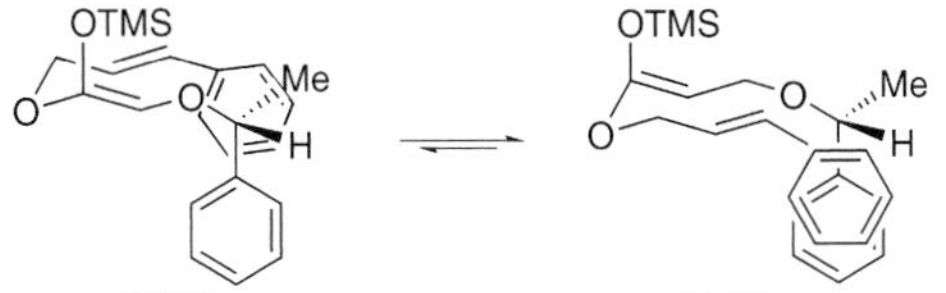

Scheme 7.53

with the auxiliary is possible and transition state **7.114** is preferred because of the steric interaction between the methyl group of the auxiliary and the cinnamyl phenyl ring in transition state **7.113**.[51]

Corey and Lee reported an enantioselective version of the Ireland–Claisen rearrangement which utilised the chiral boron reagent **7.115**. The intermediate chiral boron enolate **7.117** species derived from the simple allylic ester **7.116** and the chiral bromoborane **7.115** rearranges in a highly stereoselective manner to give the *syn*-acid **7.118** in exceptionally high e.e. (Scheme 7.54).[52] This

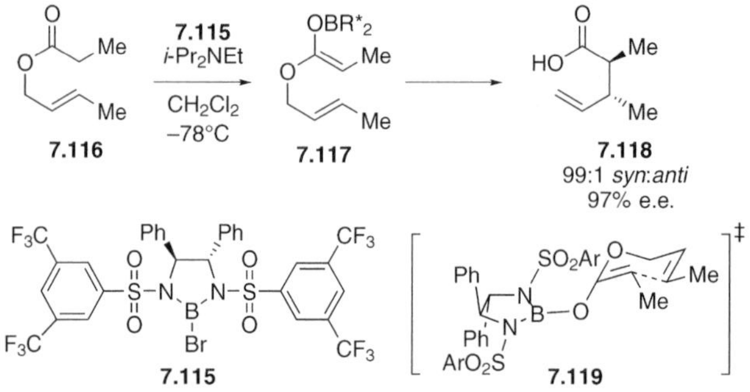

Scheme 7.54

result is in agreement with the sterically more favourable chair-like transition state **7.119**.

7.4　Asymmetric [2,3]-Wittig rearrangements

[2,3]-Wittig sigmatropic rearrangements can be generalised as an isomerisation that proceeds through a six-electron five-membered cyclic transition state.[53] The most common [2,3]-Wittig rearrangement utilised in asymmetric synthesis is the oxycarbanion type depicted in Scheme 7.55. The G substituent is an

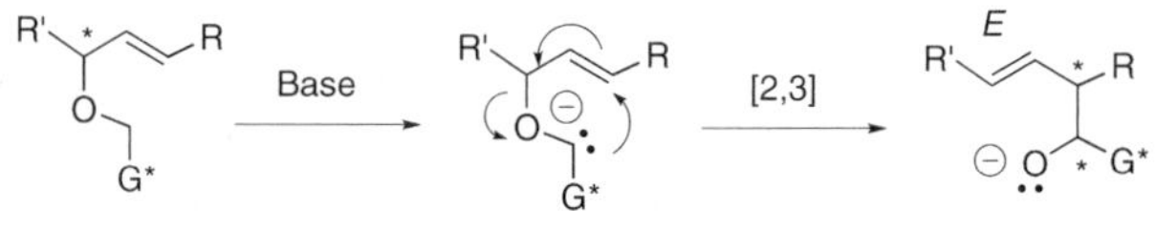

Scheme 7.55

attached functional group that can stabilise an adjacent carbanion to a degree, and, after rearrangement, the charge is transferred to the oxygen atom as shown. Asymmetry can be transferred from an existing stereogenic centre or from the G substituent to the two new centres formed in the rearrangement.

1,3-Chirality transfer in the [2,3]-Wittig process can be highly selective. The use of an (E)-configured allylic ether **7.120** with asymmetry at the 1-position usually produces the *anti* configured product **7.122** after rearrangement (Scheme 7.56).[53] The five-membered transition state can be depicted as structure **7.121** which is in the preferred 'folded-envelope' conformation analogous to a cyclopentane ring. The G substituent prefers to remain in the pseudoequatorial position and the R group is in the *exo*-orientation to minimise steric interactions. The alternative (Z)-allylic ether **7.123** rearranges via the five-membered

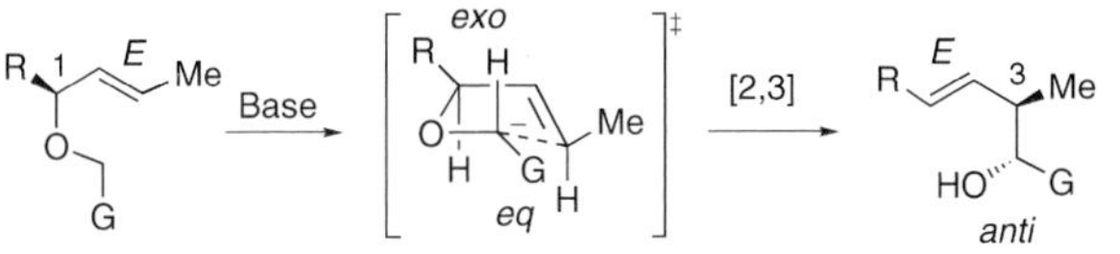

7.120 7.121 7.122

7.123 7.124 7.125

Scheme 7.56

transition state **7.124**, which again possesses *exo* R and equatorial G substituents, to provide the *syn* product **7.125**.

As shown in Scheme 7.57, treatment of the optically active propargyl ether **7.126** with base at low temperature for several hours gave the *syn*-propargylic

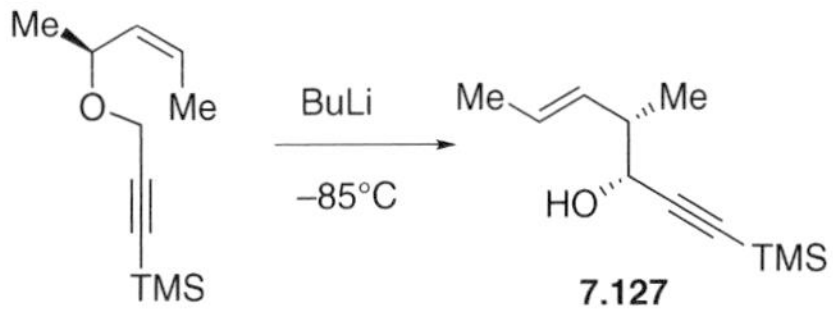

7.126 7.127

Scheme 7.57

alcohol **7.127** as the major product with high diastereoisomeric and geometric selectivity (99% d.e. and 99%*E*).[54] This result suggests that the reaction proceeds via the transition state **7.128** where the alkyne is pseudoequatorial and the methyl group is orientated *exo*.

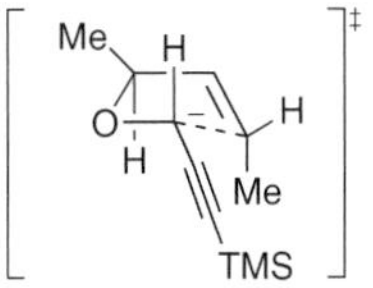

7.128

Chiral auxiliary G substituents have also been effective for asymmetric [2,3]-Wittig rearrangements. A *trans*-2,5-bis(methoxyhydroxymethyl)pyrrolidine moiety served as the chiral inducing agent in the rearrangement of allylic ether **7.129**. Generation of the zirconium enolate by deprotonation with butyllithium followed by transmetallation with zirconium provided the *syn* isomer **7.130** with excellent asymmetric induction (Scheme 7.58).[55] It was suggested that transition

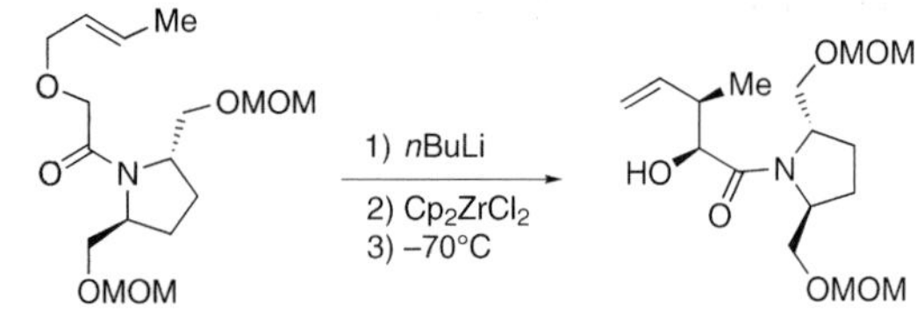

7.129 **7.130**

Scheme 7.58

state conformer **7.131** was lower in energy than the alternative conformer **7.132** owing to the indicated developing steric interaction between the metal ion and vinyllic hydrogen atom.

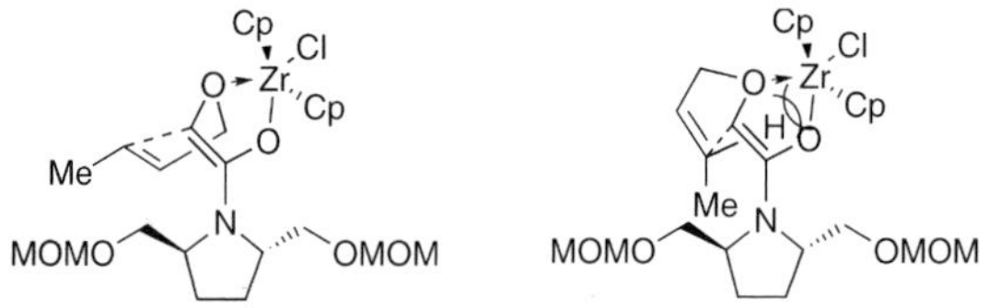

7.131 **7.132**

The 8-phenylmenthol-derived chiral auxiliary has also proven effective in controlling asymmetry in the [2,3]-Wittig rearrangement. Treatment of ester **7.133** with base in the presence of a polar solvent additive (HMPA) gave the *syn* isomer **7.134** with 9:1 selectivity in 97% d.e. (Scheme 7.59).[56] In this case, the *Si* face of the enolate is sterically less congested whereas the *Re* face back-side

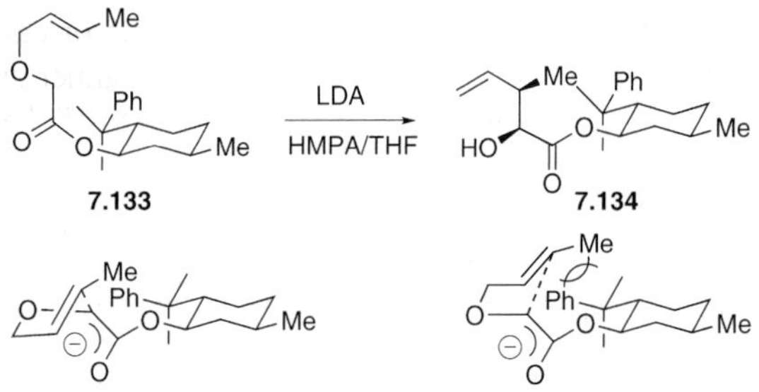

7.133 **7.134**

Scheme 7.59

rearrangement suffers from a steric interaction between the phenyl group and the allylic moiety.

Chiral bases can provide useful control in asymmetric [2,3]-rearrangements, and an example reported by Marshall and Lebreton is depicted in Scheme 7.60.[57]

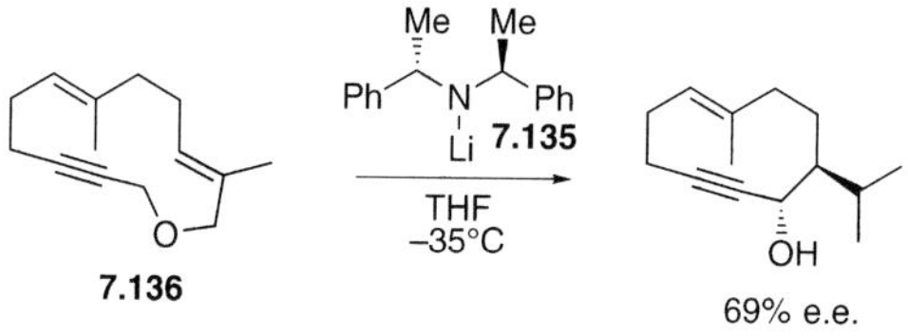

Scheme 7.60

The chiral base **7.135** effected deprotonation of achiral cyclic ether **7.136**, and subsequent rearrangement of the anion proceeded in fair e.e. The relatively complex preferred transition state **7.137** was postulated for this reaction where

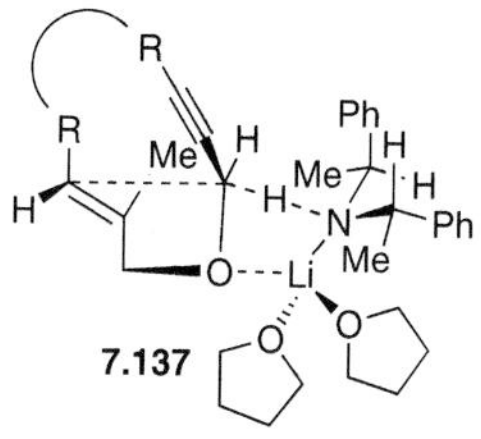

a tight chiral chelate is formed between the lithium ion and the oxygen and nitrogen atoms of both substrate and solvent.

7.5   Asymmetric ene reactions

The ene reaction is a pericyclic process where an alkene or carbonyl system (enophile) reacts with an allylic compound (ene) in a manner similar to both a cycloaddition and a [1,5]-sigmatropic shift of hydrogen (Scheme 7.61). In

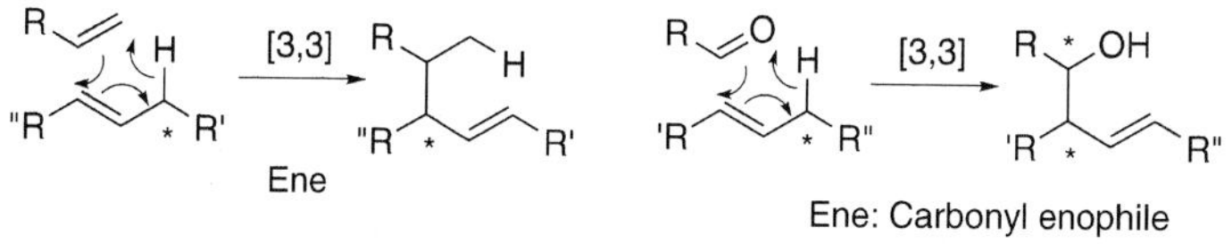

Scheme 7.61

the case of an alkene enophile, high temperatures are often required, but for carbonyl compounds the reaction can be accelerated by Lewis acids which lower the LUMO of the enophile just as in the Diels–Alder reaction. In this way, asymmetric ene reactions between alkenes and carbonyl compounds can be carried out at lower temperatures, and high selectivities can be achieved.

Relative stereocontrol in the ene reaction of carbonyl enophiles can be achieved by the type of Lewis acid promoter utilised. Treatment of methyl glyoxylate with (Z)-2-butene in the presence of tin(IV) chloride generates the *anti* isomer **7.138** as the preferred product (Scheme 7.62).[58] The *syn* isomer **7.139** is obtained almost exclusively when the Lewis acid is dimethyl aluminium chloride.

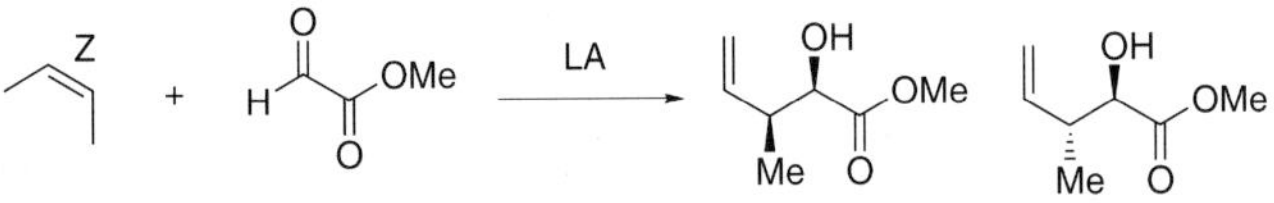

LA	syn:anti
SnCl$_4$	28:72
Me$_2$AlCl	99:1

Scheme 7.62

The tin-based Lewis acid probably binds to the glyoxylate in a *syn* fashion as shown in **7.140** and **7.141**; however, transition state **7.140** is preferred owing to the steric interaction between the SnCl$_4$ and the *cis*-methyl group in transition state **7.141**. The aluminium-based Lewis acid binds to the glyoxylate in an *anti* manner and there is a steric interaction between the methyl and aluminium in transition state **7.143**. The reaction therefore could proceed via transition state **7.142** to give the *syn* isomer.

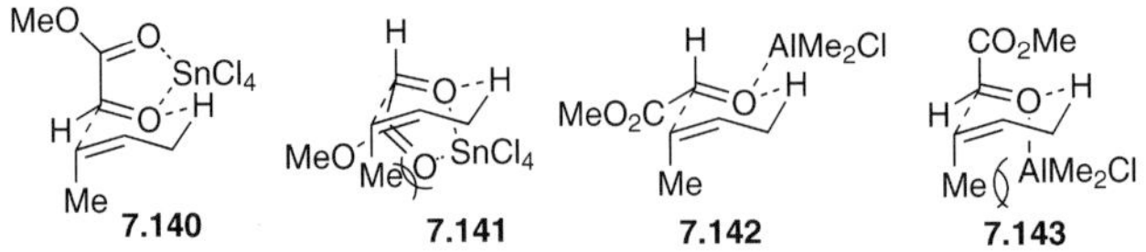

The 8-phenylmenthol chiral auxiliary has also been successfully applied in the asymmetric ene reaction of glyoxylate esters (Scheme 7.63). The ester **7.144** reacts with hexene in the presence of tin(IV) chloride to provide the hydroxy ester **7.145** in very high d.e.[59] Several alkenes have been used in this reaction and similar results were achieved.[59,60] Chiral auxiliaries were also surveyed but the best still proved to be 8-phenylmenthol.[61]

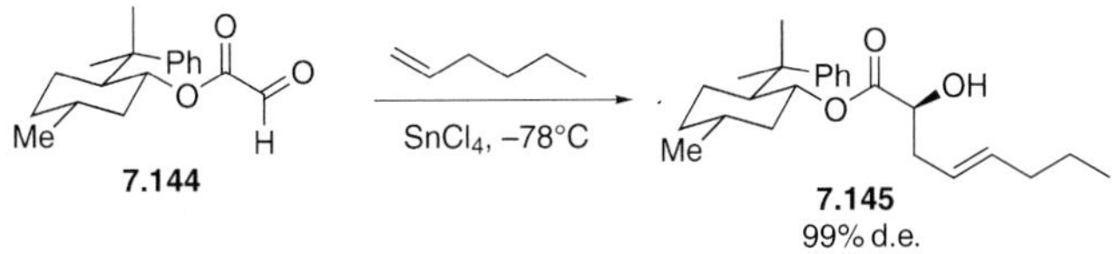

Scheme 7.63

Remote steric effects have also been useful in controlling asymmetric ene reactions. The Lewis-acid-mediated ene reaction between the chiral silyloxy ether **7.146** and methyl glyoxylate provided the 1,4-*anti*-4,5-*anti* isomer **7.147** in high d.e. (Scheme 7.64).[62] Interestingly, it was found that a 4-pentenyl silyl

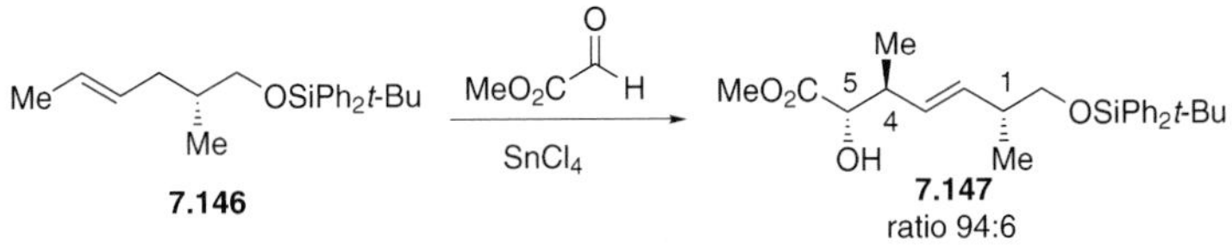

Scheme 7.64

ether was more reactive than 1-undecene. The proposed folded transition state conformers **7.148** and **7.149** anchored by the large silyloxy group can be utilised to explain the remote 1,4-stereocontrol observed. The indicated steric inter-action in **7.149** causes a preference for the reaction to proceed by transition state **7.148**.

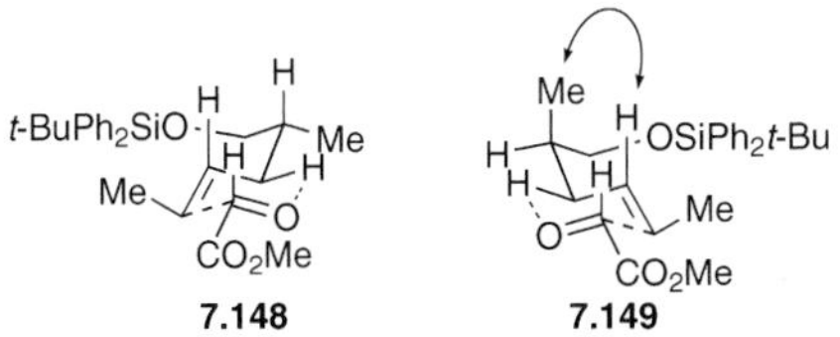

References

1. Oppolzer, W., in *Comprehensive Organic Synthesis*, Ed. C.H. Heathcock, Pergamon, Oxford, **1991**, vol. 5, p. 316.
2. Fleming, I., *Frontier Orbitals and Organic Chemical Reactions*, John Wiley, New York, **1976**.
3. Oppolzer, W., *Angew. Chem., Int. Ed. Engl.*, **1984**, *23*, 876.
4. Paquette, L.A., in *Asymmetric Synthesis*, Ed. J.D. Morrison, Academic Press, Orlando, FL, **1984**, vol. 3, pp. 455-501.
5. Choy, W., Reed, L.A. and Masamune, S., *J. Org. Chem.*, **1983**, *48*, 1137.

6. Walborsky, H.M., Barash, L. and Davis, T.C., *J. Org. Chem.*, **1961**, *26*, 4778.
7. Furuta, K., Iwanaga, K. and Yamamoto, H., *Tetrahedron Lett.*, **1986**, *27*, 4507.
8. Corey, E.J. and Ensley, H.E., *J. Am. Chem. Soc.*, **1975**, *97*, 6908.
9. Oppolzer, W., Kurth, M., Reichlin, D. and Moffatt, F., *Tetrahedron Lett.*, **1981**, *22*, 2545.
10. Oppolzer, W., Chapuis, C., Dao, G.M., Reichlin, D. and Godel, T., *Tetrahedron Lett.*, **1982**, *23*, 4781.
11. Poll, T., Sobczak, A., Hartmann, H. and Helmchen, G., *Tetrahedron Lett.*, **1985**, *26*, 3095.
12. Poll, T., Metter, J.O. and Helmchen, G., *Angew. Chem., Int. Ed. Engl.*, **1985**, *97*, 116.
13. Oppolzer, W., Chapuis, C. and Bernardinelli, G., *Helv. Chim. Acta*, **1984**, *67*, 1397.
14. Evans, D.A., Chapman, K.T. and Bisaha, J., *J. Am. Chem. Soc.*, **1988**, *110*, 1238.
15. Roush, W.R., Gillis, H.R. and Ko, A.I., *J. Am. Chem. Soc.*, **1982**, *104*, 2269.
16. Alonso, I., Carretero, J.C. and Ruano, J.L.G., *J. Org. Chem.*, **1993**, *58*, 3231.
17. Trost, B.M., O'Krongly, D. and Belleître, J.L., *J. Am. Chem. Soc.*, **1980**, *102*, 7595.
18. Siegel, C. and Thornton, E.R., *Tetrahedron Lett.*, **1988**, *29*, 5225.
19. Masamune, S., Reed, L.A., Davis, J.T. and Choy, W., *J. Org. Chem.*, **1983**, *48*, 4441.
20. Gupta, R.C., Harland, P.A. and Stoodley, R.J., *J. Chem. Soc., Chem. Commun.*, **1983**, 754.
21. Rieger, R. and Breitmaier, E., *Synthesis*, **1990**, 697.
22. Padwa, A., in *Comprehensive Organic Synthesis*, Ed. B.M. Trost, Pergamon, Oxford, **1991**, vol. 4, p. 1069.
23. Gothelf, K.V. and Jørgensen, A., *Chem. Rev.*, **1988**, *98*, 863.
24. Belzecki, C. and Panfil, I., *J. Org. Chem.*, **1979**, *44*, 1212.
25. DeShong, P., Dicken, C.M., Leginus, J.M. and Whittle, R.R., *J. Am. Chem. Soc.*, **1984**, *106*, 5598.
26. Ito, M., Maeda, M. and Kibayashi, C., *Tetrahedron Lett.*, **1992**, *33*, 3765.
27. Carruthers, W., Coggins, P. and Weston, J.B., *J. Chem. Soc., Chem. Commun.*, **1991**, 117.
28. Murahashi, S., Imada, Y., Kohno, M. and Kawakami, T., *Synlett*, **1993**, 395.
29. Deprez, P., Royer, J. and Husson, H.-P., *Tetrahedron Asymm.*, **1991**, *2*, 1189.
30. Garner, P. and Ho, W.B., *J. Org. Chem.*, **1990**, *55*, 3973.
31. Garner, P., Ho, W.B. and Shin, H., *J. Am. Chem. Soc.*, **1993**, *115*, 10742.
32. Curran, D.P., Kim, B.H., Daugherty, J. and Heffner, T.A., *Tetrahedron Lett.*, **1988**, *29*, 3555.
33. Kim, Y.H., Kim, S.H. and Park, D.H., *Tetrahedron Lett.*, **1993**, *34*, 6063.
34. Hill, R.K. and Gilman, N.W., *J. Chem. Soc., Chem. Commun.*, **1967**, 619.
35. Evans, D.A. and Golob, A.M., *J. Am. Chem. Soc.*, **1975**, 97.
36. Paquette, L.A., *Tetrahedron*, **1997**, *53*, 13971.
37. Paquette, L.A. and Maynard, G.D., *Angew. Chem., Int. Ed. Engl.*, **1991**, *30*, 1368.
38. Paquette, L.A. and Maynard, G.D., *J. Am. Chem. Soc.*, **1992**, *114*, 5018.
39. Chan, K.-K., Cohen, N., Noble, J.P.D., Specian, A.C. and Saucy, G., *J. Org. Chem.*, **1976**, *41*, 3497.
40. Denmark, S.E. and Marlin, J.E., *J. Org. Chem.*, **1987**, *52*, 5742.
41. Ireland, R.E. and Mueller, R.H., *J. Am. Chem. Soc.*, **1972**, *94*, 5897.
42. Pereira, S. and Srebnik, M., *Aldrichimica Acta*, **1993**, *26*, 17.
43. Ireland, R.E. and Willard, A.K., *Tetrahedron Lett.*, **1975**, 3975.
44. Ireland, R.E., Wipf, P. and Armstrong III, J.D., *J. Org. Chem.*, **1991**, *56*, 650.
45. Burke, S.D., Fobare, W.F. and Pacofsky, G.J., *J. Org. Chem.*, **1983**, *48*, 5221.
46. Sato, T., Tajima, K. and Fujisawa, T., *Tetrahedron Lett.*, **1983**, *24*, 729.
47. Ireland, R.E., Wipf, P. and Xiang, J.-N., *J. Org. Chem.*, **1991**, *56*, 3572.
48. Paterson, I., Hulme, A.N. and Wallace, D.J., *Tetrahedron Lett.*, **1991**, *32*, 7601.
49. Paterson, I. and Hulme, A.N., *J. Org. Chem.*, **1995**, *60*, 3288.
50. Di Florio, R. and Rizzacasa, M.A., *J. Org. Chem.*, **1998**, *63*, 8595.
51. Kallmerten, J. and Gould, T.J., *J. Org. Chem.*, **1986**, *51*, 1152.
52. Corey, E.J. and Lee, D.-H., *J. Am. Chem. Soc.*, **1991**, *113*, 4026.
53. Nakai, T. and Mikami, K., *Chem. Rev.*, **1986**, *86*, 885.

54. Sayo, N., Azuma, K., Mikami, K. and Nakai, T., *Tetrahedron Lett.*, **1984**, *25*, 565.
55. Uchikawa, M., Hanamoto, T., Katsuki, T. and Yamaguchi, M., *Tetrahedron Lett.*, **1986**, *27*, 4577.
56. Takahashi, O., Mikami, K. and Nakai, T., *Chem. Lett.*, **1987**, 66.
57. Marshall, J.A. and Lebreton, J., *J. Am. Chem. Soc.*, **1988**, *110*, 2925.
58. Mikami, K., Loh, T.-P. and Nakai, T., *Tetrahedron Lett.*, **1988**, *29*, 6305.
59. Whitesell, J.K., Bhattacharya, A., Buchanan, C.M., Chen, H.H., Deyo, D., James, D., Liu, C.-L. and Minton, M.A., *Tetrahedron*, **1986**, *42*, 2993.
60. Whitesell, J.K. and Minton, M.A., *J. Am. Chem. Soc.*, **1986**, *108*, 6802.
61. Whitesell, J.K., Lawrence, R.M. and Chem, H.-H., *J. Org. Chem.*, **1986**, *51*, 4779.
62. Shimizu, M. and Mikami, K., *J. Org. Chem.*, **1992**, *57*, 6105.

8   Reactions of alkenes

8.1   Asymmetric hydroboration

8.1.1   Background

Hydroboration of alkenes is a highly stereo- and regioselective reaction where a boron atom is delivered to the least hindered position and hydrogen to the more hindered carbon atom of the alkene in a concerted process to provide the *syn* products. Oxidation and base hydrolysis of the intermediate borane then provides the corresponding alcohol, with retention of configuration. The asymmetric version of this reaction can therefore proceed with high selectivity. Both chiral alkenes and chiral boranes can be utilised effectively (Scheme 8.1).[1]

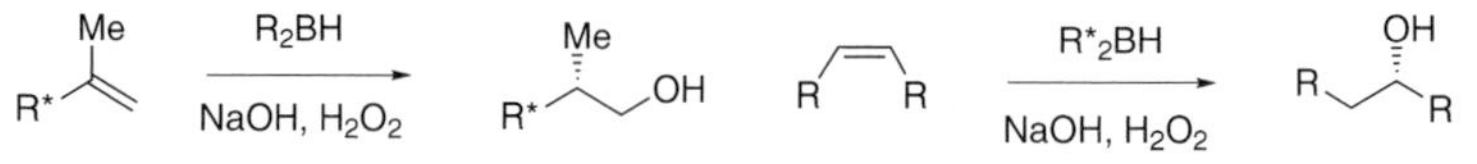

Scheme 8.1

8.1.2   Asymmetric hydroboration: chiral alkenes

Hydroboration of alkenes can be directed effectively by the pre-existing asym-metric centres in the substrate. Allylic stereocentres are able to control stereo-chemistry, and hydroboration of allylic alcohols is most effective. Still and Barrish have reported that hydroboration of the simple chiral allylic alcohol **8.1** proceeds with high stereocontrol (Scheme 8.2).[2] The selectivity of this reaction can be rationalised by considering that the alkene adopts the preferred conformation **8.2** where the hydrogen atom and double bond are *syn*-coplanar.

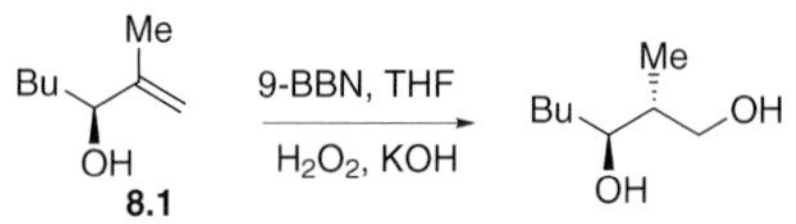

Scheme 8.2

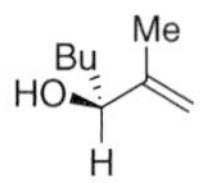

Hydroboration then occurs from the face opposite the butyl group, as shown in Figure 8.1.

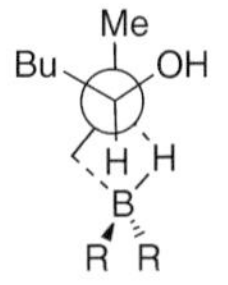

Figure 8.1

A more elaborate example is presented in Scheme 8.3. Double hydroboration of the symmetrical dienol **8.3** by using thexylborane and then borane itself to

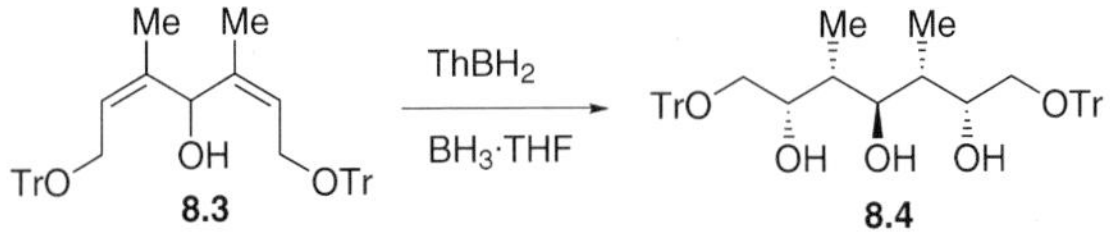

Scheme 8.3

drive the reaction to completion proceeds in a stereoselective manner giving the *meso* triol **8.4** as the major product along with undesired isomers in a 5:1 ratio.[2]

In the total synthesis of the polyether ionophore monensin, Kishi *et al.* utilised a directed hydroboration to introduce a propionate system.[3] Hydroboration of the alkene **8.5** followed by oxidation gave the alcohol **8.6** in a ratio of 8:1 (Scheme 8.4). In this case, the preferred conformer **8.7** is again the one

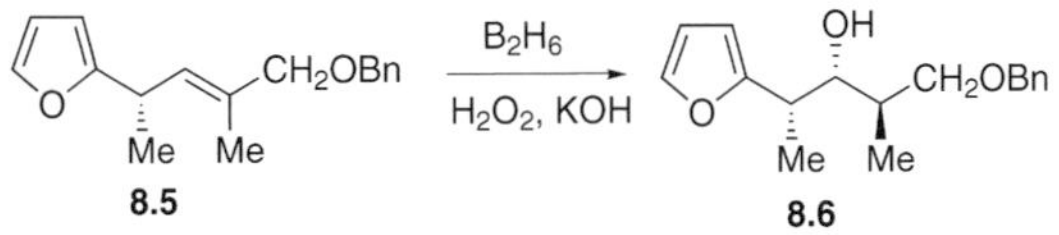

Scheme 8.4

where the hydrogen and alkene double bond are *syn*-coplanar as depicted in Figure 8.2. Borane then approaches from the face opposite the furan ring to give the major product.

In a more complex example, shown in Scheme 8.5, hydroboration of the alkene **8.8** afforded the diol **8.9** and the alternative diastereoisomer in a higher 12:1 ratio.[3] In this manner, a series of consecutive asymmetric centres were introduced stereoselectively by directed hydroboration.

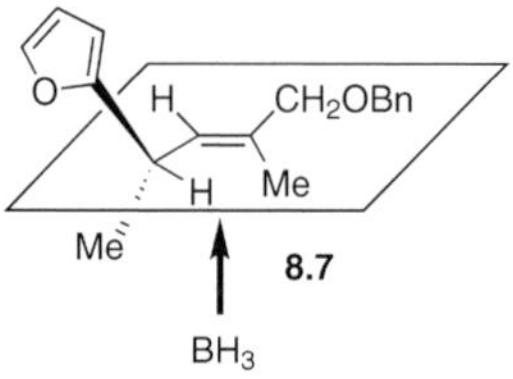

Figure 8.2

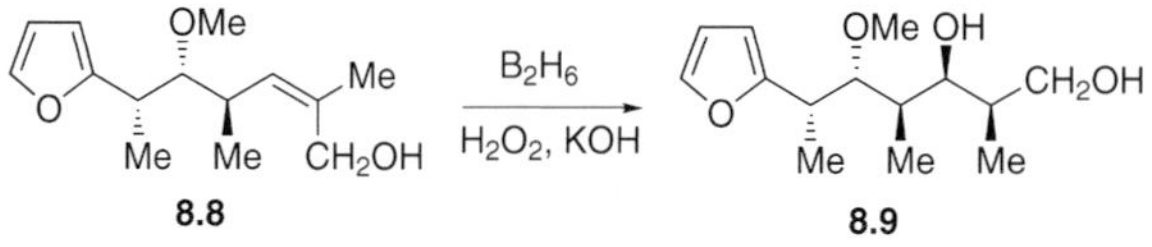

Scheme 8.5

8.1.3 Asymmetric hydroboration: chiral boranes

Chiral organoboranes are easily synthesised and transfer their chirality to the new asymmetric centre formed on hydroboration.[1] Brown and Zweifel first reported diisopinocampheylborane (Ipc$_2$BH) **8.10** as an asymmetric hydroborating reagent in 1961[4] and reported that hydroboration of (Z)-2-butene gave (R)-2-butanol in 87% e.e.[5] This was the first example of a nonenzymatic asymmetric synthesis. Later, the synthesis of Ipc$_2$BH from commercially available α-pinene was improved such that pinene of lower e.e. could be utilised. Hydroboration of (+)-α-pinene **8.11** (94.7% e.e.) with borane dimethylsulfide complex afforded Ipc$_2$BH of similar e.e. (Scheme 8.6). However, it was discovered that

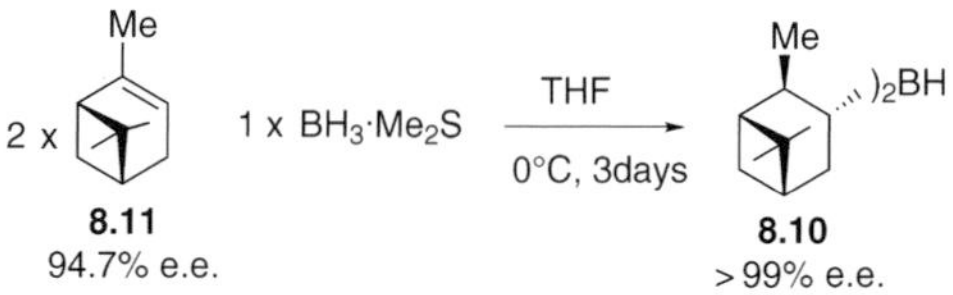

Scheme 8.6

equilibration with excess (+)-α-pinene over 3 days provided crystalline Ipc$_2$BH in 99% e.e. or better.[6,7]

Hydroboration of (Z)-2-butene with **8.10** proceeded in high e.e., and other alkenes also gave good selectivities (Scheme 8.7). Cyclic alkenes are also hydroborated in high e.e. with **8.10** as shown in Scheme 8.7 but (E)- or trisubstituted alkenes give lower selectivities. In these cases, IpcBH$_2$ (**8.12**) can be utilised to effect asymmetric hydroboration to provide products in higher e.e. (Scheme 8.8).

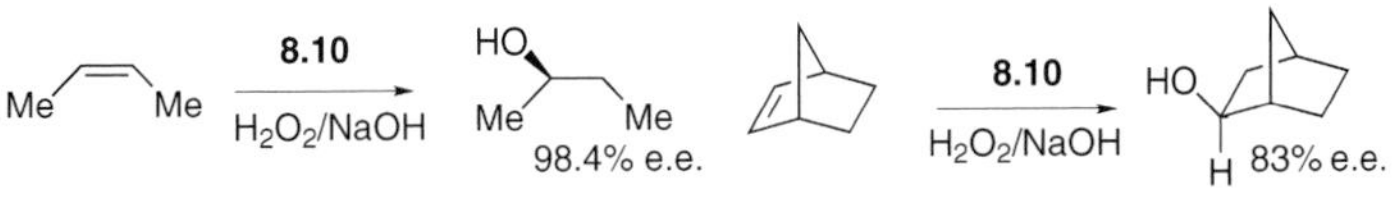

Scheme 8.7

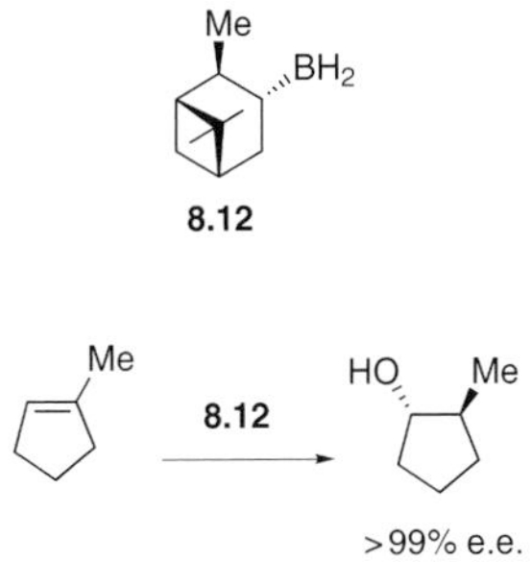

8.12

Scheme 8.8

The preparation of **8.12** by direct hydroboration of (+)-α-pinene **8.11** is difficult as the reaction cannot be controlled and proceeds rapidly to give mainly Ipc$_2$BH (**8.10**). However, **8.12** can be prepared from Ipc$_2$BH (**8.10**) by treatment with *N,N,N,N*-tetramethylenediamine (TMED) which gives the 1:2 adduct TMED · 2BH$_2$Ipc **8.13** and 1 equivalent of (+)-α-pinene (Scheme 8.9).[8] Borane

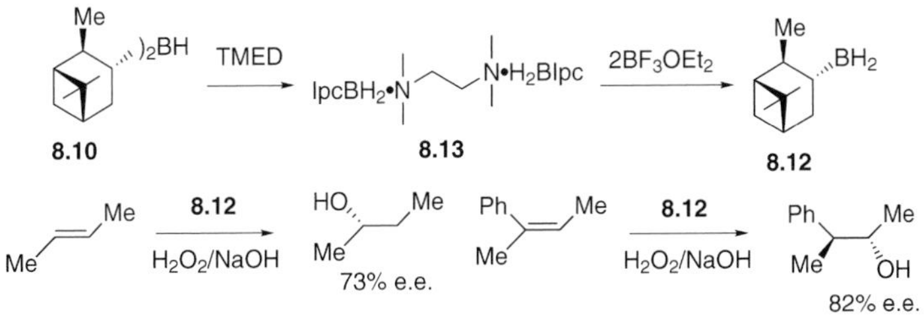

Scheme 8.9

8.12 is then liberated by exposure of the complex to BF$_3$ · OEt$_2$. Asymmetric hydroboration–oxidation of (*E*)-2-butene and trisubstituted alkenes with borane **8.12** then gives alcohols in high e.e. as shown by the example in Scheme 8.9.[9]

The chiral borane **8.10** can also be utilised in double asymmetric synthesis. Masamune *et al.* found that hydroboration of alkene **8.14** with 9-BBN gave a 2:1 mixture of the desired alcohol **8.15** and the other possible diastereoisomer (Scheme 8.10).[10] However, hydroboration with Ipc$_2$BH (**8.10**) gave compound **8.15** in greater than 50:1 selectivity.

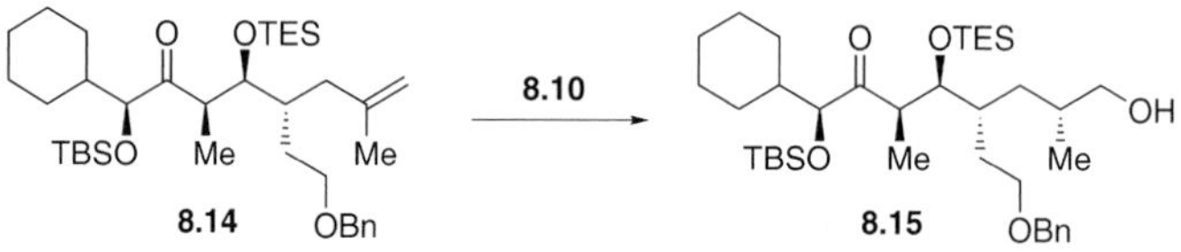

Scheme 8.10

An alternative to chiral boranes derived from pinene has been reported by Masamune *et al.*[11] The chiral C_2 symmetric borolane **8.16** hydroborates both *E* and *Z* alkenes in high e.e. (Scheme 8.11). However, the synthesis of **8.16**,

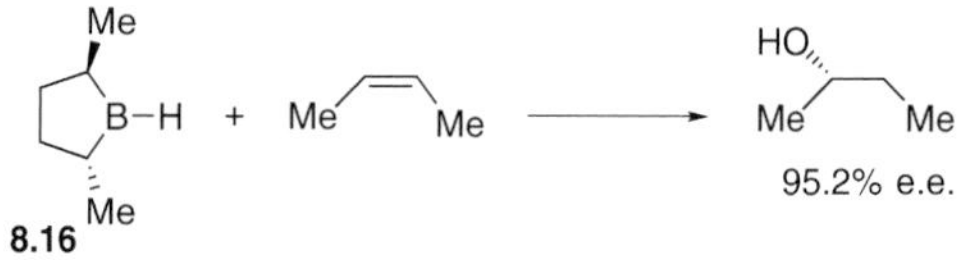

Scheme 8.11

which involves a resolution step of a racemic borolane precursor, is more difficult and costly than that of the pinene derived borane **8.10**. Furthermore, the chiral auxiliary is not recovered after the reaction.

8.2   Asymmetric conjugate addition

8.2.1   Background

Conjugate (Michael) addition of organometallic reagents to α,β-unsaturated carbonyl compounds and their nitrogen counterparts is a useful method for the formation of a carbon–carbon bond in an asymmetric manner (Scheme 8.12).[12,13] Typically, α,β-unsaturated carboxylic acid derivatives are utilised where the chiral auxiliary is attached to the carbonyl group. Another version of asymmetric conjugate addition to α,β-unsaturated carbonyl compounds is where the chirality is attached to one of the alkene carbons. Nitrogen analogues, such

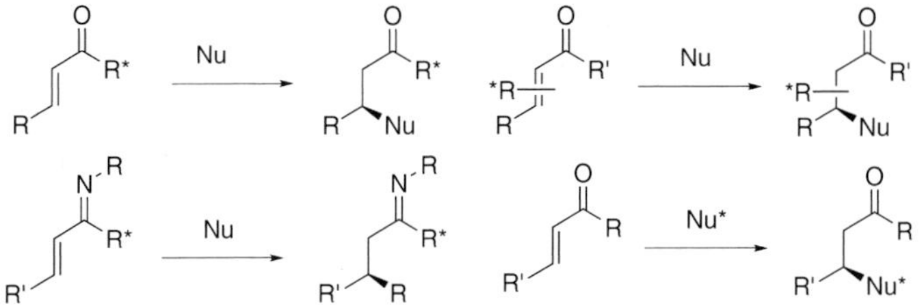

Scheme 8.12

as α,β-unsaturated chiral aldimines and oxazolines, also undergo asymmetric Michael addition with useful selectivities.

8.2.2 *Chiral esters*

High asymmetric induction in conjugate additions can be achieved by use of chiral ester derivatives. Oppolzer and co-workers reported that chiral enoate esters derived from (−)-8-phenylmenthol react with copper-based organometallic reagents in a highly diastereoselective manner.[14,15]

BF$_3$-mediated 1,4-conjugate addition of butyl copper to the chiral enoate **8.17** provided the product **8.18** in high d.e. and yield (Scheme 8.13).[14] It

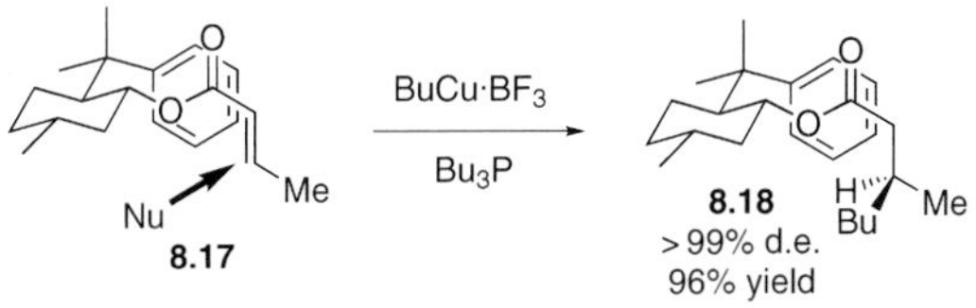

Scheme 8.13

was proposed that addition occurs from the face of the enoate as shown in Scheme 8.13. Lewis acid complexation causes the C=C/C=O of the enoate to be orientated antiperiplanar and the phenyl moiety effectively shields one enantiotopic face. The conformational rigidity as well as the π,π-orbital overlap is enhanced by coordination of the carbonyl group with the Lewis acid.

Camphor-derived esters are also effective substrates for asymmetric conjugate addition. Addition of methyl copper to the enoate **8.19** derived from (−)-camphor proceeded in high d.e. to provide ester **8.20** in good yield (Scheme 8.14).[15] Hydrolysis then gave (*S*)-(−)-citronellic acid **8.21** along with the regenerated chiral auxiliary. It was found that the instability and reactivity of methyl copper could be attenuated by the addition of PBu$_3$, and high yields were possible without the use of a large excess of this reagent. The C=C/C=O

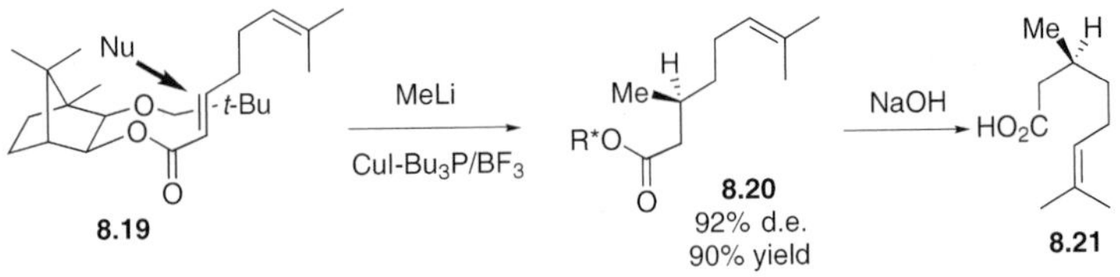

Scheme 8.14

arrangement is again antiperiplanar and the nucleophile approaches from the less-hindered face as shown.

8.2.3 Chiral amides

In addition to chiral esters, chiral amide derivatives of α,β-unsaturated acids also undergo conjugate addition in a stereoselective manner. An early example was reported by Mukaiyama and Iwasawa and involved the use of a chiral amide derived from L-ephedrine.[16] Treatment of the chiral enamide **8.22** with a variety of Grignard reagents in ether solvent provided a chiral saturated amide **8.23** which upon acid hydrolysis gave a number of β-substituted alkanoic acids **8.24**, each in good e.e. (Scheme 8.15). The highly selective conjugate addition

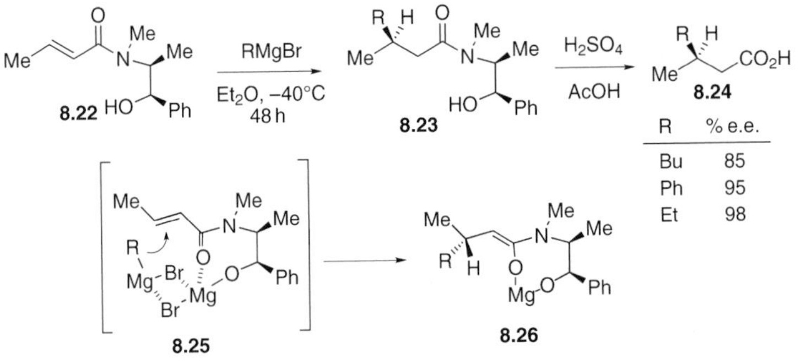

Scheme 8.15

step probably proceeds via a highly organised chelated intermediate such as **8.25** where the nucleophile is delivered to the β-carbon on the face opposite to the phenyl and methyl groups of the auxiliary. The resultant stable magnesium enolate **8.26** resists further side reactions which are sometimes observed in the conjugate addition of Grignard reagents to α,β-unsaturated amides.

Camphor-derived N-enoylsultams, which are highly selective reagents in asymmetric Diels–Alder reactions, are also useful as substrates for asymmetric conjugate additions. Oppolzer et al. reported that the β-sily enoyl sultam **8.27** undergoes conjugate addition with copper-based nucleophiles in the presence of a Lewis acid to provide a number of adducts in high d.e. (Scheme 8.16).[17] The silyl group can then be transformed into a hydroxy group with retention

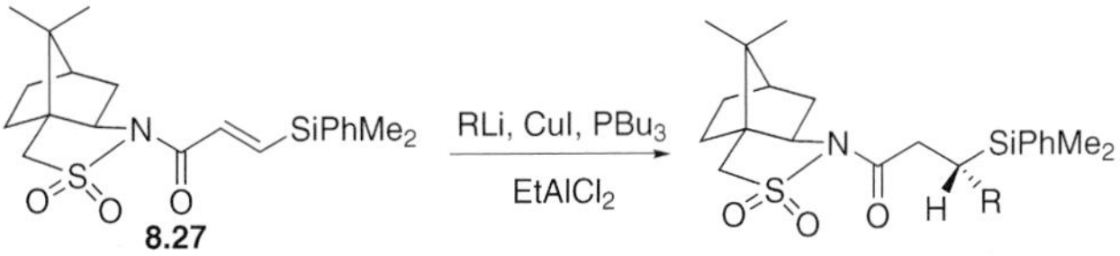

R	% d.e.
Vinyl	90
Me	86
Et	86
Ph	94.8

Scheme 8.16

of configuration. Interestingly when BF_3 was utilised as the Lewis acid, products resulting from an opposite mode of attack of the alkene were obtained.

The selectivities observed for this reaction can be attributed to the mode of coordination of the particular Lewis acid utilised. The aluminium-based Lewis acid coordinates to the carbonyl and sulfonyl oxygens to give a complex such as **8.28** which causes the nucleophile to attack the *Re* face of the enamide (Figure 8.3). The BF_3 Lewis acid coordinates to the carbonyl oxygen only and

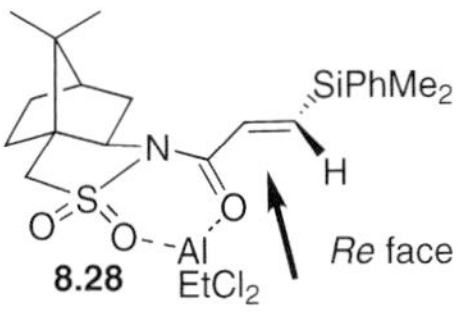

Figure 8.3

causes the SO_2 and C=O groups to be *anti* disposed and the nucleophile then approaches from the *Si* face.

The enolate derived from asymmetric addition of a Grignard reagent to a chiral *N*-enoylsultam can react with various electrophiles to introduce a second stereogenic centre with a high degree of control. Addition of propylmagnesium chloride to the sultam **8.29** followed by protonation of the derived enolate at low temperature provided the *anti* product **8.30** in high d.e. (Scheme 8.17).[18] The initial conjugate addition proceeds in a stereocontrolled manner possibly via the chelated intermediate **8.31**. Protonation of the resultant (*Z*)-magnesium enolate **8.32** then occurs stereoselectively. It was suggested that the magnesium

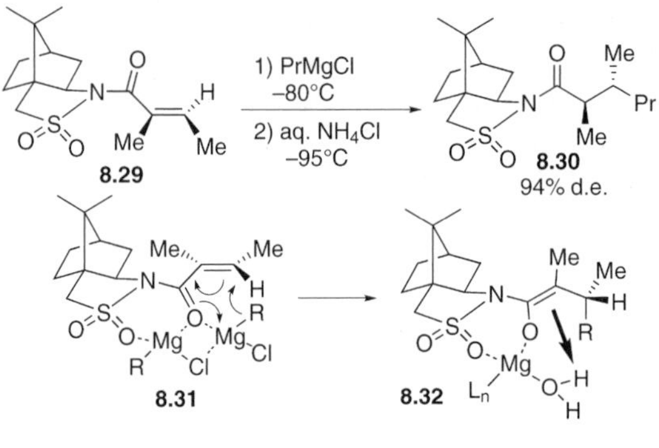

Scheme 8.17

may also be associated with an H_2O molecule which delivers the proton from the *Si* face.

8.2.4   Chiral ketones

One of the most effective methods for the synthesis of optically enriched 3-substituted cycloalkenones has been developed by Posner and co-workers.[19,20] The protocol involves the use of 2-*p*-(anisylsufinyl)-2-cycloalkenones which are readily synthesised from the corresponding 2-bromocycloalkenones. Cuprates sometimes react sluggishly but Grignard reagents in the presence of chelating agents add in a stereoselective conjugate fashion effectively to provide chiral 3-substituted cycloalkanones after reductive removal of the chiral sulfoxide auxiliary.

Treatment of the enantiomerically pure sulfoxide **8.33** with the Grignard reagent **8.34** in the presence of $ZnBr_2$ followed by reductive removal of the sufinyl group provided cyclopentanone **8.35** in good e.e. (Scheme 8.18).[19]

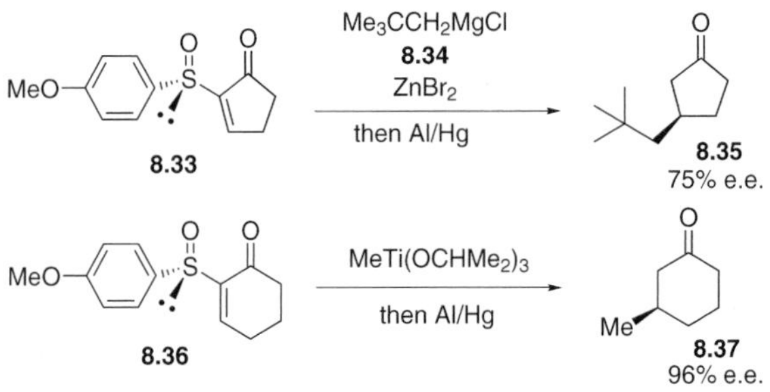

Scheme 8.18

Almost complete asymmetric induction was observed during conjugate addition of methyltitanium triisopropoxide to cyclohexenone **8.36**. (*R*)-Cyclohexanone

8.37 was obtained after desulfurisation in nearly 100% e.e. Additions of Grignard reagents to **8.36** were not as stereoselective.

Normally 2-(arylsufinyl)-2-cyclohexenones exist in a conformation where the sufinyl sulfur–oxygen bond dipole is *anti* to the carbonyl C=O bond as shown in conformer **8.38** (Figure 8.4). Addition of a nucleophile to the β-carbon

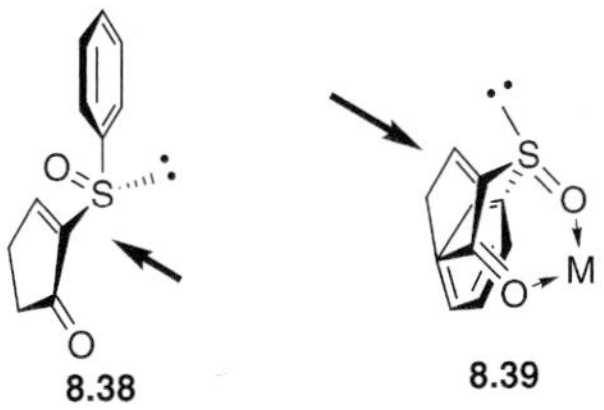

Figure 8.4

atom then occurs from the direction indicated by the arrow. However, upon addition of a metal salt, chelation of the sufinyl and carbonyl oxygens occurs to give the complex **8.39**. Conjugate addition then occurs from the opposite face as shown, to give the alkanone product in high d.e. The one problem with this method is the fact that the chiral auxiliary is destroyed upon its removal.

Recently, Funk and Yang have reported that menthol derivatives of 2-cyclohexenones give similar results to the corresponding 2-(arylsufinyl)-2-cyclohexenones with cleavage/recovery of the chiral auxiliary in a one-pot reaction.[21] Addition of a higher order phenyl cuprate to chiral cyclohexenone **8.40** gave the cyclohexanone **8.41** in excellent e.e. along with the auxiliary-derived menthone **8.42** (Scheme 8.19). The mechanism of this reaction

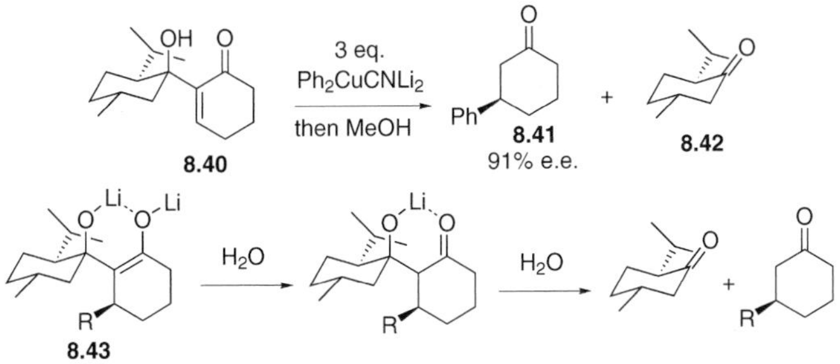

Scheme 8.19

apparently involves sequential protonation of the dianion **8.43**, which results from treatment of **8.40** with excess nucleophile, followed by a facile retro-aldol reaction to give the ketone products.

8.2.5   Chiral imines

Chiral α,β-unsaturated imines and their analogues are also useful in asymmetric Michael additions. For example, chiral aldimines can be derived from the corresponding amines, and the auxiliary is easily removed after conjugate addition by hydrolysis to yield scalemic β-substituted aldehydes. Koga *et al.* reported that condensation of an optically active amino acid ester with an α,β-unsaturated aldehyde gives a chiral imine that undergoes effective asymmetric conjugate addition.[22] Treatment of (*E*)-crotonaldehyde with **8.44** gave imine **8.45** which reacted with butylmagnesium bromide to give aldehyde **8.46** after hydrolysis of the imine (Scheme 8.20).

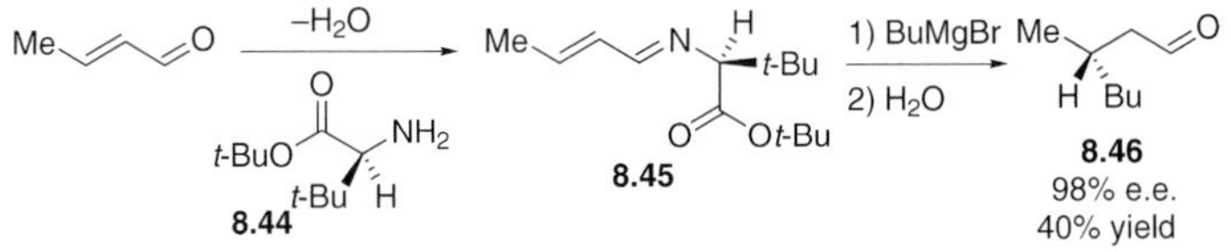

Scheme 8.20

The stereoselectivity in this reaction might arise from the initial cyclic complex **8.47**. Addition of the chelated Grignard reagent to the β-carbon occurs from the less-hindered side of the α,β-unsaturated aldimine (which is in the *s*-*cis* conformation) as shown in Scheme 8.21. Protonation of the resultant magnesium amide then gave the β-substituted aldehyde product.

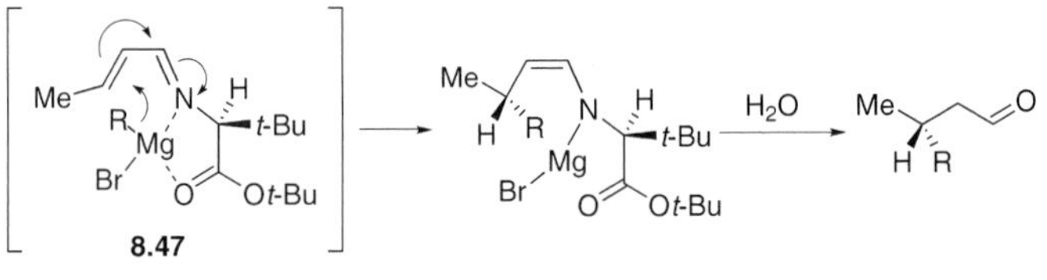

Scheme 8.21

8.2.6   Chiral oxazolines

In addition to simple imines, chiral oxazolines are also effective auxiliaries for diastereoselective conjugate addition. Optically active oxazolines are easily synthesised from a carboxylic acid and chiral amino alcohol derived from an amino acid. Meyers *et al.* first reported that conjugate addition to α,β-unsaturated chiral oxazolines proceeds with excellent stereocontrol.[23] Addition of butyllithium to the oxazoline **8.48** followed by acid hydrolysis of the oxazoline moiety provided the β-substituted acid **8.49** in excellent e.e.

(Scheme 8.22).[24] It was proposed that the mechanism of addition involves a cyclic chelated intermediate and internal delivery of the nucleophile to the

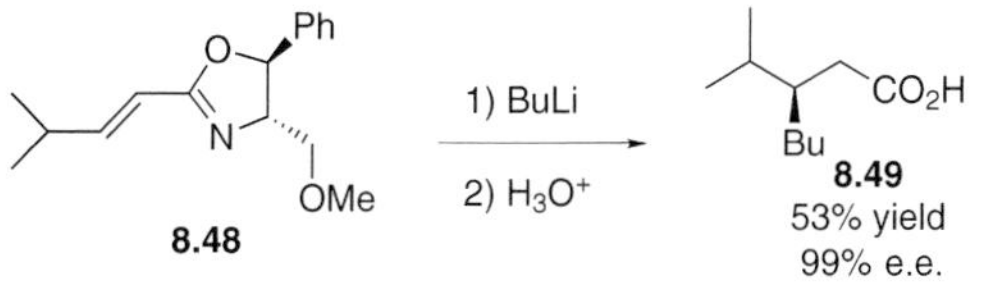

Scheme 8.22

β-carbon occurs in a stereoselective manner to give an intermediate aza-enolate (Scheme 8.23).[24]

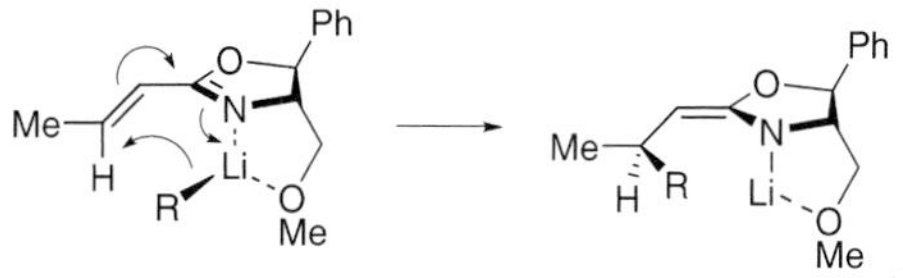

Scheme 8.23

Chiral naphthyloxazolines also undergo asymmetric conjugate addition to afford optically active dihydronaphthylenes.[25,26] An example of this interesting reaction is presented in Scheme 8.24. Conjugate addition to the naphthyloxazoline **8.50** followed by subsequent alkylation of the resulting intermediate

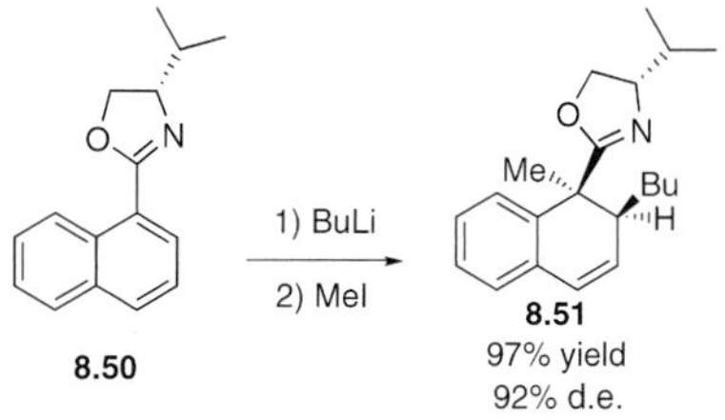

Scheme 8.24

aza-enolate gives the oxazoline **8.51** in high yield and d.e. The selectivity can be explained by the sequence shown in Scheme 8.25. Initial addition of the complexed organolithium to the naphthyloxazoline occurs from the β-face opposite to the oxazoline substituent to provide intermediate **8.52** which then reacts with the electrophile from the more accessible α-face.[26]

Meyers and Lutomski have also reported a synthesis of chiral biaryls which involves nucleophilic aromatic displacement of an *o*-methoxy group activated

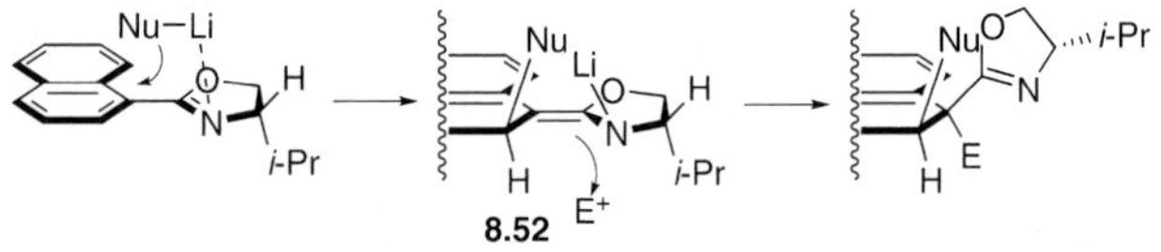

Scheme 8.25

by a chiral oxazoline. This reaction initially involves an asymmetric conjugate addition reaction, similar to that described above, to give an aza-enolate intermediate which upon elimination of methoxide affords a biaryl. For example, treatment of the chiral oxazoline **8.53** with the Grignard reagent **8.54** provides the biaryl **8.55** in reasonable d.e. (Scheme 8.26).[27]

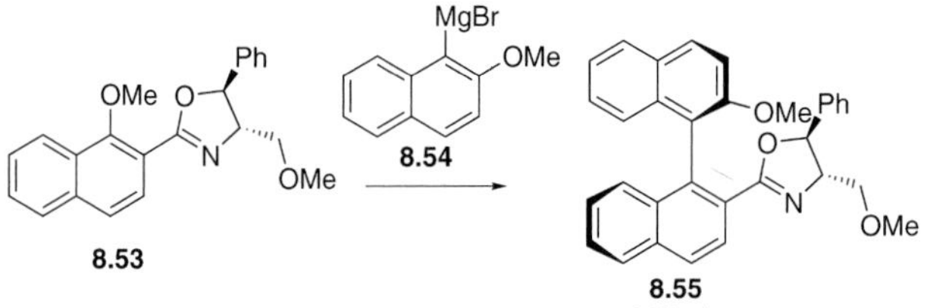

Scheme 8.26

The proposed mechanism involves initial asymmetric conjugate addition of the naphthyl Grignard reagent to give an intermediate which can exist as two conformers **8.56A** and **8.56B** (Scheme 8.27). There is a preference for

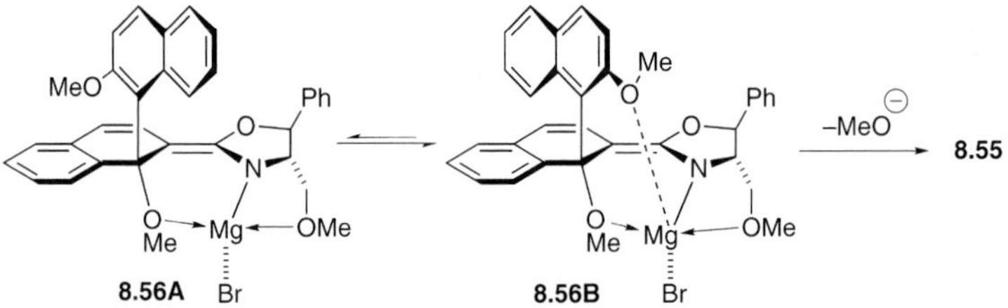

Scheme 8.27

conformer **8.56B** as the methoxy group on the Grignard reagent is able to chelate to the magnesium, and collapse of this rotomer would then give biaryl **8.55** as the major diastereoisomer.

8.2.7   Chiral nucleophiles: enolates

Addition of chiral nucleophiles to achiral α,β-unsaturated carbonyl compounds usually involves the use of chiral enolates as the source of asymmetric induction.

Heathcock and Oare have studied the relative diastereoselection of the conjugate addition of enolates to α,β-unsaturated ketones and have showed that the outcome is dependent on the enolate geometry. (Z)-O-enolates give the *anti* products whereas (E)-O-enolates provide the *syn* isomers (Scheme 8.28).[28] This

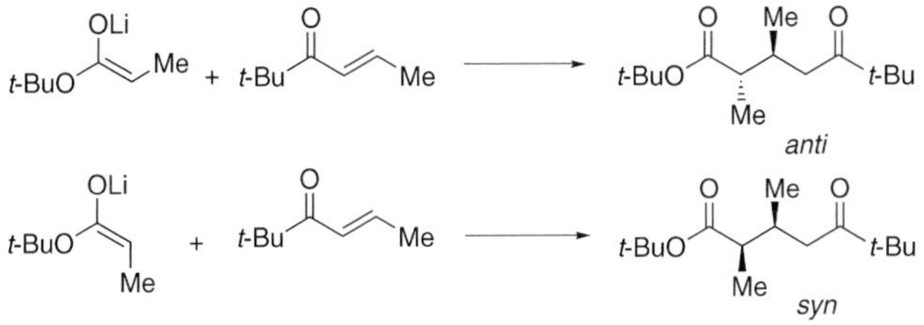

Scheme 8.28

simple diastereoselection probably involves cyclic synclinal transition states rather than the antiperiplanar transition states depicted in Scheme 8.29 and this

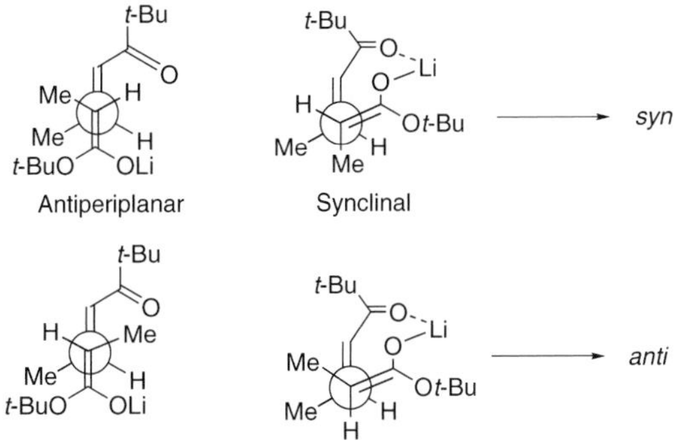

Scheme 8.29

proposal is supported by studies on Michael additions to nitroolefins; however, both transition states are possible.[29]

Corey and Peterson have reported the stereoselective addition of enolates of phenylmenthol esters to crotonates.[30] Good diastereoselectivities were observed and the relative stereochemistry of the products were again dependent on enolate geometry. Exposure of methyl crotonate to the (E)-O-enolate **8.57** gave the *syn* product **8.58** along with the *anti* product in a 9:1 ratio (Scheme 8.30). The alternative *syn* diastereoisomer was also obtained and the ratio of these isomers was 95:5. The outcome for this reaction was rationalised by considering the synclinal closed transition state **8.59** where the lithium ion is bound to the ester carbonyl oxygen and enolate oxygen.

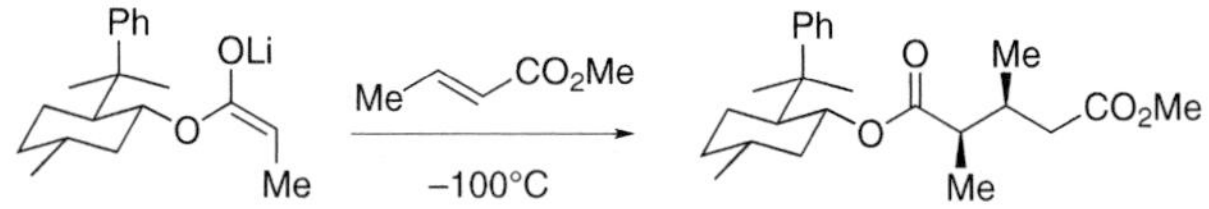

Scheme 8.30

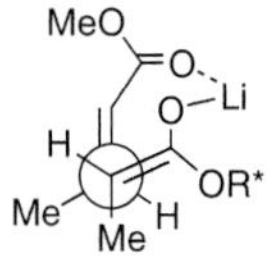

8.59

Chlorotitanium enolates derived from chiral *N*-acyloxazolidinones add to various electrophilic olefins in a diastereoselective manner to provide Michael adducts in high d.e. Addition of the chiral titanium enolate derived from oxazolidinone **8.60** to methyl acrylate gives the imide **8.61** in high yield and d.e. (Scheme 8.31).[31] The scope of these reactions, however, does not extend to β-substituted α,β-unsaturated esters or nitriles.

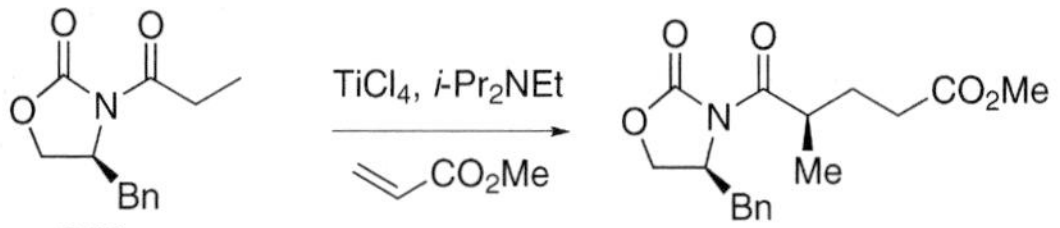

Scheme 8.31

8.2.8   Chiral nucleophiles: cuprates

A large number of examples of addition of chiral nucleophiles to prochiral α,β-unsaturated carbonyl compounds have been reported and most involve the use of a chiral mixed cuprate derived from an amino alcohol or amine. Corey *et al.* reported that the chiral organo(alkoxo)cuprate (derived from ephedrine) provides high selectivities for conjugate additions to cyclohexenone (Scheme 8.32).[32] The reagent is formed by treatment of the diamino alcohol **8.62** with an alkyllithium and copper iodide. When old solutions of alkyllithium were utilised it was found that small amounts of methyl iodide had to be added as low enantioselectivities were observed presumably because of small amounts of alkoxide impurities present.

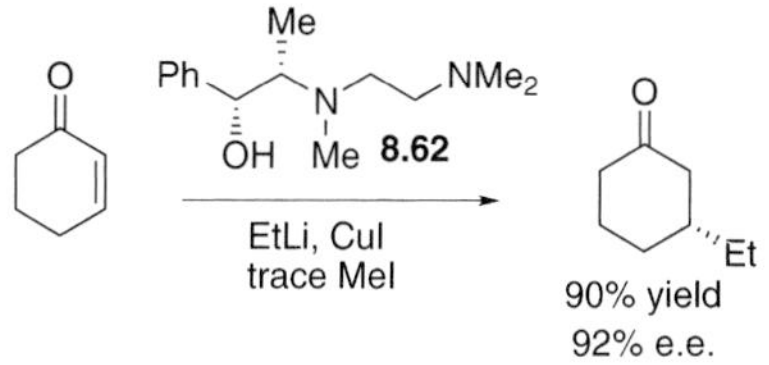

Scheme 8.32

The results using the ligand **8.62** can be understood in terms of the model shown in Scheme 8.33. Association of the tridentate lithium cuprate–aminoalcohol complex with the enone gives a complex such as **8.63** where the

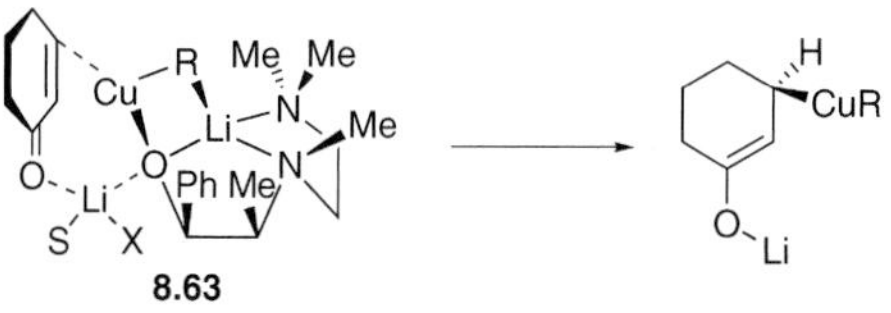

Scheme 8.33

nucleophilic copper species interacts with the *Re* face of C(3) of the 2-cyclohexenone and delivers the copper from that side. Electrophilic lithium is important as a second lithium coordinates with the carbonyl oxygen and alkoxy group of the ligand to form the complex shown. The alternative pathway involving attack of the copper on the *Si* face of C(3) is less favourable because of steric interactions with the coordinated ligand. This model also explains why alkoxide impurities reduce the enantioselectivity.

The final example of a chiral cuprate is that derived from the ligand **8.64** and lithium dimethylcuprate. The resultant enantioselective reagent added to cyclopentadec-2-enone **8.65** to provide (*R*)-(−)-muscone **8.66** in high yield and 100% e.e. (Scheme 8.34).[33]

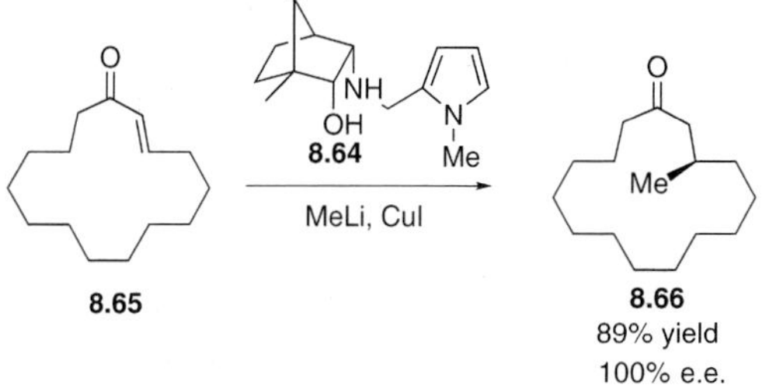

Scheme 8.34

References

1. Brown, H.C. and Singaram, B., *Acc. Chem. Res.*, **1988**, *21*, 287.
2. Still, W.C. and Barrish, J.C., *J. Am. Chem. Soc.*, **1983**, *105*, 2487.
3. Schmid, G., Fukuyama, T., Akasaka, K. and Kishi, Y., *J. Am. Chem. Soc.*, **1979**, *101*, 259.
4. Brown, H.C. and Zweifel, G., *J. Am. Chem. Soc.*, **1961**, *83*, 486.
5. Brown, H.C. and Zweifel, G., *J. Am. Chem. Soc.*, **1964**, *86*, 393.
6. Brown, H.C., Jadhav, P.K. and Desai, M.C., *J. Org. Chem.*, **1982**, *47*, 5065.
7. Brown, H.C. and Singaram, B., *J. Org. Chem.*, **1984**, *49*, 945.
8. Brown, H.C., Schwier, J.R. and Singaram, B., *J. Org. Chem.*, **1978**, *43*, 4395.
9. Brown, H.C., Jadhav, P.K. and Mandal, A.K., *J. Org. Chem.*, **1982**, *47*, 5074.
10. Masamune, S., Lu, L.D.-L., Jackson, W.P., Kaiho, T. and Toyoda, T., *J. Am. Chem. Soc.*, **1982**, *104*, 5523.
11. Masamune, S., Kim, B.M., Peterson, J.S., Sato, T., Veenstra, S.J. and Imai, T., *J. Am. Chem. Soc.*, **1985**, *107*, 4549.
12. Schmalz, H.-G., in *Comprehensive Organic Synthesis*, Ed. C.H. Heathcock, Pergamon, Oxford, **1991**, vol. 2, p. 199.
13. Rossiter, B.E. and Swingle, N.M., *Chem. Rev.*, **1992**, *92*, 771.
14. Oppolzer, W. and Löher, H.J., *Helv. Chim. Acta*, **1981**, *64*, 2808.
15. Oppolzer, W., Moretti, R., Godel, T., Meunier, A. and Löher, H., *Tetrahedron Lett.*, **1983**, *24*, 4971.
16. Mukaiyama, T. and Iwasawa, N., *Chem. Lett.*, **1981**, 913.
17. Oppolzer, W., Mills, R.J., Pachinger, W. and Stevenson, T., *Helv. Chim. Acta*, **1986**, *69*, 1542.
18. Oppolzer, W., Poli, G., Kingma, A.J., Strkemann, C. and Bernardinelli, G., *Helv. Chim. Acta*, **1987**, *70*, 2201.
19. Posner, G.H., Frye, L.L. and Hulce, M., *Tetrahedron*, **1984**, *40*, 1401.
20. Posner, G.H., *Acc. Chem. Res.*, **1987**, *20*, 72.
21. Funk, R.L. and Yang, G., *Tetrahedron Lett.*, **1999**, *40*, 1073.
22. Hashimoto, S.-I., Yamada, S.-I. and Koga, K., *J. Chem. Soc.*, **1976**, *98*, 7450.
23. Meyers, A.I. and Whitten, C.E., *J. Am. Chem. Soc.*, **1975**, *97*, 6266.
24. Meyers, A.I., Smith, R.K. and Whitten, C.E., *J. Org. Chem.*, **1979**, *44*, 2250.
25. Meyers, A.I. and Barner, B.A., *J. Org. Chem.*, **1986**, *51*, 120.
26. Rawson, D.J. and Meyers, A.I., *J. Org. Chem.*, **1991**, *56*, 2292.
27. Meyers, A.I. and Lutomski, K.A., *J. Am. Chem. Soc.*, **1982**, *104*, 879.
28. Heathcock, C.H. and Oare, D.A., *J. Org. Chem.*, **1985**, *50*, 3022.
29. Seebach, D. and Golinski, J., *Helv. Chim. Acta*, **1981**, *64*, 1413.
30. Corey, E.J. and Peterson, R.T., *Tetrahedron Lett.*, **1985**, *26*, 5025.
31. Evans, D.A., Bilodeau, M.T., Somers, T.C., Clardy, J., Cherry, D. and Kato, Y., *J. Org. Chem.*, **1991**, *56*, 5750.
32. Corey, E.J., Naef, R. and Hannon, F.J., *J. Am. Chem. Soc.*, **1986**, *108*, 7114.
33. Tanaka, K. and Suzuki, H., *J. Chem. Soc., Chem. Commun.*, **1991**, 101.

9 Free radical processes

9.1 Introduction and background

When one considers a potential strategy for the stereoselective synthesis of a target molecule, the reputed lack of selectivity observed in some free radical reactions has often meant that this whole class of reactions has been overlooked. However, with the modern methods available for the directed formation of radicals, and a greater understanding of the factors influencing selectivity in the reactions of these radicals, the number of applications of radicals in stereoselective synthesis is increasing. That is not to say that radical reactions are the choice for all situations, but the situations in which the high reactivity of radicals can best be exploited are becoming clear. For example, the formation of highly congested tertiary centres via radical reactions is proving invaluable.

A number of books[1-3] and reviews[4-11] have been published on modern radical chemistry, discussing this everincreasing field of research. This chapter will endeavour to give an overview of the factors effecting selectivity in radical reactions and it will highlight some examples where high levels of stereoselectivity have been achieved in the formation of carbon–carbon bonds.

9.1.1 Radical chain reactions

The majority of radical reactions used in synthesis are chain processes, and one must consider the initiation, propagation and termination steps for carbon–carbon bond-forming processes.

Radical reactions are generally initiated by thermal or photochemical means. In this process, a weak bond in a suitable initiating molecule is cleaved either thermally or photochemically, giving a radical which can begin the chain reaction. Radical initiators which are often used include di-*t*-butyl peroxide, dibenzoyl peroxide and azobisisobutyronitrile (AIBN) (Scheme 9.1).

The key to a successful radical reaction in synthesis is a highly efficient propagating chain reaction. In modern radical chemistry the use of efficient radical chain carriers has facilitated this, and one of the most used reagents for this has proved to be tributyltin hydride. For example, the radical addition of cyclohexane to acrylonitrile can be carried out successfully by using a thermally initiated (via AIBN) radical chain reaction exploiting tributyltin hydride as the hydrogen atom source and radical chain carrier (Scheme 9.2).

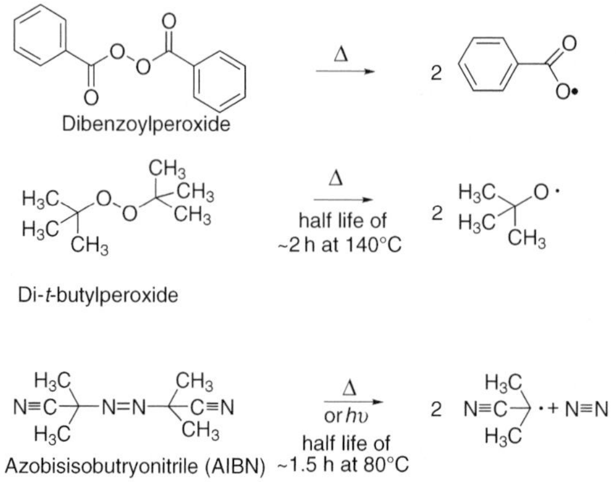
Δ
2
Dibenzoylperoxide
Δ
half life of
~2 h at 140°C
2
Di-t-butylperoxide
Δ
or hυ
half life of
~1.5 h at 80°C
2
Azobisisobutryonitrile (AIBN)

Scheme 9.1

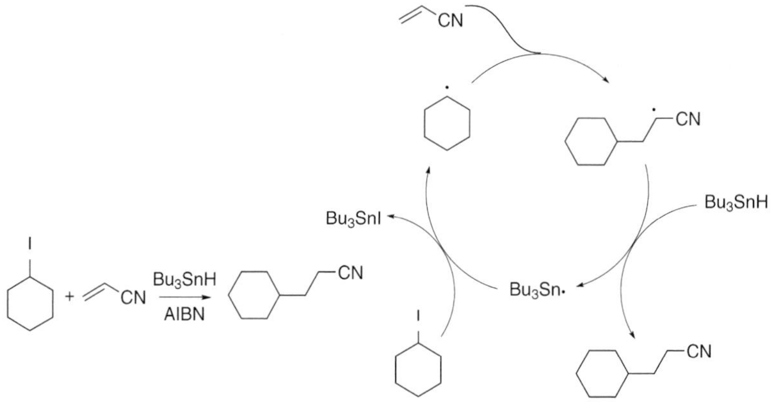
Bu3SnH
AIBN
Bu3SnI
Bu3Sn·
Bu3SnH

Scheme 9.2

9.2 Factors affecting the selectivity of radical reactions

9.2.1 Introduction

Radical reactions are generally regarded as showing high chemoselectivity and regioselectivity but rather less stereoselectivity and enantioselectivity. This is to some extent true, but examples of high stereoselectivity and enantioselectivity are increasing. The factors affecting the selectivity are the thermochemistry, steric effects, stereoelectronic effects and polar effects.

The underlying selectivities in radical reactions are governed by the thermochemistry. This is based on the assumption that activation energies reflect activation enthalpies, that is, endothermic reactions will have high activation energies and the more exothermic a reaction the lower its activation energy. This leads to the conclusion that the relative rates of radical reactions can be determined from bond dissociation energies.

Although the basic assumption that the thermodynamics of the process will always be the underlying influence on the rates of radical reactions, other factors may contribute to or even dominate the rates of radical reactions. The effects which must be considered are the steric effects, stereoelectronic effects and polar effects. The steric effects are determined by consideration of the contribution of van der Waals repulsions (and attractions) to transition structure energies. The stereoelectronic effect is based on the assumption that the energy of the transition structure depends on the extent of overlap of frontier molecular orbitals. The consideration of polar effects acknowledges the possibility that radical transition structures may be dipolar and reflect the relative electronegativities of constituent atoms.

9.2.2   Thermodynamic control of radical reactions

An example of thermodynamic control is seen in the simple free radical halogenation of 2-methylpropane, where it is observed experimentally that chlorination gives a mixture of the primary and tertiary halides, whereas bromination gives almost exclusively the tertiary halide (Scheme 9.3).[1]

$$
\underset{\overset{|}{CH_3}}{\overset{\overset{CH_3}{|}}{H_3C-C-H}} \quad \xrightarrow[h\upsilon]{X_2} \quad \underset{\overset{|}{\underset{X}{CH_2}}}{\overset{\overset{CH_3}{|}}{H_3C-C-H}} \qquad \underset{\overset{|}{CH_3}}{\overset{\overset{CH_3}{|}}{H_3C-C-X}}
$$

X = Cl	ca 40%	ca 60%
X = Br	< 1%	> 99%

Scheme 9.3

When one takes into account the statistical weight of the methyl hydrogens the selectivity of the bromination is significant. This selectivity can be explained by considering that the selection is occurring in the hydrogen atom abstraction by the halogen radical and in the difference in the energy of the bond being formed. For chlorination the bond being formed in this reaction (H—Cl) has a homolytic bond dissociation energy of 103 kcal/mol whereas the bond being broken has a bond energy of 101 kcal/mol, making the abstraction slightly exothermic. Abstraction of the tertiary hydrogen (bond dissociation energy 95 kcal/mol) by the chlorine radical is 6 kcal/mol more exothermic. For both these processes one can invoke the Hammond postulate and assume the transition states will be *reactant*-like and thus not very different in energy (perhaps 1 kcal/mol).

Thus one would not expect chlorination to be very selective. On the other hand, when the bromine radical abstracts the methyl hydrogen the bond formed (H—Br) is significantly weaker, at 87.5 kcal/mol, making the abstraction rather endothermic. Abstraction of the tertiary hydrogen by the bromine radical is 6 kcal/mol less in energy but still rather endothermic. Again, by invoking the Hammond postulate we can say that these endothermic processes will have product-like transition states and thus be rather different in energy. This significant difference in energy results in the selectivity observed for the bromination reaction (Figure 9.1).

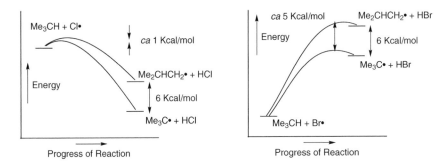

Figure 9.1

9.2.3   Steric effects in radical reactions

Free radicals add to the less-substituted carbon of an alkene, and examples of simple steric effects for radical reactions can be seen in the addition of radicals to alkenes of varying substitution. For example, the attack of a radical on 2-methylpropene **9.1** is very much faster than its attack on 2-methylbut-2-ene **9.2** (Scheme 9.4).[1]

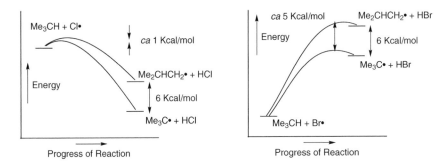

Scheme 9.4

However, the issue of regioselectivity is often not simple when considering substituents which stabilise the intermediate radicals, as are present in the common radical acceptors such as acrylates and α,β-unsaturated nitriles.

In these cases the α and β effects of all the substituents on the double bond should be considered, but the predominant cause of addition to the less-substituted end of the alkene is still primarily a *steric* one and not the difference in stability of the forming radical.[10] This is seen in the addition of cyclohexyl radicals to different substituted methyl acrylates. When the acrylate is unsubstituted addition occurs almost exclusively (99.8%) β to the CO_2Me giving product **9.3**, but as the substituent becomes significantly larger than the CO_2Me the selectivity is reversed, giving **9.4** as the major product (Scheme 9.5).[10]

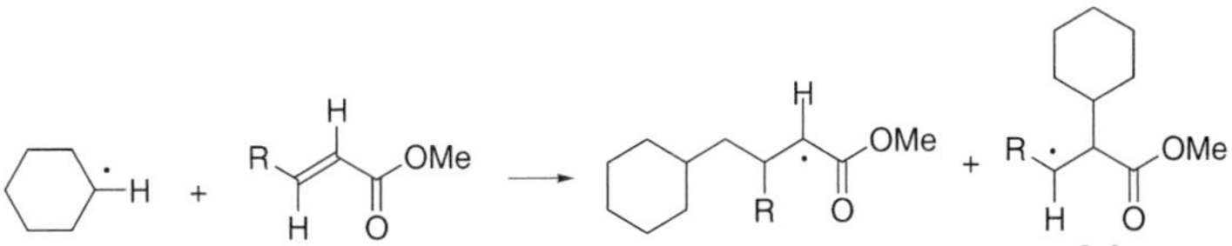

R	% **9.3**	% **9.4**
H	99.8	0.2
CH$_3$	92	8
Et	88	12
^{i}Pr	75	25
^{t}Bu	20	80

Scheme 9.5

9.2.4 *Stereoelectronic effects in radical reactions*

The observation that the 5-hexenyl radical cyclised predominantly to the less thermodynamically stable cyclopentylmethyl radical **9.5** rather than the isomeric cyclohexyl radical **9.6** (Scheme 9.6) could not be explained on thermodynamic

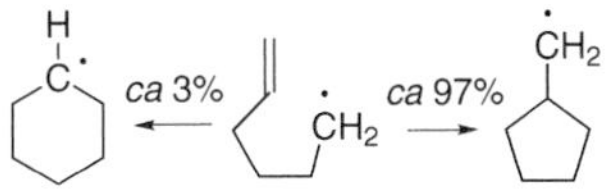

Scheme 9.6

grounds. This must be an example of a radical reaction where the kinetic barrier to the more stable product was higher than the barrier to an alternative less-stable product (Figure 9.2).

This reaction is occurring under stereoelectronic control where the controlling factor is the efficient overlap of the singly occupied molecular orbital (SOMO) of the radical and the π^* orbital of the double bond. Molecular orbital theory has been used to calculate that the addition of a methyl radical to ethene

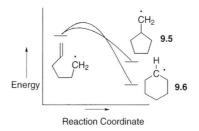

Figure 9.2

passes through a transition state where the angle of attack on the π system is 108° and the distance about 2.27 Å (Figure 9.3).[12]

Figure 9.3

The preferred formation of the five-membered ring product can be explained by considering that it is much easier to position the radical centre so that it interacts with the π-system at C(5) with the appropriate angle (Figure 9.3).[13] This formation of five-membered rings involves a chair-like transition state where orbital overlap is maximised and nonbonded interactions minimised. This cyclisation has gained considerable synthetic importance in radical chemistry and has received detailed theoretical analysis.[13–16] In some cases a boat-like transition state (ts) is also accessible (Figure 9.4).

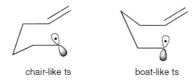

chair-like ts   boat-like ts

Figure 9.4

The postulate of chair-like transition states is supported by the result of cyclisation of the different positional isomers of the methyl-6-bromohex-1-enes. It has been observed experimentally[17] that cyclisation of the 4- and 5-methyl-substituted compounds under radical chain reaction conditions with Bu_3SnH

gives the *cis* and *trans*-1,3-dimethylcyclopentane, respectively, as the major product. The 3- and 6-substituted isomers on the other hand give *trans* and *cis*-1,2-dimethylcyclopentane, respectively, as the major product (Scheme 9.7).

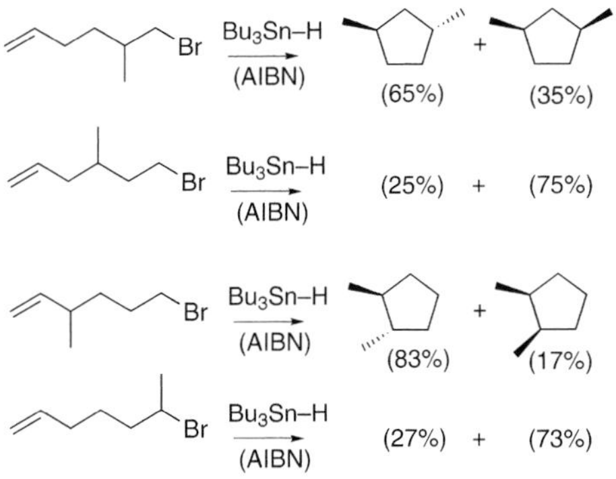

Scheme 9.7

These results are consistent with the reactions occurring through a chair transition state where the methyl group is either *pseudo*equatorial or *pseudo*axial. For example, reaction of the 4-methyl-6-bromohex-1-ene will give more of the *cis* isomer via a transition state where the methyl is *pseudo*equatorial (Scheme 9.8).

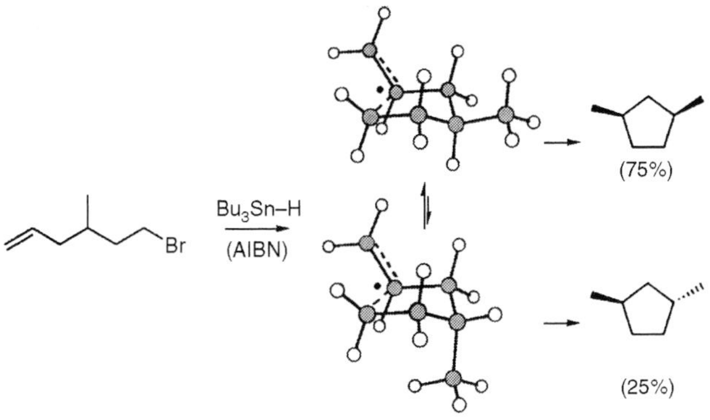

Scheme 9.8

A similar stereoelectronic control argument can be used to rationalise the observations for the opening of cyclopropane rings in suitably substituted radicals. β-Fission of the cyclopropyl radical to give the allyl radical is a highly exothermic process as a result of the release of the ring-strain energy and the formation of a resonance-stabilised radical, but the rate of the reaction is at least six orders of magnitude slower than for the mildly exothermic ring-opening of the cyclopropylcarbinyl radical ($k \approx 1.3 \times 10^8$ s^{-1} at 25°C).[18] These results can be explained by considering that the transition state for β-fission involves overlap of the SOMO with the σ* orbital of the bond undergoing fission. This is readily achieved in the cyclopropylcarbinyl radical, but not in the cyclopropyl radical (Scheme 9.9).

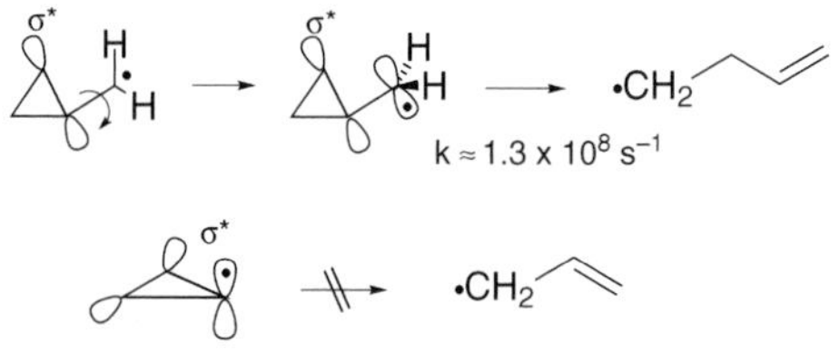

Scheme 9.9

9.2.5 Polar effects in radical reactions

Another effect which can moderate the energy of the transition state for a radical reaction is the possibility that the movement of electrons is not totally synchronous with the bond-breaking and bond-forming steps. If this is the case then some *polarisation* of the transition state can occur, and, if suitable substituents are present, it can favour the transition state leading to the cleavage of a stronger bond over a weaker one. For example, this is seen in the hydrogen atom abstraction from γ-butyrolactone where the —CH$_2$O— hydrogens are selectively abstracted by the oxygen-centred *t*-butoxy radical giving radical **9.7** in preference to the hydrogens (—CH$_2$C=O—) α to the carbonyl, for which the bond strength is about 5 kcal/mol weaker. This system has been studied by using electron spin resonance (ESR) by Roberts *et al.* (Scheme 9.10).[19]

This preference is rationalised by the pictorial representation of the transition state for the abstraction of the —CH$_2$O— hydrogens (Scheme 9.10), where partial charge separation leads to the electronegative oxygen bearing a partial negative charge and thus to a reduction in energy of the transition state. On the other hand, the polar transition state for the abstraction of the hydrogens α to the carbonyl group, with a partial negative charge on the electronegative oxygen, results in a partial positive charge on the carbon adjacent to the electron-withdrawing carbonyl group and thus would not be expected to be stablised.

This idea of polar transition states in radical reactions is supported by the observation that carrying out the experiment in the presence of borane–trimethylamine complex results in a switch in the selectivity of hydrogen atom abstraction: now the hydrogens α to the carbonyl are abstracted, giving

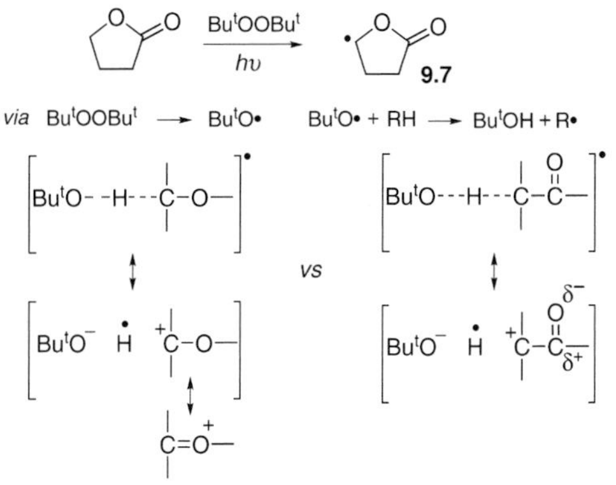

Scheme 9.10

radical **9.8**. This can be rationalised by the idea that the borane–trimethylamine acts as a radical chain carrier and that a boron-centred radical is now abstracting the hydrogen atom (Scheme 9.11).

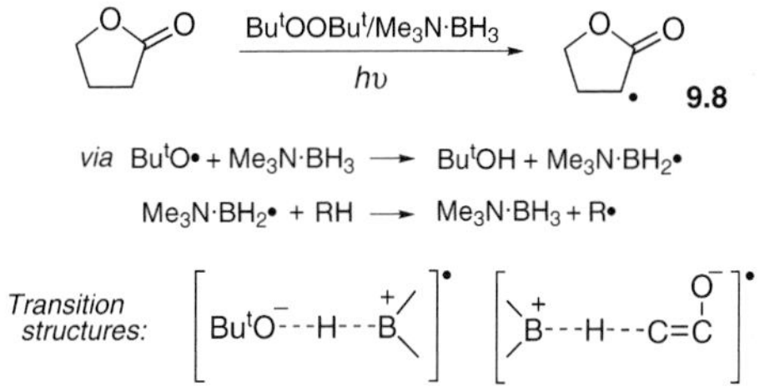

Scheme 9.11

Now the hydrogen abstraction is taking place by the electropositive boron, which takes on a partial positive charge in the transition state, and the hydrogens α to the carbonyl are preferred, with the carbonyl stabilising the negative charge which now develops on the α carbon.

9.3   Stereocontrol in the formation of cyclic systems

9.3.1   Formation of five-membered rings

By far the most widely used carbon–carbon bond-forming radical reactions in synthesis have been those involving the cyclisation of hexenyl radicals to give five-membered rings. These reactions, as described in section 9.2.4, generally proceed through chair-like transition states under stereoelectronic control.

These reactions can occur in a cascade manner where the radical formed from the first cyclisation is positioned to undergo a further cyclisation leading to the formation of a second ring. An example of this is seen in the synthesis of hirsutane reported by Curran and Kuo.[20,21] In this process two *cis*-fused five-membered rings are formed (Scheme 9.12). The second cyclisation to give

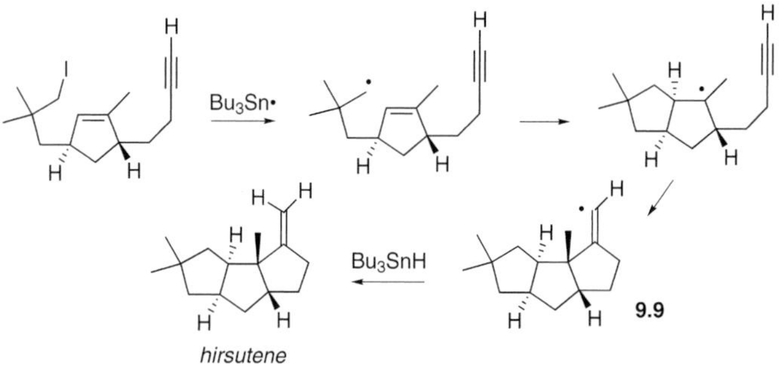

Scheme 9.12

intermediate **9.9** is favoured by formation of a σ-bond and a vinyl radical at the expense of a tertiary radical and a π-bond.

In developing this general approach to the triquinanes, Curran's group have synthesised a number of natural products, and one example of this general approach is seen in the synthesis of capnellene[22] as shown in Scheme 9.13.

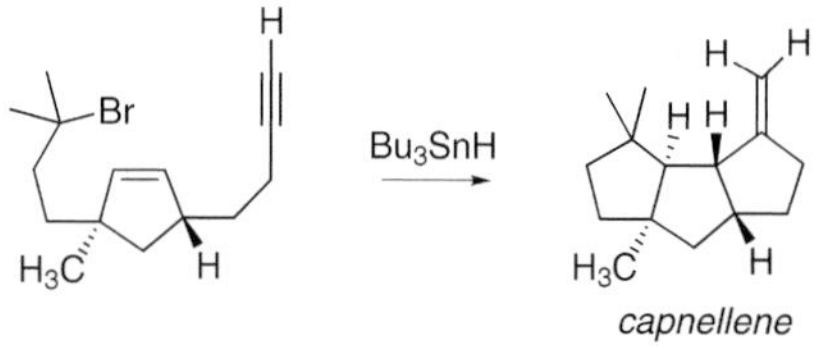

Scheme 9.13

The synthesis of siphiperfol-6-ene, an angular triquinane, has been carried out in racemic form by Curran and Kuo[20,21] and in optically active form by Meyers and Lefker[23] (Scheme 9.14). In this sequence the initially formed vinyl radical **9.10** produced by bromine-atom abstraction is not configurationally stable and undergoes isomerisation to the correct geometry before adding to the suitably positioned π system (despite the very fast cyclisation of vinyl radicals). The radical **9.11** then cyclises onto the pendant alkene, but this second cyclisation is not totally selective, giving a 1:2.5 mixture of epimers at the newly formed stereocentre.

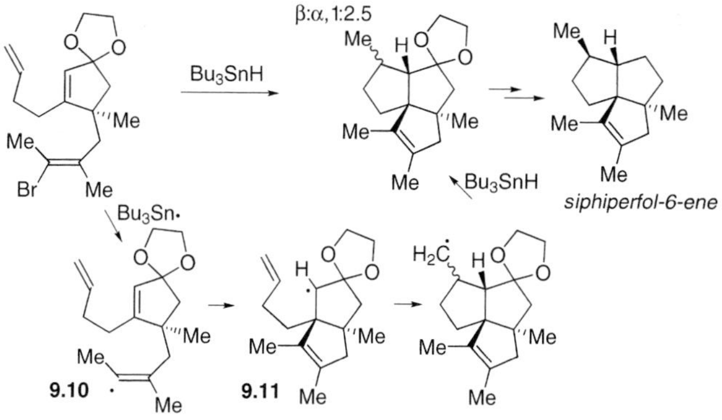

Scheme 9.14

Formation of a mixture of epimers at this stereocentre can be rationalised by a situation where a boat-like transition state competes with the chair-like transition state (Scheme 9.15). In this case the major product (the α isomer required for

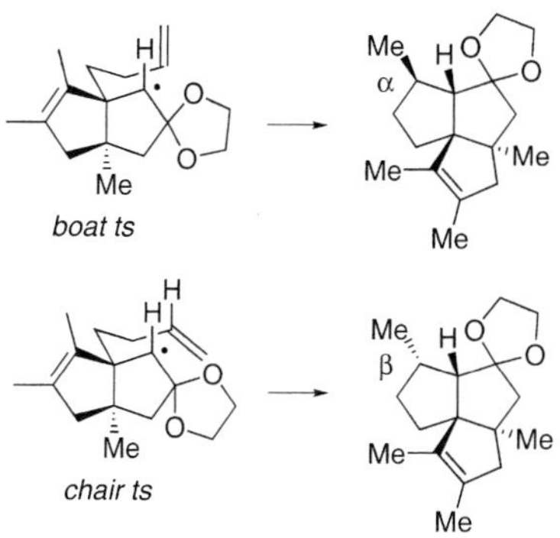

Scheme 9.15

the total synthesis) is formed via the boat-like transition state. It should be noted that when the same reaction is carried out on the compound which does not have the carbonyl protected as the bulky acetal the reaction proceeds predominantly through the chair-like transition state, giving the β isomer as the major product (α:β, 1:3).

9.3.2 Formation of six-membered rings

Although formation of five-membered rings is by far the most common of radical cyclisations, other sized rings can be formed. An example of cyclisation to form a six-membered ring is seen in Ladlow and Pattenden's construction of the neopentyl quaternary centre in alliacolide.[24,25] In this approach an activated α,β-unsaturated lactone receptor is used to facilitate the addition of the radical to the fully substituted terminus of the double bond and thus give the six-membered ring (Scheme 9.16).

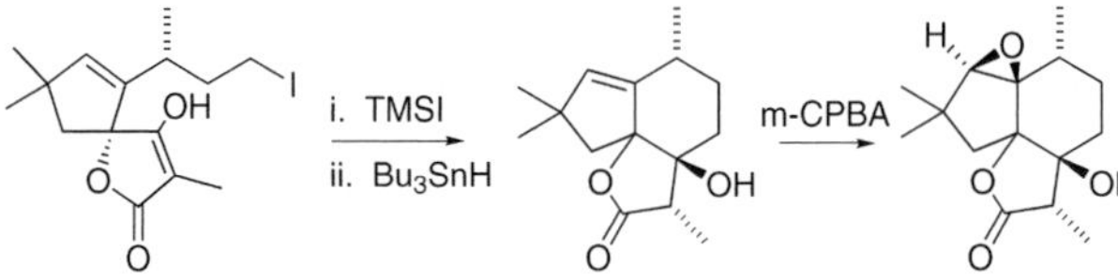

Scheme 9.16

9.4   Stereocontrol in the formation of acyclic systems

9.4.1   Stereocontrol in additions of chiral radicals to alkenes

The quest for effective acyclic stereocontrol in radical reactions is ongoing, and only in recent years has significant progress been made in this pursuit. When a radical is generated α to a stereocentre it is often observed that this stereocentre controls the facial selectivity of subsequent radical reaction via a 1,2 induction process.   An early example[26] is seen in Scheme 9.17 where the addition of

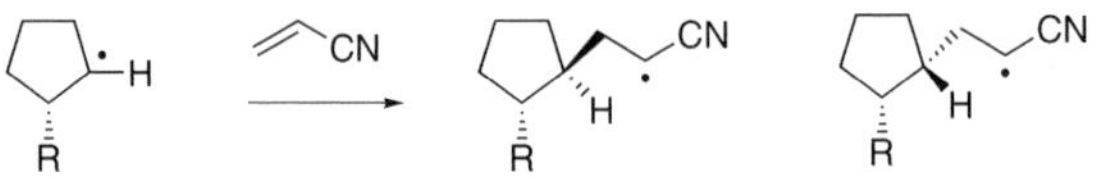

R	9.13:9.14
Me	92:8
OEt	77:23
O^{t}Bu	80:20

Scheme 9.17

2-substituted cyclopentyl radicals **9.12** to acrylonitrile results in the formation of predominantly the *anti* product **9.13** over the *syn* product **9.14**. The selectivity for R = Me is somewhat higher (92:8) than that observed for R = OEt (77:23) or R = O^{t}Bu (80:20). Similar selectivity is observed for the related acetal **9.15**, which gives a ratio of 86:14 as shown in Scheme 9.18.

If more than one substituent is present on the ring this can result in either an increase or a decrease in selectivity, depending on the spatial arrangement of

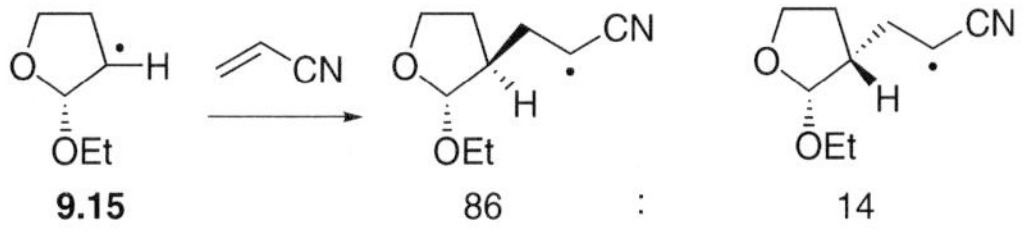

Scheme 9.18

these substituents. For example the compounds **9.16–9.18** show exclusive *anti* attack whereas radicals **9.19** and **9.20** show reduced selectivity (Figure 9.5) in additions to acrylonitrile.[27–29]

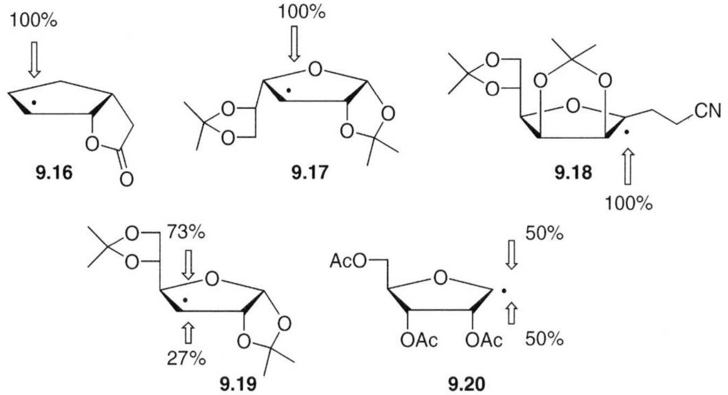

Figure 9.5

A multiply substituted cyclopentane ring has been used to control the stereochemistry of a radical addition in Stork *et al.*'s approach to prostaglandin synthesis.[30] In this case the radical intermediate **9.21** is trapped with *t*-butyl isocyanide (Scheme 9.19).

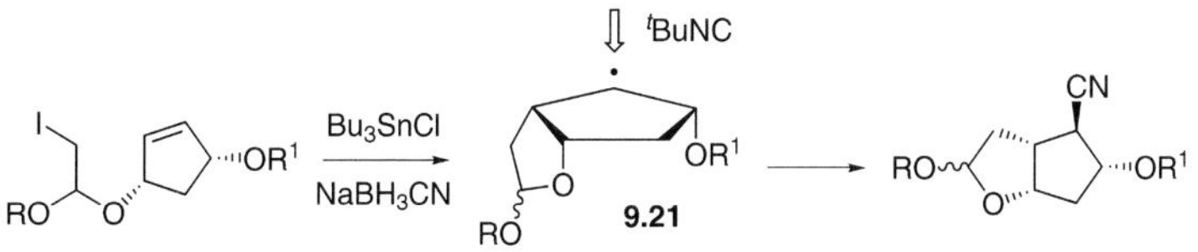

Scheme 9.19

In sugar systems radicals may be generated at the anomeric carbon, which reacts to give predominantly α-substituted products. The selectivity of these glucosyl radicals is independent of whether the radical precursor is the α or β anomer. The glucosyl radical precursor **9.22** undergoes a radical addition to acrylonitrile with high selectivity, as shown in Scheme 9.20.[31,32]

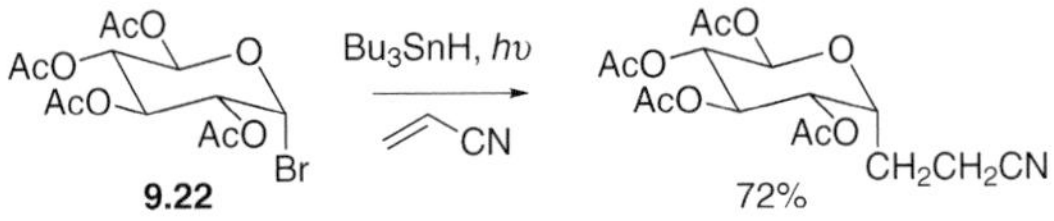

Scheme 9.20

Recently there have been reports of auxiliary-based systems in which high levels of stereocontrol have been achieved by using 1,4 induction.[33,34] The Oppolzer camphorsultam derivative **9.23** reacts with reasonable selectivity at 80°C (12:1) but the selectivity increases to 25:1 at −25°C (Scheme 9.21).

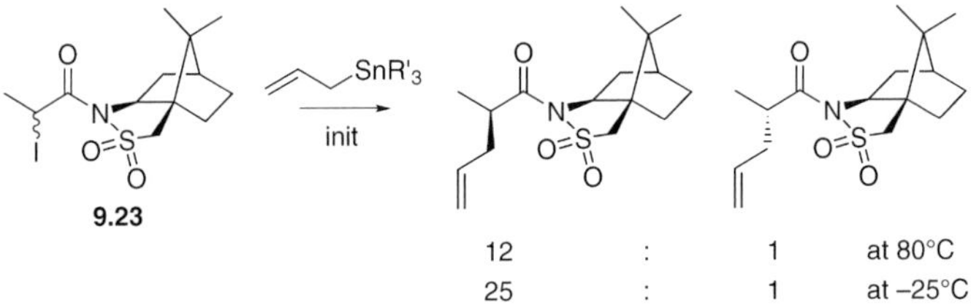

12	:	1	at 80°C
25	:	1	at −25°C

Scheme 9.21

Another example of 1,4 stereoinduction using an auxiliary-based system is seen in the addition of radical **9.24**, formed from the Barton ester **9.25** to ethyl acrylate in the presence of tributyl tin hydride.[35] Under the reaction conditions the 1:1 adduct **9.26** is formed with good stereocontrol but the 1:2 adduct **9.27** is a mixture of epimers at the second-formed stereocentre (Scheme 9.22).

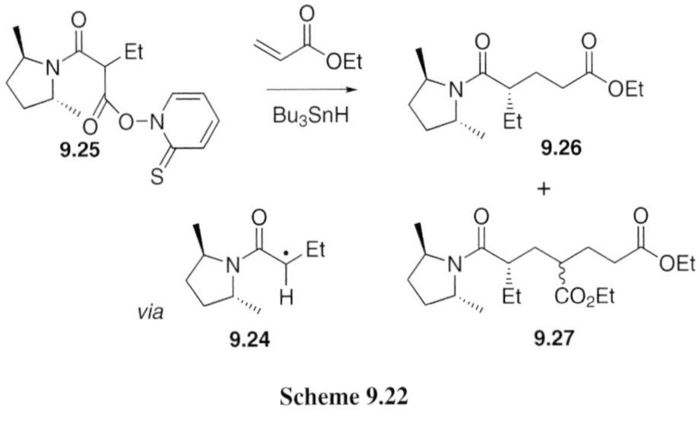

Scheme 9.22

The selectivity observed for both the Oppolzer sultam and the dimethyl-pyrrolidine amide can be rationalised by simple models (Figure 9.6).[8] The

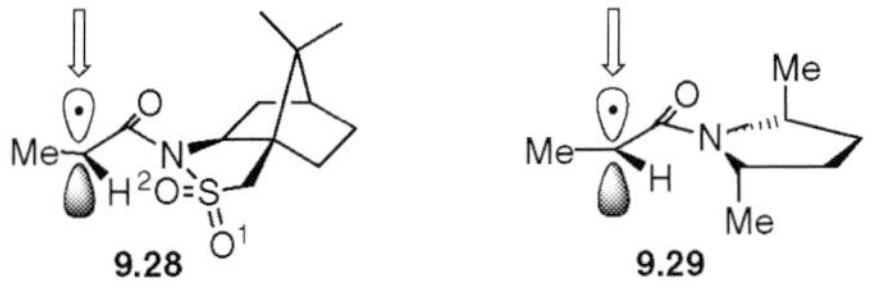

Figure 9.6

Oppolzer sultam model **9.28** has a number of important features. First, the low-energy conformation has the amide and sultam oxygens *anti* to minimise dipole repulsion, the radical has Z geometry (minimising $A^{1,3}$ strain), and O^1 shields one face much more effectively than O^2 shields the other face. For the dimethyl-pyrrolidine **9.29** similar considerations exist: the C2 symmetry eliminates C—N rotamer problems, the radical again has Z geometry, and the *cis* methyl group shields one face more effectively than the *trans* methyl shields the other face.

An oxazolidine-based auxiliary has been introduced by Sibi and Ji.[36] In this system the intermediate radical **9.30** is generated by radical addition to the enone **9.31** or halogen abstraction of a bromide precursor **9.32**. The selectivity of this reaction depends on the presence of a suitable Lewis acid. In the absence of Lewis acid the ratio of products **9.33** to **9.34** is only 1:1.8, but when an excess of MgI_2 or $MgBr_2$ is used under the same reaction conditions the ratio of products increases to greater than 100:1 (Scheme 9.23).

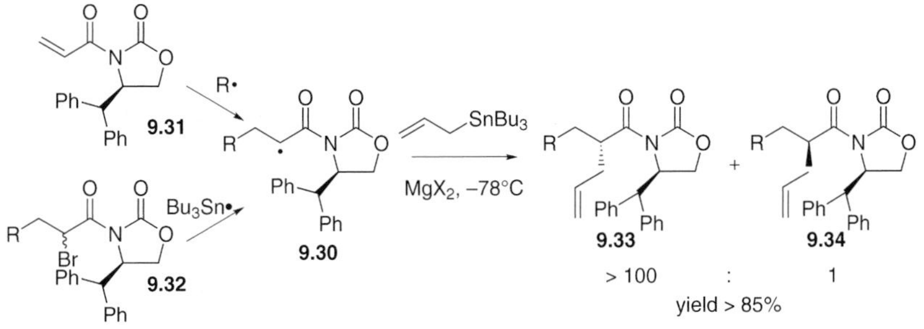

Scheme 9.23

The high selectivity for this reaction can be rationalised by considering the model where the Lewis acid coordinates to both carbonyl oxygens, the Z conformation of the radical minimises $A^{1,3}$ strain, and the diphenylmethyl substituent effectively shields one face from attack (Figure 9.7).

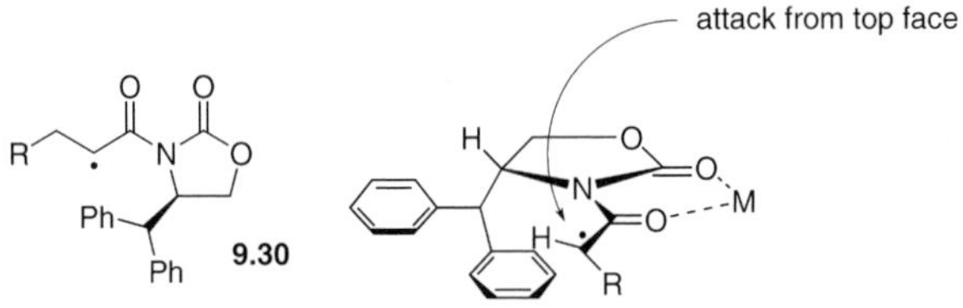

Figure 9.7

9.4.2   Stereocontrol in atom abstraction by chiral radicals

We have seen a number of cases where selectivity has been observed during addition of a radical to a π system; now we consider the formation of a new stereocentre by selective atom abstraction processes. An early example of this is seen in the addition of radical to methylmaleic anhydride **9.35**.[37] In this system addition of a cyclohexyl radical to the double bond generated a stereocentre, and hydrogen atom abstraction by this radical gave predominantly the *cis*-fused product **9.36**, as seen in Scheme 9.24. (Thus attack occurs on the face opposite the cyclohexyl group.)

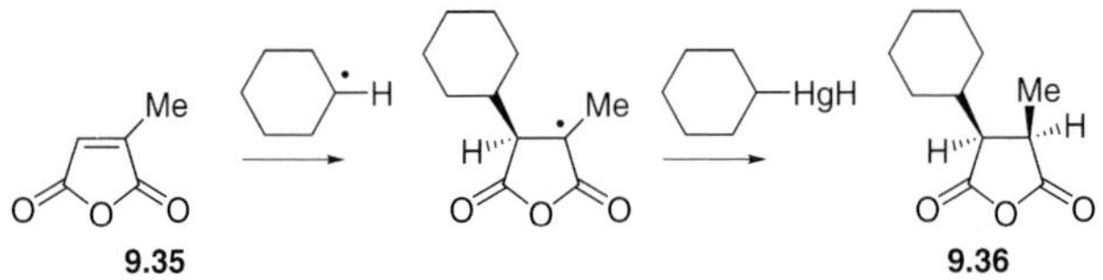

Scheme 9.24

Acyclic systems have been studied in some detail and it has generally been found that the *threo* products **9.37** predominate when R^2 is larger than X, whereas the *erythro* isomers **9.38** are the main products when X is larger than R^2. If R^2 is ^{t}Bu and X is CO_2Me a 98:2 mixture of **9.37:9.38** is formed, whereas $R^2 = CH_3$ and $X = OSiPh_2{}^t$Bu the selectivity is reversed giving a 3:97 mixture of **9.37:9.38** (Scheme 9.25).

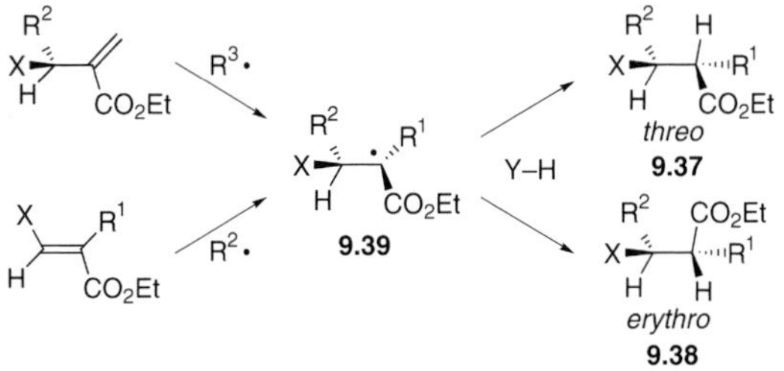

Scheme 9.25

It appears that the radical **9.39** adopts a conformation in which the substituents R^2 and X shield opposite faces of the adjacent radical centre and attack occurs opposite the substituent which is sterically larger (Figure 9.8).

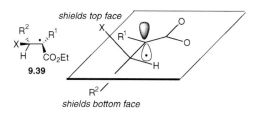

Figure 9.8

An example of auxiliary-controlled atom abstraction is seen in the halogen atom abstraction by the chiral radical **9.24** generated from the Barton ester **9.25** by a photochemical reaction to give **9.40** and **9.41** initiated reaction (Scheme 9.26). The abstraction of bromine is slightly more selective than iodine at 25°C and the selectivity increases if the temperature is lowered to 0°C.[38]

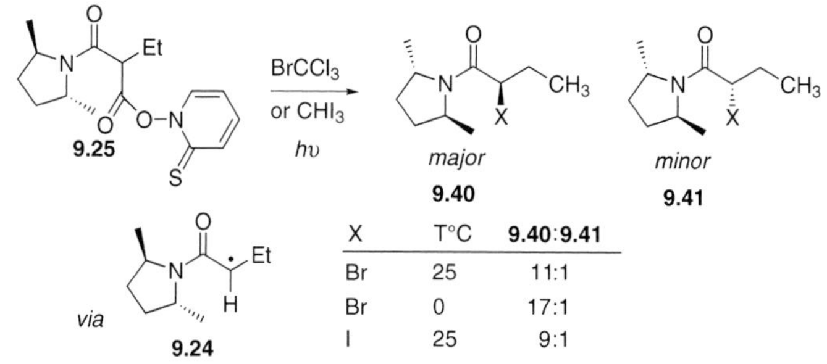

X	T°C	9.40:9.41
Br	25	11:1
Br	0	17:1
I	25	9:1

Scheme 9.26

9.4.3 Stereocontrol in radical additions to chiral alkenes

There are few examples of intermolecular additions of achiral radicals to acyclic chiral alkenes that give selectivities higher than 3:1. The case for terminal alkenes is worse, with no known cases of high selectivity.

A significant advance was made with the use of chiral α,β-unsaturated amides as the radical acceptors. The first report of high selectivity in radical addition with use of an amide of this type was in the synthesis of the large-ring macrocycle muscone, noted by Porter and co-workers.[39,40] In this synthesis the *exo:endo* product ratio is 8:1 and the diastereoselectivity of the two *endo* products 14:1 at 80°C (Scheme 9.27).

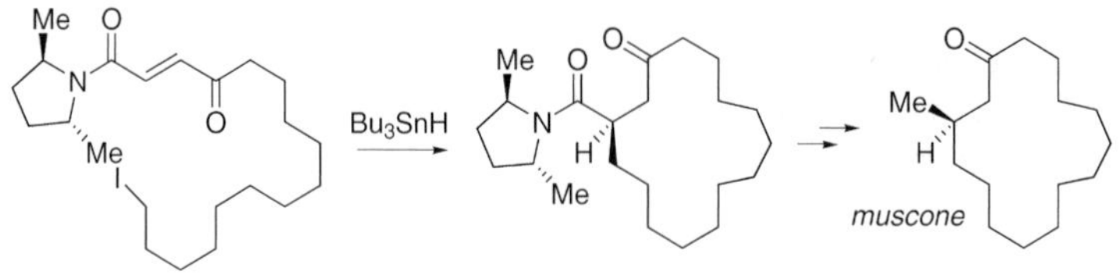

Scheme 9.27

However, there was no diastereoselectivity observed for the *exo* product and this has proved to be typical of these C(2) symmetric amide auxiliaries, where the auxiliary is only effective when it is attached to the α carbon and is not effective when it is attached to the β carbon. The problem of regiochemistry in the addition of the radical has been overcome by the use of the symmetric diamide derived from fumaric acid. Addition of *t*-butyl radical to this alkene gives very predominantly one stereoisomer at room temperature (Scheme 9.28).[41]

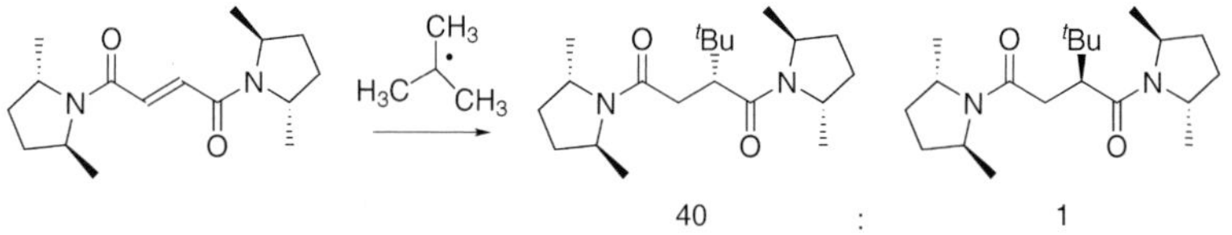

40   :   1

Scheme 9.28

The effect of 1,2 induction compared with 1,4 induction has been studied in this system where the intermediate radical **9.42** abstracts deuterium, giving the product with a d.s. of 93% as shown in Scheme 9.29.[42] This result is consistent

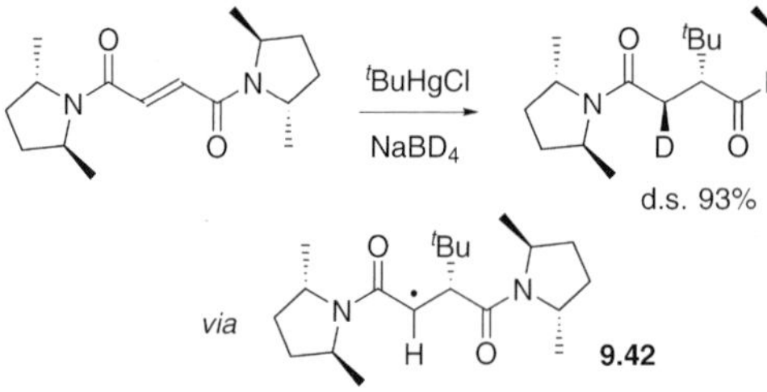

Scheme 9.29

with the major influence arising from the 1,2 induction from the stereocentre bearing the *t*-butyl group in accord with the model presented in Figure 9.8.

A highly reactive trisubstituted alkene which reacts with high regioselectivity and stereoselectivity is shown in Scheme 9.30. The reaction with cyclohexyl radicals gives exceptionally high stereoselectivity (> 125:1 at −80°C).[43,44]

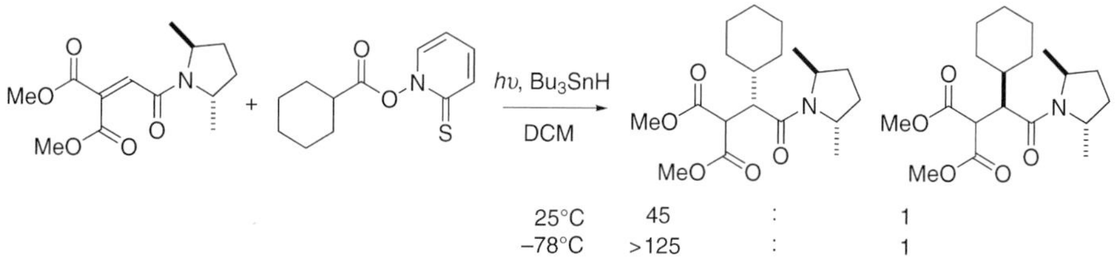

25°C	45 :	1
−78°C	>125 :	1

Scheme 9.30

An auxiliary derived from Kemp's acid has been reported where control of stereochemistry from attack on the β carbon has been achieved (Scheme 9.31).[45]

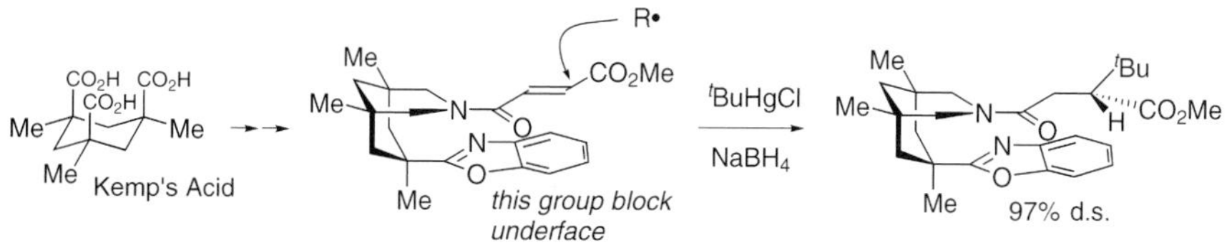

Scheme 9.31

Attack of the *t*-butyl radicals occurs exclusively at the β carbon and has a stereoselectivity of 97:3.

One other example of stereocontrol for addition at the β carbon has been reported in which an oxazolidine-based auxiliary **9.43** complexed to a Lewis acid was used.[46] In this case *iso*-propyl radicals add selectively to the β carbon with very high stereoselectivity to give the major product **9.44** with very small amounts of **9.45**, as shown in Scheme 9.32.

In a direct analogy to the selectivity observed for the related radical **9.30**, the high selectivity in addition of radicals to **9.43** can be rationalised by considering the model where the Lewis acid coordinates to both carbonyl oxygens and the diphenylmethyl substituent effectively shields one face from attack (Figure 9.9).

9.4.4 Enantioselective radical reactions

The highest degree of difficulty in a radical reaction is to achieve the production of one enantiomer by using a chiral reagent. This has only very recently

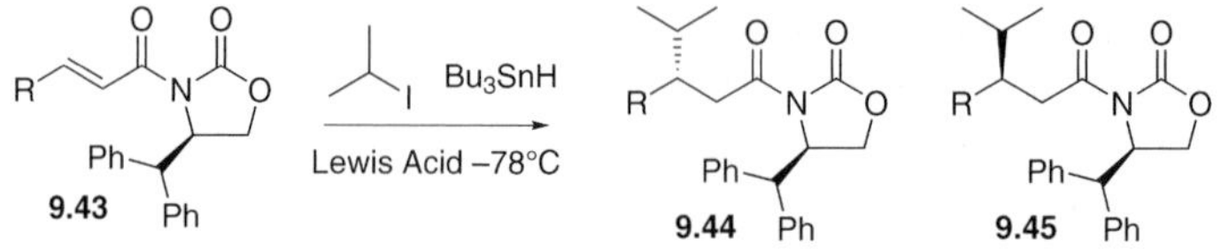

R	Lewis Acid	% Yield	9.44:9.45
Me	Yb(OTf)$_3$	> 90	25:1
Ph	Yb(OTf)$_3$	> 90	45:1
COOEt	Er(OTf)$_3$	> 90	53:1

Scheme 9.32

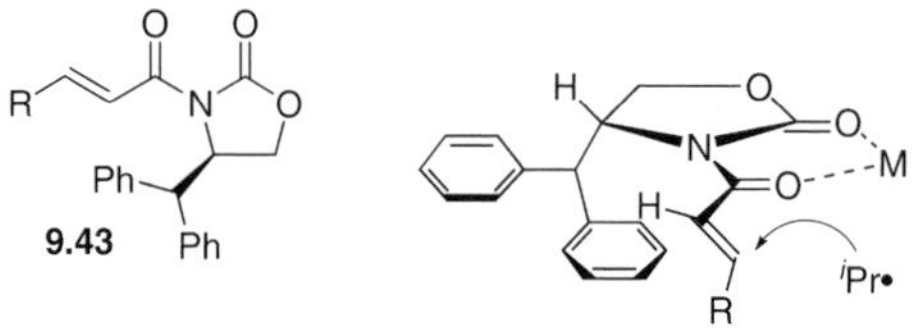

Figure 9.9

been achieved.[11] In most cases this enantioselectivity has been achieved by the complexation of the radical acceptor to a chiral Lewis acid. In some cases these chiral Lewis acids can be used catalytically (and thus go beyond the scope of this book), but we will highlight examples where the use of a stoichiometric amount of Lewis acid leads to high enantioselectivity.

The first reported carbon-to-carbon bond-forming radical reaction that proceeded with high enantioselectivity employed stoichiometric amounts of the bisoxazoline ligand **9.46** and Zn(OTf)$_2$ and is shown in Scheme 9.33.[47] The

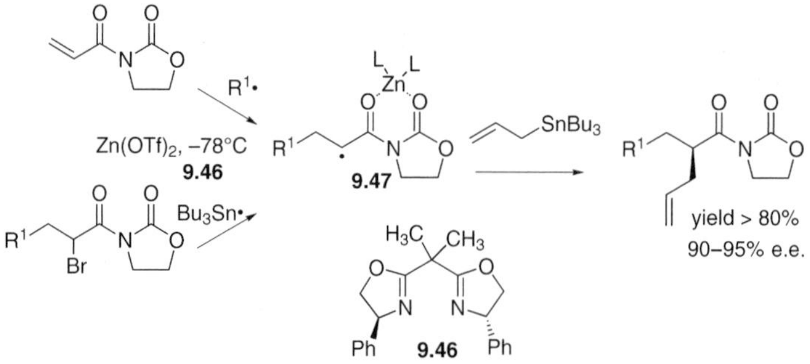

Scheme 9.33

ligand complexes to both oxygens of the achiral oxazolidinone auxiliary, producing a chiral complex **9.47** which is then selectively allylated.

A second system has been investigated[48] where the addition to the β carbon is enantioselective (Scheme 9.34). In this study three different chiral ligands were

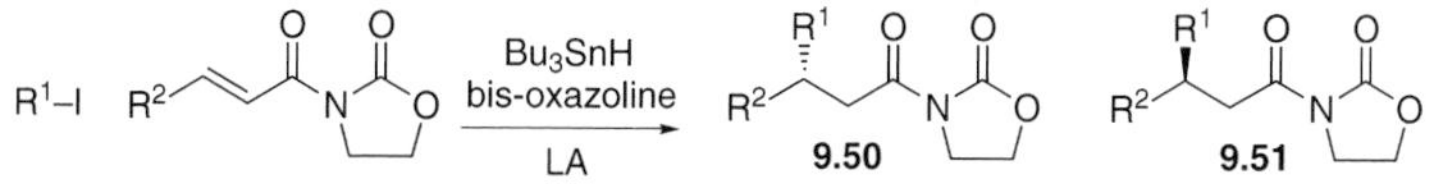

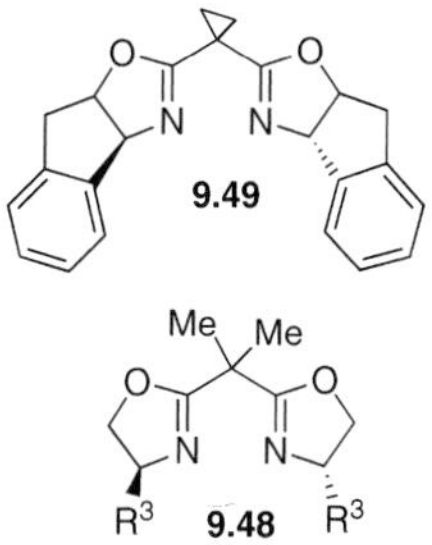

$$R^1\text{–I} \quad R^2\overset{O}{\diagup}\overset{O}{\diagdown}N\diagdown O \xrightarrow[\text{LA}]{\substack{Bu_3SnH \\ \text{bis-oxazoline}}} R^2 \overset{R^1}{\underset{9.50}{\diagup}}\overset{O}{\diagdown}\overset{O}{\diagdown}N\diagdown O \quad R^2\overset{R^1}{\underset{9.51}{\diagup}}\overset{O}{\diagdown}\overset{O}{\diagdown}N\diagdown O$$

9.49

9.48

R¹	R²	R³ (eq.)	LA (eq.)	Yield (%)	9.50:9.51
^{t}Bu	Me	**9.48** Ph (1.0)	Zn(OTf)$_2$ (1.0)	65	10:1
^{t}Bu	Me	**9.48** ^{t}Bu (1.0)	MgBr$_2$ (1.0)	90	1:10
^{t}Bu	Me	**9.49**	MgI$_2$ (1.0)	85	1:99
^{t}Bu	Me	**9.49** (0.2)	MgI$_2$ (0.2)	75	1:10
^{i}Pr	Ph	**9.49**	MgI$_2$ (1.0)	85	1:98

Scheme 9.34

tested. The ligand **9.49** compared with **9.48** gave the best selectivity (98–99:1, **9.51**:**9.50**) when used in a stoichiometric amount. Use of a catalytic amount (20 mol%) reduced the selectivity to 10:1.

9.5 Conclusions

We have seen sufficient examples of stereoselective radical reactions to conclude that radical reactions should be considered a viable method for the generation of stereocentres in the synthesis of natural products.

In this discussion only limited analysis of stereochemical models for predicting or rationalising the stereochemical outcome of these radical reactions has been presented. Interested readers are directed towards Smadja's very detailed analysis of the factors controlling the stereochemical outcome in the whole range of acyclic radical reactions.[6]

References

1. Perkins, M.J., *Radical Chemistry*, Ellis Horwood, New York, **1994**.
2. Giese, B., *Radicals in Organic Synthesis:Formation of Carbon–Carbon Bonds*, 1st edn, Pergamon, Oxford and New York, **1986**.
3. Fossey, J., Lefort, D. and Sorba, J., *Free Radicals in Organic Chemistry*, John Wiley, Masson, Chichester, New York and Paris, **1995**.

4. Jasper, C.P., Curran, D.P. and Fevig, T.L., *Chem. Reviews*, **1991**, *91*, 1237.

5. Beckwith, A.L.J., *Chem. Soc. Rev.*, **1993**, 143.

6. Smadja, W., *Synlett*, **1994**, 1.

7. Curran, D.P., *Synthesis*, **1988**, 489.

8. Porter, N.A., Giese, B. and Curran, D.P., *Acc. Chem. Res.*, **1991**, *24*, 296.

9. Giese, B., *Angew. Chem., Int. Ed. Engl.*, **1989**, *28*, 969.

10. Giese, B., *Angew. Chem., Int. Ed. Engl.*, **1983**, *22*, 753.

11. Sibi, M.P. and Porter, N.A., *Acc. Chem. Res.*, **1999**, *32*, 163.

12. Houk, K.N., Paddon-Row, M.N., Spellmeyer, D.C., Rondan, N.G. and Nagase, S., *J. Org. Chem.*, **1986**, *51*, 2874.

13. Beckwith, A.L.J. and Schiesser, C., *Tetrahedron*, **1985**, *41*, 3925.

14. Spellmeyer, D.C. and Houk, K.N., *J. Org. Chem.*, **1987**, *52*, 959.

15. Beckwith, A.L.J. and Zimmermann, J., *J. Org. Chem.*, **1991**, *56*, 5791.

16. RajanBabu, T.V., *Acc. Chem. Res.*, **1991**, *24*, 139.

17. Beckwith, A.L.J., Easton, C.J., Lawrence, T. and Serelis, A.K., *Aust. J. Chem.*, **1983**, *36*, 545.

18. Bowry, V.W., Lusztyk, J. and Ingold, K.U., *J. Am. Chem. Soc.*, **1991**, *113*, 5687.

19. Kaushal, P., Mok, P.L.H. and Roberts, B.P., *J. Chem. Soc., Perkin Trans.*, **1990**, *2*, 1663.

20. Curran, D.P. and Kuo, S.-C., *Tetrahedron*, **1987**, *43*, 5653.

21. Curran, D.P. and Kuo, S.-C., *J. Am. Chem. Soc.*, **1986**, *108*, 1106.

22. Curran, D.P. and Chen, M.-H., *Tetrahedron Lett.*, **1985**, *26*, 4991.

23. Meyers, A.I. and Lefker, B.A., *Tetrahedron*, **1987**, *43*, 5663.

24. Ladlow, M. and Pattenden, G., *Tetrahedron Lett.*, **1985**, *26*, 4413.

25. Ladlow, M. and Pattenden, G., *J. Chem. Soc., Perkin Trans.*, **1988**, *1*, 1107.

26. Giese, B., Heuck, K., Lehardt, H. and Lüning, U., *Chem. Ber.*, **1984**, *117*, 3132.

27. Burke, S.D., Fobare, W.F. and Armistead, D.H., *J. Org. Chem.*, **1982**, *47*, 3348.

28. Giese, B., Gónzalez-Gómez, J.A. and Witzel, T., *Angew. Chem., Int. Ed. Engl.*, **1984**, *23*, 69.

29. Dupuis, J., Giese, B., Hartung, J., Leising, M., Korth, H.-G. and Sustmann, R., *J. Am. Chem. Soc.*, **1985**, *107*, 4332.

30. Stork, G., Sher, P.M. and Chen, H.L., *J. Am. Chem. Soc.*, **1986**, *108*, 6384.

31. Giese, B. and Dupuis, J., *Angew. Chem., Int. Ed. Engl.*, **1983**, *22*, 622.

32. Giese, B., Dupuis, J. and Nix, M., *Org. Synth.*, **1987**, *65*, 622.

33. Curran, D.P. and Heffner, T.A., *J. Org. Chem.*, **1990**, *55*, 4585.

34. Curran, D.P., Kim, B.H., Daugherty, J. and Heffner, T.A., *Tetrahedron Lett.*, **1988**, *29*, 3555.

35. Porter, N.A., Swann, E., Nally, J. and McPhail, A.T., *J. Am. Chem. Soc.*, **1990**, *112*, 6740.

36. Sibi, M.P. and Ji, J., *Angew. Chem., Int. Ed. Engl.*, **1996**, *35*, 190.

37. Giese, B. and Meixner, J., *Tetrahedron Lett.*, **1977**, *18*, 2783.

38. Porter, N.A., Breyer, R., Swann, E., Nally, J., Pradhan, J., Allen, T. and McPhail, A.T., *J. Am. Chem. Soc.*, **1991**, *113*, 7002.

39. Porter, N.A., Chang, V.H.-T., Magnin, D.R. and Wright, B.T., *J. Am. Chem. Soc.*, **1988**, *110*, 3554.

40. Porter, N.A., Lachner, B., Chang, V.H.-T. and Magnin, D.R., *J. Am. Chem. Soc.*, **1989**, *111*, 8309.

41. Porter, N.A., Scott, D.M., Lacher, B., Giese, B., Zeitz, H.G. and Linder, H.J., *J. Am. Chem. Soc.*, **1989**, *111*, 8311.

42. Giese, B., Zehnder, M., Roth, M. and Zeitz, H.-G., *J. Am. Chem. Soc.*, **1990**, *112*, 6741.

43. Scott, D.M., McPhail, A.T. and Porter, N.A., *Tetrahedron Lett.*, **1990**, *31*, 1679.

44. Porter, N.A., Wu, W.-X. and McPhail, A.T., *Tetrahedron Lett.*, **1991**, *32*, 707.

45. Stack, J.G., Curran, D.P., Rebek, J.J. and Ballester, P., *J. Am. Chem. Soc.*, **1991**, *113*, 5918.

46. Sibi, M.P. and Jasperse, C.P., *J. Am. Chem. Soc.*, **1995**, 10779.

47. Wu, J.H., Radinov, R. and Porter, N.A., *J. Am. Chem. Soc.*, **1995**, *117*, 11029.

48. Sibi, M.P., Ji, J., Wu, J.H., Gürtler, S. and Porter, N.A., *J. Am. Chem. Soc.*, **1996**, *118*, 9200.

Index